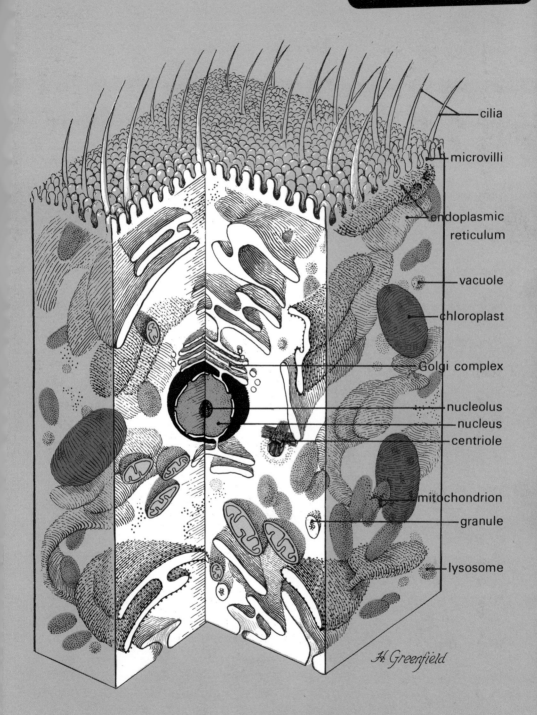

Cell Biology

Cell Biology

JACK D. BURKE, Ph.D.

Professor of Anatomy,
Medical College of Virginia,
Health Sciences Division,
Virginia Commonwealth University,
Richmond, Virginia

The Williams & Wilkins Company
Baltimore/1970

Copyright ©, 1970
The Williams & Wilkins Company
428 E. Preston Street
Baltimore, Maryland 21202 U.S.A.

Made in the United States of America

Library of Congress Catalog Card Number 70-108276 SBN 683-01212-6

All rights reserved. This book is protected by copyright. No part of this book may be reproduced in any form or by any means, including photocopying, or utilized by any information storage and retrieval system without written permission from the copyright owner.

Cover: The trilaminar appearance exhibited by cell membranes studied by electron microscopy leads one to think of the ubiquitous unit membrane as a universal concept. This electron micrograph of a human red blood cell was made and kindly furnished by Dr. J. D. Robertson of the Duke University School of Medicine.

Composed and Printed at
Waverly Press, Inc.
Mount Royal and Guilford Avenues
Baltimore, Maryland 21202 U.S.A.

Preface

Cell Biology has been written for students who are entering their first phase of professional or graduate training and for those in their last phase of undergraduate education. I believe that the subject matter is organized to provide a direction for a student in his learning about cell structure and function and that the text material which has been chosen for discussion will give him a better insight into "the most important invention in nature."

We are now entering an academic era in which medical, dental, and graduate education is undergoing a drastic change. In many instances old curricula are being redesigned, and new curricula are being formulated. The subject matter of various disciplines is gradually being correlated, synthesized, and integrated. As a result, one may probe deeper into a cell and find interdisciplinary thinking regarding its structure and function. Emphasis of this kind on cell study has resulted in many departments offering more courses concerned with the biology of cells.

This book is about the biology of cells. As a discipline, cell biology is a study of how reproducing, living cells maintain themselves in a steady state relative to their environment. It deals with both structure and function of the organelles in the cell and the matrix which supports them. The text of *Cell Biology* has been organized into three sections. The first is concerned with the anatomy and physiology of cells, the second with the metabolism of cells, and the third section with the nucleocytoplasmic relations of cells.

In the first section the anatomy and physiology of the cell are discussed in relation to microscopy, organelles, matrix, and transfer of substances across the cell membrane. The anatomy of the cell, as it is known today, is based mostly on information obtained from light and electron microscopy. As the evolution of microscopy progressed, detailed descriptions of organelles improved. However, limitations of microscopical analysis stimulated development of sophisticated instrumentation and cytochemical techniques. These clarified many points concerned with structure and function in organelles as well as the physical and chemical properties of the colloidal matrix. The factors affecting movement of waste products and foodstuff across the cell membrane to maintain the steady state are also discussed.

In the second section cell metabolism is discussed in relation to bioenergetics, enzymes and cofactors, biological oxidations and mitochondria, cellular respiration, and intermediary metabolism. Flow sheets are used to illustrate important pathways in metabolic processes. It is indicated in the discussions that the steady state of the cell is maintained as energy flows through it in enzymatically controlled reactions. The conserved energy in these reactions is made available to drive many cellular processes. It is also shown that most foodstuff is metabolized because of its conversion to acetyl coenzyme A or an intermediate of the tricarboxylic acid cycle. Continuing discussions show that during subsequent oxidations of the intermediate compounds in the tricarboxylic acid cycle, energy-rich adenosine triphosphate is ultimately produced by oxidative phosphorylation, and carbon dioxide and water are formed as waste products of the reactions. Not only are the metabolic pathways for the metabolism of macromolecules described, but the important aspects of cellular respiration are also discussed.

Section three is concerned with nucleocytoplasmic relations. One of the most exciting areas for research in cell biology is concerned with the theory of the gene. Discussions here are involved with chromosome morphology, deoxyribonucleic acid (DNA) replication and cell division, ribosomes and protein synthesis, the genetic code, and the regulation of enzymes. The morphological features of the nuclear envelope aid the diffusion process in which substances are exchanged between the nucleus and cytoplasm. The structure of the chromosome is discussed in relation to DNA, the chemical basis of inheritance. The DNA serves as a template for messenger ribonucleic acid which carries the genetic message to the cytoplasmic ribosomes; here protein synthesis occurs. The current concept of protein synthesis is discussed and illustrated in relation to the central dogma of protein synthesis. The relationship of the genetic code and enzyme regulation to the central dogma is also discussed. The importance of the effects of radiation on cells has not been overlooked. Although different cells can have varying degrees of radiosensitivity, all cells can have their genetic stability affected by high doses of ionizing radiation. Discussions relate the effects of radiation on chromosomes, DNA, and cell division, as well as the gross effects on whole organisms.

Acknowledgments: Indeed, I am grateful to those colleagues whose stimulating discussions have clarified certain vague points for me; but all errors, omissions, and corrections needed are solely my responsibility. Quite frankly, this book would not have been written without the aid of my wife, who collaborated with me at nearly every phase; she served as graphic-arts illustrator, typed the manuscript, and aided with the index. Again, I thank her. I also want to express thanks to Mrs. Ranice Crosby, Director of the Department of Art as Applied to Medicine at the Johns Hopkins Medical School, for her advice on the artwork of the cell shown on the inside covers. Especially am I indebted to Mrs. Harriet Greenfield of the same department for conveying my ideas of the cell into such a superb and artful reality. It would be remissful of me not to acknowledge the kindness and helpfulness the library staff of the Medical College of Virginia showed me in every instance. The same kind of generosity was shown me by the MCV Visual Education staff. In particular I want to thank Mrs. Louise W. Morton, Reference Librarian, and Mr. Melvin C. Shaffer, Director of the Visual Education Department, for their aid. I could not have asked for any more cooperation and consideration than that which was given me by the staff of The Williams & Wilkins Company; their encouragement and confidence, as well as their many courtesies, were most valuable.

JACK D. BURKE
January, 1970

TO MY WIFE AND CHILDREN

Contents

Section I.

Cell Anatomy and Physiology

1.
Historical Microscopy 3
   A. *Light Microscopy* 5
   B. *Electron Microscopy* 10
   C. *Phase Contrast Microscopy* 16
   D. *Measurements of Cells* 19

2.
Organelles: Structure and Function 22
   A. *A Representative Cell* 22
   B. *Nucleoplasm* 22
      1. Nucleolus 22
      2. Chromatin and Chromosomes 25
      3. Nuclear Envelope 26
   C. *Cytoplasm* 33
      1. Endoplasmic Reticulum 33
      2. Golgi Complex 34
      3. Ribosomes 36
      4. Lysosomes 39
      5. Mitochondria 43
      6. Chloroplasts 49
      7. Cilia and Flagella 54
      8. Centrioles and Basal Bodies 55
      9. Microtubules 57

10.	Tonofibrils	61
11.	Cell Membrane	11
12.	Modifications of the Cell Surface	67
13.	Cell Wall	68

3.
Matrix: Physical and Chemical Characteristics 75

- A. *Properties of Aqueous Solutions* 76
  1. Solvent 76
  2. Solutes 78
  3. Molecular and Ionic Solutions 78
  4. Colloids 78
- B. *Macromolecules* 79
  1. Carbohydrates 81
     a. Classification 81
     b. Monosaccharides 82
     c. Derived Monosaccharides 85
     d. Oligosaccharides 85
     e. Polysaccharides 87
  2. Proteins 88
     a. Classification 89
     b. Properties 90
     c. Amino Acids and Peptides 90
  3. Nucleoproteins and Nucleic Acids 99
- C. *Porphyrins* 103
- D. *Lipids* 105
  1. Simple Lipids 105
  2. Compound Lipids 108
- E. *Vitamins* 111
  1. Fat-soluble Vitamins 112
  2. Water-soluble Vitamins 113
- F. *Minerals* 115
- G. *Acids, Bases, and Salts* 116
- H. *pH and Buffers* 117

4.
Transport Across Membranes 123

- A. *Permeability of Membranes* 126
- B. *Osmosis and Diffusion* 126
- C. *Active Transport* 128

    D. *Donnan Equilibrium* 131
    E. *Pinocytosis and Phagocytosis* 132
    F. *The Nerve Impulse* 134
    G. *Impulse-Transmission in Muscle Cells* 136

Section **II**.
Cell Metabolism

   5.
   Bioenergetics, Enzymes, and Cofactors 143
       A. *Bioenergetics* 143
           1. Free Energy 144
           2. Redox Reactions 147
           3. High-energy Compounds 148
       B. *Enzymes* 151
           1. Theory of Action 152
           2. Kinetics 153
           3. Inhibitors and Antimetabolites 157
           4. Nomenclature 159
           5. Isozymes 162
       C. *Cofactors* 162

   6.
   Biological Oxidations and Mitochondria 168
       A. *Mitochondrial Pathways* 168
       B. *The Tricarboxylic Acid Cycle* 173
       C. *Electron Transport and Oxidative Phosphorylation* 177

   7.
   Cellular Respiration 183
       A. *Calorimetry and Respiratory Quotient* 183
       B. *Hemoglobins and Gas Transport* 195
       C. *Carbohydrate Formation and Conversions* 211
           1. Light Reactions 215
           2. Dark Reactions 218

   8.
   Intermediary Metabolism 225
       A. *Carbohydrate Metabolism* 227
       B. *Lipid Metabolism* 235

CONTENTS

  C. *Protein Metabolism* 244
  D. *Nucleic Acids Metabolism* 251
   1. Pyrimidines 252
   2. Purines 254
   3. RNA 256
   4. DNA 258

Section **III.**
Nucleocytoplasmic Relations

9.
Theory of the Gene 263

  A. *Chromosome Morphology* 263
  B. *DNA Replication and Cell Division* 275
  C. *Ribosomes and Protein Synthesis* 285
  D. *Genetic Code* 293
  E. *Enzyme Regulation* 298

10.
Cell Radiation Biology 310

  A. *Physical Aspects of Radiation* 310
   1. Historical 310
   2. Ionizing Radiations 313
   3. Measurements of Radioactivity 317
   4. Theories of Action 318
  B. *Radiosensitivity* 320
   1. Influence of Environmental Factors 320
   2. Cellular Conditions 321
  C. *Radiation Effects on the Cell* 322
   1. DNA 322
   2. Chromosome Structure 324
   3. Mitotic Apparatus 325
   4. Influencing Factors 326
  D. *Gross Effects of Radiation* 327
   1. Viruses 327
   2. Microorganisms 327
   3. Plants 328
   4. Animals 328
   5. Mammalian systems 329

Index 335

"The living cell is the most important invention in nature. In its capacity to maintain the constant equilibrium in which life is manifest, it must be the continuous wonder of any thoughtful person."

Sir Rudolph Peters
British Medical Bulletin,
24:99, 1968

Section I.
Cell Anatomy and Physiology

Chapter One

Historical Microscopy

A. LIGHT MICROSCOPY

B. ELECTRON MICROSCOPY

C. PHASE CONTRAST MICROSCOPY

D. MEASUREMENTS OF CELLS

During the past 15 years there has been an astonishing increase in the number of papers appearing in the literature concerned with cell structure and cell function. This increment seems to be directly correlated with new and improved instrumentation and techniques. Among these can be mentioned the electron microscope and ultramicrotome, scintillation monitors, ultracentrifuge, amino acid analyzer, x-ray diffraction units, double-beam spectrophotometers, and chromatographic and electrophoretic apparatuses; of course there are others. The impetus in research has been aided by grants and contracts from private industries and governmental agencies. As more information about structure and function appears, cell research becomes more detailed—and specialization results. When this happens, structure and function tend to become separated. Upon listening to experts present papers at scientific meetings, one is hard-pressed to realize that the speaker is describing a structure or function which relates to a cell. Be that as it may, cell structure and cell function are related. The scope of this book about cell biology is to place the relation of cell structure to cell function in its proper perspective by showing where and how this relation exists.

The study of cell biology has been inseparable from the evolution of the microscope. In the latter part of the 15th century, da Vinci stressed the importance of using lenses to view small objects. During the 17th century, as the development of light microscopy progressed, so did new discoveries concerned with the biology of cells. These new discoveries were made mostly by van Leeuwenhoek, a name familiar to nearly every student who has taken a course in general biology. He made his own microscopes and ground his own lenses. In fact, his estate disposed of 247 microscopes. Included in the 450 letters which van Leeuwenhoek published were descriptions of Protozoa, bacteria and spermatozoa, red blood cells and nuclei of red cells in fish. There were also cellular observations and histological descriptions of muscle, nerve, skin structure, and teeth as well as ant and aphid development and metamorphosis of a flea.

Hooke used a microscope similar to van Leeuwenhoek's to study cells from a leaf surface and the cell walls (and cavities) of cork cells for the first time as the structural unit in an organism. Although Adams in 1780 was able to cut thin sections of tissue cells about $1/2000$ inch thick, the 18th century saw very little contribution to either the development of the light microscope or new discoveries about the cell. Instead, most 18th century investigators placed their research emphasis on taxonomy and natural history of both plants and animals.

The 19th century was an era in which the cell was investigated extensively with the microscope. It must be kept in mind, however, that many different investigators were using microscopes in a study of the biology of cells. Sometimes one or more cell structures would be observed and incorrectly identified by today's terminology. It also happened that later another worker would describe the structure again, but in this instance name it correctly, and describe its function. Then the problem of priority arose—to whom should the credit be given for original description or elucidation. The cell theory furnishes an example of such a difference between science history and the commonly stated view.

In 1824 Dutrochet stated that "all organic tissues are actually globular cells of exceeding smallness, which appear to be united only by simple adhesive forces; thus all tissues, all animal and plant organs, are actually only a cellular tissue variously modified." This seems to be the first clear statement that all living organisms are composed of cells. It was not until 1838 that Schleiden, who is often credited with this discovery, emphasized that "cells are organisms and entire animals and plants are aggregations of these organisms arranged according to definite laws." A year later Schwann stated "We have seen that all organisms are composed of essentially like parts, namely, of cells." And Virchow in 1858 reported "Where a cell exists there must have been a pre-existing cell, just as the animal arises only from an animal and the plant only from a plant." Thus Dutrochet and Virchow made the two statements which mark the emergence of the cell theory in its definite form. The cell theory ranks with the theory of evolution proposed by Wallace and Darwin (1858) and supported with firm evidence by Darwin in 1859, and the genetic theory discovered by Mendel in 1866 as the three great foundations of modern biology. It is of importance that Dutrochet recognized that organismal growth resulted from the addition of new cells as well as an increase in the volume of cells. It is also noteworthy that Virchow reported that disease symptoms in an organism reflect an impairment at the cell level which was the beginning of the science of pathology.

Microscopic details of the cell became more apparent as the microscope was improved in design. Acromatic lenses appeared about 1830 and in the 1870's immersion lenses were available for higher magnification. In 1878 Abbé constructed the first modern microscope with the subsequent design of a substage condenser as the last important improvement in the ordinary laboratory microscope.

Improvements in light microscopy engendered and enlivened interest in the 19th century so that cell structure was studied with vigor by investigators in both animals and plants. As a result nearly all of the cell structures which can be identified today with a light microscope were described in the 19th century. Shown in Table 1-1 are the names, dates, and contributions which the 19th century workers made to cell biology.

The Nobel laureates introduced the 20th century. The names of those prize winners and their contributions are listed in Table 1-2. Through their efforts in chemistry, physiology and medicine, and physics our knowledge of the cell has continued to expand.

A. Light Microscopy

Microscopes are important in the study of cells because they possess the characteristics of magnification and resolving power. Important in magnification is the ability of the microscope to resolve, or distinguish, between two objects lying close together. The lesser the distance that can be discerned (measured) between two points in a microscope field, the greater its resolving power. The capacity of the microscope both

TABLE 1-1
19th Century Contributors to Cell Biology

Date	Name	Contribution
1817	Pelletier and Caventou	Named chlorophyll
1824	Dutrochet	All plants and animals are composed of cells
1826	Turpin	Reported the occurrence of cell division
1827	Dutrochet	Coined the terms endosmose; exosmose.
1828	Brown	Described the characteristic dancing of cell particles which is now referred to as Brownial movement
1831	Brown	Also named nuclei from plant cells although van Leeuwenhoek observed nuclei in fish red blood cells (1674)
1832	Dutrochet	Showed that stomata communicate with intercellular space within a leaf although Malpighi (1675) described stomata
1838	Schleiden	Described nucleoli although first noted by Fontana (1781)
1838	Mulder	Named proteins
1845	von Siebold	Recognized Protozoa as unicellular animals (although van Leeuwenhoek described Protozoa and bacteria in 1676, Goldfuss named Protozoa in 1817)
1845	Kölliker	Reported that spermatozoa (described by van Leeuwenhoek, 1679) and ova are single cell products
1846	von Mohl	First used the term protoplasm, but described the cell area of of protoplasm which is now called cytoplasm
1855	Pringsheim	Described the entrance of a ciliated spermatozoid into a female (receptive) cell in a freshwater alga
1858	Virchow	Stated that all cells arise from pre-existing cells
1861	Schultze	Said that the cell is a living substance possessing a nucleus and cell membrane; he used the term protoplasm to refer to the living substance of a cell. In 1863 Schultze stated that protoplasm is the "physical basis of life"; this term was chosen by Huxley (1869) for the title of a book. Schultze (1864) described protoplasmic bridges
1863	Waldeyer	Reported the use of the now common hematoxylin to stain tissue cells in which he described chromosomes
1865	Sachs	Stated that chlorophyll is located in special bodies, that chlorophyll is formed only in light, and that sunlight determines the activity of the special bodies in absorbing carbon dioxide
1866	Mendel	Discovered the fundamental principles of genetics
1866	Haeckel	Named plastids
1867	L. St. George	Discovered what was later called the Golgi complex
1870	His	Developed the microtome for cutting serial sections of tissues for cell study. Preservation of tissues dates from Boyle (1663) who used alcohol as a preservative for specimens
1871	Miescher	Discovered nucleic acid (nuclein)
1879	Fol	Observed a spermatazoan penetration of an ovum

TABLE 1-1—*Continued*

Date	Name	Contribution
1879	Flemming	Introduced the term chromatin and described the splitting of chromosomes. In 1882 he described cell division in animal cells by the term mitosis, and named the aster (1892). He suggested a correlation between nucleic acid and chromatin
1882	Strasburger	Described cell division in plant cells, and introduced the modern usage of the terms cytoplasm and nucleoplasm.
1883	Schimper	Named chloroplasts, the special bodies of Sachs (1865), and the green granules of Comparetti (1791)
1883	Metchnikoff	Observed and named the process of phagocytosis (pinocytosis was named by Lewis, 1931).
1887	van Benedin	Demonstrated that sperm and egg cells are haploid in a horse roundworm, and that diploidy (2 n=4) is restored at fertilization
1888	Waldeyer	Introduced the term chromosome
1888	Boveri	Named the centrosome, and in 1892 he published the still current diagrams illustrating spermatogenesis and oogenesis
1893	Hertwig	Cytology as a science emerged from his work on cells and tissues. Owen (1844) had already introduced the term histology
1894	Strasburger	Demonstrated that chromosomal reduction occurred in flowering plants similar to the process in animal cells described by van Benedin (1887)
1894	Sachs	Reported that starch appeared in chloroplasts following immediate absorption of carbon dioxide, but that starch did not appear in the parts of leaves coated with wax which occluded the stomata
1898	Benda	Named the mitochondrion
1898	Golgi	Described the Golgi complex as an internal reticular apparatus
1897	Garnier	Named and described ergastoplasm

to magnify and resolve depends on the kind of illumination used. Because of the nature of light, microscopes using light as a source of illumination cannot resolve objects less than one-half the wave length of that light as Abbé proved in 1878. The average wave length of white light is about 0.55 μ. Therefore, since white light is ordinarily used in laboratory microscopes, light microscopes using white light cannot resolve objects less than about 0.2 μ. A lens must be achromatic and perfectly ground to attain such a high resolving power.

In a compound light microscope the lens for resolving between objects is called the objective lens, and it lies closest to the specimen being examined. The ocular lens is the uppermost lens and it can

TABLE 1-2
List of Nobel Laureates Whose Discoveries Have Provided Insight into Cell Biology

Date	Name	Discovery
1901	J. H. van't Hoff (N)*	Formulation of the laws of chemical dynamics and of osmotic pressure
1902	H. E. Fischer (G)	Synthesis of purines and sugars
1903	S. A. Arrhenius (S)	Theory of electrolytic dissociation
1906	C. Golgi (I)	Structure of the nervous system
	S. Ramon y Cajal (SP)	
1910	A. Kossel (G)	Chemistry of cell nucleus
1915	R. M. Willstätler (G)	Research on chlorophyll
1920	W. H. Nernst (G)	Research in thermochemistry
1922	A. V. Hill (GB)	Production of heat in muscles
	O. F. Meyerhof (G)	Fixed relation between oxygen consumption and metabolism of lactic acid in muscle
1923	F. Pregl (A)	Method of microanalysis of organic substances
1925	R. A. Zsigmondy (G)	Nature of colloid solutions
1926	T. Svedberg (S)	Ultracentrifuge
1928	A. O. R. Windaus (G)	Relation of sterols to vitamins.
1929	C. Eijkman (N)	Discovery of vitamin B_1
	F. A. Hopkins (B)	Discovery of the growth-stimulating vitamins
1930	H. Fischer (G)	Synthesis of hemin
1931	O. H. Warburg (G)	Action of the respiratory enzyme
1932	C. S. Sherrington (GB)	Functions of neurons
	E. D. Adrian (GB)	
1933	T. H. Morgan (USA)	Function of chromosomes in transmission of heredity
1934	H. C. Urey (USA)	Discovery of heavy hydrogen
1935	H. Spemann (G)	Organizer effect in embryonic development
1936	P. J. W. Debye (N)	Study of molecular structure and electric polarity in molecules
1936	H. H. Dale (GB)	Chemical transmission of nerve impulses
	O. Loewi (A)	
1937	W. N. Haworth (GB)	Constituency of carbohydrates and vitamin C
	P. Karrer (SW)	Constituency of carotenoids, flavins, vitamins A and B
1937	A. Szent-Györgyi von Nagyrapoct (H)	Biological condensation processes related especially to vitamin C and catalysis of fumaric acid
1938	R. Kuhn (G)	Research on carotenoids and vitamins
1943	G. de Hevesy (H)	Isotopes as tracer elements in chemical processes
1943	E. A. Doisy (USA)	Vitamin K
	H. C. P. Dam (D)	
1944	J. Erlanger (USA)	Functions of single nerve fibers
	H. J. Gasser (USA)	
1946	H. J. Muller (USA)	Production of mutations by x-ray irradiations
1946	J. B. Sumner (USA)	Crystallized the first enzymes
	J. H. Northrop (USA)	Preparation of enzymes and virus proteins in a pure form
	W. M. Stancey (USA)	

*(A) Austria; (AR) Argentina; (AU) Australia; (D) Denmark; (F) France; (GR) Germany; (GB) Great Britain; (H) Hungary; (I) Italy; (N) The Netherlands, (S) Sweden; (SP) Spain; (SW) Switzerland; (R) Russia; (USA) United States of America.

TABLE 1-2—*Continued*

Date	Name	Discovery
1947	C. F. Cori (USA) G. T. Cori (USA)	Catalytic conversion of glycogen
	B. A. Houssay (AR)	Relation of the anterior pituitary hormone to the metabolism of sugar
1948	A. W. K. Tiselius (S)	Electrophoresis and adsorption analysis
1950	E. C. Kendall (USA) P. S. Hench (USA) T. Reichstein (SW)	Structure and biological effects of the suprarenal cortex hormones
1952	A. J. P. Martin (GB) R. L. M. Synge (GB)	Partition chromatography
1953	F. Zernike (N)	Phase contrast microscopy
1953	H. Staudinger (G)	Chemistry of macromolecular substances
1953	H. A. Krebs (GB)	Citric acid cycle
	F. A. Lipman (USA)	Coenzyme A and its importance in intermediary metabolism
1954	L. C. Pauling (USA)	Nature of the chemical bond
1955	V. du Vigneaud (USA)	First synthesis of a polypeptide hormone
1955	A. H. T. Theorell (S)	Nature and action of oxidizing enzymes
1956	C. N. Hinshelwood (GB) N. N. Semenov (R)	Mechanisms of chemical reactions
1957	A. R. Todd (GB)	Nucleotides and nucleotidic coenzymes
1958	F. Sanger (GB)	Structure of insulin
1958	G. W. Beadle (USA) E. L. Tatum (USA)	One gene regulates one definite chemical process
	J. Lederberg (USA)	Genetic recombination and organization of the genetic apparatus of bacteria
1959	S. Ochoa (USA)	*In vitro* synthesis of polyribonucleotides
	A. Kornberg (USA)	*In vitro* synthesis of polydeoxyribonucleotides
1961	M. Calvin (USA)	Work on photosynthesis
1962	J. D. Watson (USA) F. H. C. Crick (GB) M. H. F. Wilkins (GB)	Structure of DNA
1962	J C. Kendrew (GB)	Atomic structure of myoglobin
	M. F. Perutz (GB)	Atomic structure of hemoglobin
1963	J. B. Eccles (AU) A. L. Hodgkin (GB) A. F. Huxley (GB)	How a nerve impulse is produced and how it is transmitted within the nerve fiber and from one cell to another
1964	K. Bloch (USA) T. Lynen (G)	Mechanism and regulation of cholestrol and fatty acid metabolism
1965	F. Jacob (F) J. Monod (F) A. Lwoff (F)	Discovery of a class of genes which regulate the activities of other genes
1967	G. Wald (USA)	Biochemistry of perception
	R. Granit (S)	Physiology of vision
	H. K. Hartline (USA)	Interrelationships between nerve cells
1968	M. W. Nirenberg (USA) H. G. Khorana (USA)	Genetic code
	R. H. Holley (USA)	Base sequence of tRNA
1969	M. Delbrueck (USA) A. D. Hershey (USA) S. E. Luria (USA)	Reproductive pattern of viruses

magnify only what has been resolved by the objective lens. Although the human eye has no power of magnification, it can resolve points at about 25–100 μ. For comparison, the light microscope has a resolving power about 500 times that of the human eye.

Since the optical limit of microscopes was attained about the beginning of the 20th century, very little new information about cell structure appeared until a microscope was invented with a resolving power greater than 0.2 μ.

B. Electron Microscopy

Upon investigating other sources of illumination, physicists found that a beam of high speed electrons properly controlled by electromagnetic lenses (Fig. 1-1) could increase the resolving power of a microscope because the wave length of the moving electron is so much less than the wave length of light. An electron gun shooting a beam of electrons through an aperature in the body tube (Fig. 1-2) of a microscope in a vacuum with a charge of 50,000 volts has been estimated to have a wave length approximating 0.54 Angstrom unit (A). For 100,000 volts the wave length estimate attainable is 0.37 A. Beam potentials of 50–100 kv are most commonly used with the electron microscope in studying details of the cell. Theoretically, the electron microscope can resolve points to 0.018 A which is less than the diameter of the smallest atom (H = 1.06 A). But there is difficulty in constructing electromagnetic lenses so that the actual resolving power of modern electron microscopes is about 5 A in metallurgy studies, but for cell study it is about 10–20 A. Even at 100 A the electron microscope has a resolving power about 10,000 times that of the human eye (Fig. 1-3).

The development of the electron microscope has been attributed to Knoll and Ruska in 1931 as a result of their lectures and demonstrations of the electron microscope at the Technische Hochschule in Berlin. In 1932 these two German workers published the first description of an electron microscope. However, Rüdenberg in 1931 filed an application in Germany for the first patent on an electron microscope. Freundlich in 1963 has discussed the question of priorities. Certain discoveries and inventions in physics occurred before the electron microscope was developed. In 1897 Braum invented the cathode ray tube, and Thomson in the same year discovered the electron and demonstrated its relation to the cathode ray tube. That material par-

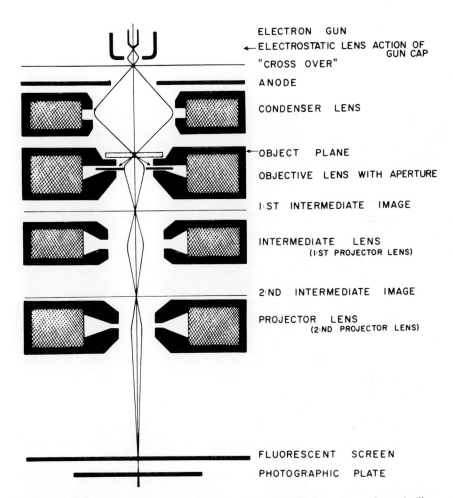

FIG. 1-1. The optical system of an electron microscope is shown in a schematic illustration. From Sjöstrand, F. S. 1967. *Electron Microscopy of Cells and Tissues.* Academic Press. New York. p. 69. Fig. III. 3.

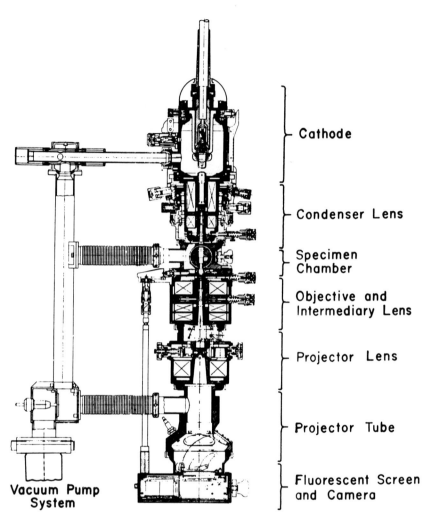

FIG. 1-2. Illustration of an electron microscope column shown in a cut-away section view. From Sjöstrand, F. S. 1967. *Electron Microscopy of Cells and Tissues*. Academic Press. New York. p. 68. Fig. III. 2.

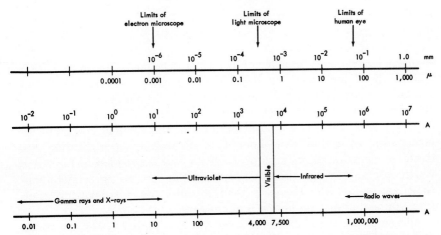

Fig. 1-3. The electromagnetic spectrum on a log scale, measured in millimeters (mm), microns (μ), and Angstrom units (A). 1 μ = 0.001 mm = 10,000 A. The approximate lower limits of resolution of the human eye, the light microscope, and the electron microscope are given. From Swanson, C. P. 1964. *The Cell*. 2nd edition. Prentice-Hall. Englewood Cliffs, N.J. p. 2. Fig. 1.1.

ticles as an electron beam have a wave motion somewhat similar to that of light was advanced by de Broglie in 1924. He derived the formula $\lambda = h/m\,v$ for the wave length of matter waves where h is Planck's constant, m is particle mass, and v is particle velocity. Busch in 1926 showed that an electron beam could be converged by a magnetic field acting as a lens, and the following year Gabor perfected a magnetic lens. In 1927, Davisson and Germer in the USA and Thompson and Reid in England verified experimentally de Broglie's hypothesis for electron beams. In 1934, Ruska showed pictures with magnifications to 12,000 diameters with a resolution greater than that of the 1934 light microscope by Driest and Müller. By 1938 von Borries and Ruska described an electron microscope capable of achieving a magnification of 20,000 diameters and a resolution of 100 A for general laboratory use. In 1939 Burton, Hillier, and Prebus in Canada described a practical laboratory magnetic electron microscope. Also in 1939 Hall built a similar practical instrument in the USA. Improvements continued, and by 1945 electron microscopes capable of achieving a resolving power of less than 20 A were developed commercially by Siemens Halske AG in Germany, and Radio Corporation of America in the USA.

As the electron microscope was improved, improvements were necessary in techniques preparing biological specimens for examination. The methods for preparing biological specimens for examination are similar in both light and electron microscopy as summarized in Table 1-3. The basic steps in each procedure require first that a tissue be obtained and fixed in a solution that will immediately coagulate the protoplasm with the least amount of distortion. After fixation, the tissue is dehydrated in organic solvents and embedded in a matrix which will support the tissue for cutting thin sections on a microtome. After sectioning, the tissue is then mounted on a supporting structure, stained, and examined in a microscope.

Although the methods for microscopy are similar there are differences in procedure, and some of these have been very difficult to over-

TABLE 1-3
Comparison of Basic Procedural Steps Common to Both Electron and Light Microscopy

Step	Electron Microscopy	Light Microscopy
Fixation	Osmium tetroxide; potassium permanganate; formalin; glutaraldehyde	Bouin's solution; formalin; Zenker's fluid
Dehydration	Increasing concentrations of ethanol (or acetone) followed by propylene oxide	Increasing concentrations of ethanol followed by benzene
Embedding	Araldite; Vestopal W; Epon 812; Maraglas; Durcopan	Paraffin
Sectioning	Usually 50–100 mμ thin sections cut with a glass or diamond knife on an ultramicrotome	Usually 6 μ sections cut with a razor blade on a microtome
Mounting	On a perforated metal disc (grid) usually covered with Formvar or Parlodian	On a glass slide with an egg albumin adhesive. Deparaffinized in xylol for staining
Staining	With salts of heavy metals such as lead acetate, lead citrate, lead hydroxide; uranyl acetate; phosphotungstic acid	Selective chromatic stains (as hematoxylin and eosin), dehydrated in ethanol series, cleared in xylol, and mounted for viewing in balsam or Permount
Viewing	Grid is placed between the condenser and objective lenses in a vacuum and the image is viewed on a phosphorescent screen	Slide is placed between the condenser and objective lenses, and the image viewed in the ocular lens

come. The tissue prepared for examination by electron microscopy must be better prepared than for light microscopy. If thinner sections are cut the electron microscope has a greater resolving power because there is less scattering of an electron beam. Chromatic stains cannot be used with an electron beam to differentiate structures in tissues since they lead to greater scattering. Specific developments have been made by various contributors to overcome these procedural differences between light and electron microscopy.

Although the fixation of some cells is better in one fixative or another, a general fixation solution for electron microscopy is osmium tetroxide. It was shown by Strangeways and Canti in 1927 that osmium tetroxide did not alter the structural organization of cells in tissue culture when they were observed by light microscopy. In 1939 Wolpers and Ruska used osmium tetroxide as a fixative and electron stain in electron microscopy. Palade in 1952 reported the use of buffered osmium tetroxide as a successful fixative for cells prepared for electron microscopy. Other chemicals which have been used in solutions for fixing cells are formalin (formalin is the correct term for formaldehyde in solution since formaldehyde is a gas), potassium permanganate, and glutaraldehyde. These chemicals may be used alone or in some other combination—as with osmium tetroxide. The cells may be treated with a fixation fluid, or the cells may be fixed by perfusion through the vascular system with a fixation solution.

After the cells are fixed in a suitable fixation solution and dehydrated, they are embedded in a supporting medium of sufficient hardness for cutting on a microtome. For observation in light microscopy, tissues are ordinarily embedded in paraffin or colloidin, but these substances do not provide the hardness that is necessary for thin sectioning with an ultramicrotome for electron microscopic studies. Consequently, it is necessary to embed tissues in plastics, but certain difficulties must be overcome. The plastic must infiltrate the tissue completely and polymerization must occur so that not only the proper hardness is attained for thin sectioning, but the cell must remain free from distortion of cell structure. The plastic also must be miscible with the dehydrating agent, and thermostable to the electron beam. The first embedding plastic used for biological electron microscopy was methacrylate. The technique was described by Newman *et al.* in 1949. Since that time there have been other plastics described and successfully employed as embedding media in preparing tissues for ultramicrotomy. Among those presently used are the epoxy resin Araldite, a polyester Vestopal W, another epoxy resin Epon 812, and the epoxy resin Maraglas as well as Durcopan.

Tissues which have been properly fixed and embedded must be cut in thin sections in order for the electron beam to penetrate the cells. Thicker sections cause a scattering of electrons and a loss in resolving power. Fortunately, Latta and Hartmann in 1950 found that plate glass could be fractured rather simply to make glass knives for adequate cuttings of thin sections of tissues. In 1953 Fernández-Morán developed the diamond knife which is particularly useful in cutting long serial sections from the same tissue block. And in the same year Porter and Blum introduced a practical microtome for ultrasectioning of tissues. Cutting tissue sections for light microscopy can be done with a steel knife, but usually razor blades are used. Sjöstrand in 1951 described a method for sharpening razor blades to cut sections of ultrathinness for electron microscopy.

Whereas chromatic stains can be used selectively to stain different structures in a cell, the morphological detail of a cell can be discerned and measured in electron micricsopy by shadow-casting as first shown in 1944 by Williams and Wyckoff. Details of structural profile are enhanced by electron stains. These stains make structures appear more or less electron dense than their surroundings as shown originally by Gibbons and Bradfield in 1956. The electron stains are solutions of heavy metals. The stains commonly used include lead acetate, lead citrate, lead hydroxide, uranyl acetate, and phosphotungstic acid. Osmium tetroxide and potassium permanganate also act as electron stains.

C. Phase Contrast Microscopy

Another type of microscope which is useful for the study of cell structure, particularly living cells in which contrast is poor, is the phase contrast microscope. The principle of phase contrast was discovered by Zernike of The Netherlands while he was working with diffraction gratings. In 1932 he applied phase contrast study to the microscope, and 10 years later a commercial model was available in Germany.

Phase contrast microscopy is possible because of the nature of light and wave motion. A light wave can be characterized by four properties which are amplitude, frequency, polarization, and phase. Amplitude is the maximum displacement of a wave from an equilibrium position. Differences in amplitude are distinguished by the eye as intensity, that is, as differences in lightness and darkness. Frequency is the number times a wave crest passes a particular point in 1 second. Wave length is related to frequency in that wave length is the distance between two wave crests. Differences in wave length register in the eye as differences in color.

Polarization refers to light that vibrates only in one direction at a right angle to the direction of wave propagation. Ordinarily, a light wave vibrates in all directions at right angles to the direction of wave propagation. The eye distinguishes polarized light from unpolarized light as a difference in intensity. When two waves are vibrating in phase with each other, the crest of one wave is coincident with the crest of the other wave. The eye does not distinguish between two light waves having the same amplitude and same frequency. Optical mechanics can be used to make one of two waves shift its frequency, and that shifted wave will arrive a point out of phase with the other wave which was not delayed, or shifted. In-phase waves, like polarized waves, make no direct impression on the eye. But out-of-phase waves are interpreted by the eye as variations in intensity. By taking advantage of the two properties of light—amplitude and frequency—impressive differences can be made directly on the eye.

In ordinary light microscopy, waves of background light illuminate the area seen in the field of view when no specimen is present on the microscope stage. When a specimen is placed in a light path optical phenomena occur such as diffraction, refraction, dispersion, and phase changes. According to the Huygens principle, every point on an advancing wave front may be regarded as a new source of further light waves. Thus, new light waves originate at the illuminated specimen. This light is called diffracted light. When two light waves such as from background and diffracted light meet out of phase, the interference pattern is one of increased amplitude which is greater than the amplitude of either of the two waves. The eye interprets this as an increase in intensity or brightness. The objective lens functions to bring the interference pattern from the specimen to a focus in the image plane, and the ocular lens magnifies the image formed.

The phase contrast microscope functions much like the ordinary light microscope except for the optical mechanics which make an increase in specimen contrast. An annulus is placed in the diaphragm between the source of illumination and the specimen on the stage.

CELL ANATOMY AND PHYSIOLOGY

The objective lens has a phase-altering plate built into it. When the specimen is illuminated, there is a separation of the background and diffracted light waves. A change occurs in the phase and amplitude of both the background and diffracted waves as they pass through the phase plate. The interference pattern thus reflects an increase in specimen contrast as a result of the increase in intensity of the interference pattern. If the background light is suppressed completely by making its amplitude zero, a specimen will be seen to appear lighter against a dark field. Comparisons are made in Figure 1-4 of the light, phase contrast, and electron microscopes.

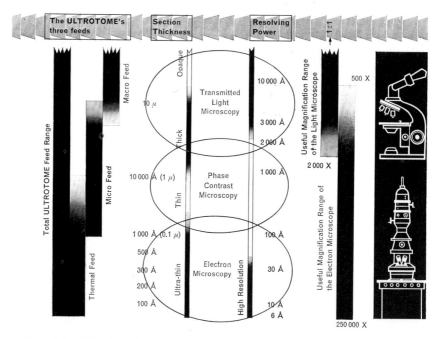

FIG. 1-4. Although strictly speaking there exists no exact correlation between the resolving power of a microscope and the thickness of the transparent section being observed, it is true to say, however, that the higher the resolving power desired, the thinner the section needed. The schematic picture illustrates the approximate relationship, and shows also the practical ranges of the light, phase, and electron microscopes. From LDB Instruments, Inc. 12221 Parklawn Drive, Rockville, Md. 20852. Brochure 4800 Eo 6, p. 9.

D. Measurements of Cells

Although cells are typically microscopic, they do vary in size. The size of a cell can be measured in a light microscope with an ocular micrometer disc and a stage micrometer. As a matter of convenience measurements of gross components in a cell are expressed ordinarily in microns. The ocular micrometer disc is usually a glass plate which is placed on the ocular diaphragm. The micrometer disc must be calibrated before measurements can be made. A calibration is made for each ocular lens, length of body tube, and objective to be used. The stage micrometer is usually a glass slide on which a scale of known intervals has been marked. When the ocular micrometer disc and the stage micrometer are graduated in the same units, the calibration of the ocular micrometer disc can be done simply. This is done by focusing on the stage micrometer, setting the zero line on it coincident with the zero line on the disc, and reading across the scale to the next two lines which are coincident with each other. The magnification factor is obtained by dividing the known value on the stage micrometer into the ocular value. Then the size of any object viewed in the light microscope will be the dimension shown on the ocular disc scale divided by the magnification factor.

Magnification is important in interpreting dimensions of cell structure in electron micrographs. All of the electron microscope photographs in this book are labeled with the degree of magnification. Therefore, it is important that determination of magnification be understood. Determining the magnification of an electron microscope can be done somewhat simply if a grating replica is used which contains a known number of lines per inch. An exposure is made on a photographic plate in the electron microscope of a field on the grating. The plate is developed and measured. A straight edge marker is placed perpendicularly across the grating lines to mark off and include in a certain length the maximum number of lines on the grating. The distance between the two outer limiting grating lines is measured precisely in millimeters. Then the total number of spaces is counted between the two limiting lines within the measured distance. The magnification can be calculated by

$$M = (XG/25400Y)(10^3)$$

where
M = magnification
X = the total distance between the limiting lines in millimeters

G = the number of lines per inch on the grating
Y = the number of spaces between the limiting lines.
It is not necessary to determine the magnification at each exposure of a specimen. A calibration curve can be made. A series of exposures are made over the magnification range of the electron microscope, and the plates measured. Calculations are made for each magnification at which exposures were made. A curve is plotted of magnification versus tap settings of the electron microscope, and the intervening settings calibrated.

Measurement of cell structure is so detailed with the electron microscope and the magnification is usually so great that the measurements are ordinarily reported in Angstrom units. Cell structures in an electron micrograph can be measured easily with a millimeter ruler when the magnification is stated by

$$\text{Size in A} = \text{size (mm)} \times 10^7/\text{magnification}.$$

Suggested Reading—Chapter 1

Barer, R. Phase contrast microscopy. Research *8:* 341, 1955.
Burton, E. F., Hillier, J., and Prebus, A. A report on the development of the electron supermicroscope at Toronto. Physical Rev. *56:* 1171, 1939.
Farber, E. *Nobel Prize Winners in Chemistry: 1901–1961*. Abelard-Schuman. New York. 1963.
Fernández-Morán, H. A diamond knife for ultrathin sectioning. Exp. Cell Res. *5:* 255, 1953.
Freeman, J. A. *Cellular Fine Structure*. McGraw-Hill Book Co. New York. 1964.
Freeman, J. A. and Spurlock, B. O. A new epoxy embedment for electron microscopy. J. Cell Biol. *13:* 437, 1962.
Friend, D. S. and Murray, M. J. Osmium impregnation of the Golgi apparatus. Amer. J. Anat. *117:* 135. 1965.
Frenster, J. H. Ultrastructural continuity between active and repressed chromatin. Nature (London) *205:* 1341, 1965.
Freundlich, M. M. Origin of the electron microscope. Science *142:* 185, 1963.
Gabriel, M. L. and Fogel, S. (editors). *Great Experiments in Biology*. Prentice-Hall. Englewood Cliffs, N. J. 1955.
Goldfischer, S., Essner, E., and Novikoff, A. B. The localization of phosphatase activities at the level of ultrastructure. J. Histochem. Cytochem. *12:* 72, 1964.
Golgi, C. Sur la structure des cellules nerveuses. Arch. Ital. Biol. *30:* 60, 1898.
Haguenau, F. The ergastoplasm; its history, ultrastructure and biochemistry. Int. Rev. Cytol. *7:* 425, 1958.
Hall, C. E. *Introduction to Electron Microscopy*. 2nd edition. McGraw-Hill Book Co. New York. 1966.
Hall, C. E. and Schoen, A. L. Application of the electron microscope to the study of photographic phenomena. J. Opt. Soc. Amer. *31:* 281, 1941.
Hooke, R. Of the schematisme or texture of cork, and of the cells and pores of some other such frothy bodies. Micrographia. London. 1665.

Latta, H. and Hartmann, J. Use of a glass edge in thin sectioning for electron microscopy. Proc. Soc. Exp. Biol. Med. *74:* 436, 1950.
Luft, J. H. Permanganate—a new fixative for electron microscopy. J. Biophys. Biochem. Cytol. *2:* 799, 1956.
Luft, J. H. Improvements in epoxy resin embedding methods. J. Biophys. Biochem. Cytol. *9:* 409, 1961.
Mellors, R. C. Microscopy. I. A review. Cancer Res. *13:* 101, 1953.
Newman, S., Borysko, E., and Swerdlow, M. New sectioning techniques for light and electron microscopy. Science. *110:* 66, 1949.
Nobel Lectures. *Physiology and Medicine.* Elsevier Publishing Co. New York. 1901–1921; 1922–1941; 1942–1962.
Palade, G. A study of fixation for electron microscopy. J. Exp. Med. *95:* 285, 1952.
Porter, K. and Blum, J. A study in microtomy for electron microscopy. Anat. Rec. *117:* 685, 1953.
Prebus, A. F. and Hillier, J. The construction of a magnetic microscope of high resolving power. Canad. J. Res. *17:* 49, 1939.
Preece, A. *A Manual for Histologic Technicians.* 2nd edition. Little, Brown and Co. Boston. 1965.
Rüdenberg, R. The early history of the electron microscope. J. Appl. Phys. *14:* 434, 1943.
Ruska, E. What is the theoretical resolution limit of the electron microscope and when will it be reached? *In Proceedings of the Fifth International Congress of Electron Microscopy,* Vol. 1, A-1, 1962.
Sabatini, D., Bensch, D., and Barnett, R. Cytochemistry and electron microscopy. The preservation of cellular ultrastructure and enzymatic activity by aldehyde fixation. J. Cell Biol. *17:* 19, 1963.
Selby, C. C. Microscopy. II. Electron microscopy: A review. Cancer Res. *13:* 753, 1953.
Simpson, G. G., Pittendrigh, C. S., and Tiffany, L. H. *Life—An Introduction to Biology.* Harcourt, Brace and Co. New York. 1957.
Singer, C. *A History of Biology.* Henry Schuman. New York. 1950.
Sjöstrand, F. S. A method for making ultrathin sections for electron microscopy at high resolution. Nature (London) *168:* 646, 1951.
Sjöstrand, F. S. *Electron Microscopy of Cells and Tissues.* Academic Press. New York. 1967.
Sjöstrand, F. S. and Andersson-Cedergren, E. The ultrastructure of the skeletal muscle myofilaments at various states of shortening. J. Ultrastruct. Res. *1:* 74, 1957.
Sourkes, T. L. *Nobel Prize Winners in Medicine and Physiology 1901–1965.* Abelard-Schuman. New York. 1966.
Swanson, C. P. *The Cell.* 2nd edition. Prentice-Hall. Englewood Cliffs, N. J. 1964.
Weiss, J. M. The ergastoplasm; its fine structure and relation to protein synthesis as studied with the electron microscope in the pancreas of the Swiss albino mouse. J. Exp. Med. *98:* 607, 1953.
Wolpers, C. and Ruska, H. Struktureuntersuchrungen zur Blutgerinnung. Klin. Wschr. *18:* 1077, 1111. 1939.
Wyckoff, R. W. G. The electron microscope in biology. Nature (London) *173:* 419, 1954.
Zernike, F. How I discovered phase contrast. Science *121:* 345, 1955.

Chapter Two

Organelles: Structure and Function

A. A REPRESENTATIVE CELL

B. NUCLEOPLASM
   1. NUCLEOLUS
   2. CHROMATIN AND CHROMO- SOMES
   3. NUCLEAR ENVELOPE

C. CYTOPLASM
   1. ENDOPLASMIC RETICULUM
   2. GOLGI COMPLEX
   3. RIBOSOMES
   4. LYSOSOMES
   5. MITOCHONDRIA
   6. CHLOROPLASTS
   7. CILIA AND FLAGELLA
   8. CENTRIOLES AND BASAL BODIES
   9. MICROTUBULES
   10. TONOFIBRILS
   11. CELL MEMBRANE
   12. MODIFICATIONS OF THE CELL SURFACE
   13. CELL WALL

A. A Representative Cell

An illustration of a representative cell is printed on the inside covers of this book showing nucleoplasmic inclusions and cytoplasmic organelles. These cell structures are represented in a relative size to each other, but some details are necessarily enlarged on the margin of the illustration for clarity. Only the best educated guess can determine the dimensions for a representative cell for both plants and animals.

Cell size varies. For example, an ostrich egg is the largest animal cell, *i.e.*, the yolk is the cell which does not include the white. The smallest cells are the pleuropneumonia-like organisms, abbreviated as PPLO. *Mycoplasma laidlaivii* has a diameter of about 0.1 μ (1000 A). Like bacteria, PPLO grow in non-living media, and like viruses, they pass through filters. Chemical analyses have shown that PPLO conduct the same biochemical processes as larger cells.

B. Nucleoplasm

1. Nucleolus

A cell which is not in the process of cell division is in an active metabolic state usually referred to as the interphase. Such a cell exhibits ordinarily one or more nucleoli and condensed portions of chromosomes called chromatin lying in the nucleoplasm. The nucleoplasm is surrounded by a nuclear envelope which consists of an inner and an outer membrane.

The nucleolus is a somewhat rounded body usually lying eccentrically in the nucleus. The nucleolus contains protein and is rich in ribonucleic acid (RNA), but it does not generally give a positive reaction with the Feulgen stain for deoxyribonucleic acid (DNA). The nucleolus can vary in size in different cells. During the prophase stage of cell division, the nucleolus begins to disappear about the same time as the nuclear envelope, and reforms in the telophase stage. It has been demonstrated that the nucleolus is associated with a specific chromosomal locus called the nucleolus organizer. This site may be one of synthesis or collection. It seems that the site contains cistrons for ribosomal RNA. If this organizer region is a length of DNA coding for ribosomal RNA and the cistrons are actually synthesizing corresponding RNA's, then nucleolar proteins may collect around the active site, although originating in some other parts of chromosomes. During the interphase stage the nucleolus probably remains attached to the organizer site, and in close contact with the DNA and histone of the organizer. This could give rise to the nucleolar associated chromatin. When a nucleolus is present in the prophase stage, the nucleolar organizer site is often marked by a secondary constriction in the following metaphase stage chromosomes.

Although in some cells the nucleolus may be diffuse or fragmented, most nucleoli of cells are compact and composed of two regions, often referred to as the pars amorpha and the nucleolonema as shown in Figure 2-1. The former appears to be a structureless component, but at high magnification it appears as a dense mass of filaments about 50 A thick. The latter is a fibrillar component which often appears as a coarse thread. This strand may form a torturous network around the pars amorpha in vertebrate cells. In plant cells, the nucleolonema can appear as a central core with the dense granules forming a peripheral localization. The fibrillar elements have a width of 50–80 A usually, but a dimension of 500 A has been reported. Often embedded in the fine matrix of the nucleolonema are ribosomal-like, electron dense granules with a diameter of about 150 A. The ribosomal granules are distributed rather irregularly throughout the nucleolus. Their staining reaction, as well as their size, has been shown to be identical with the cytoplasmic ribosomes. The granules contain both RNA and protein,

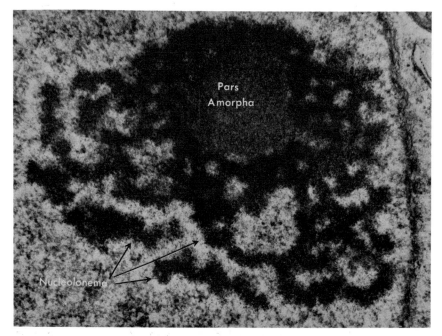

FIG. 2-1. Nucleoli from spermatogonia of opossum testis. Collidine buffered gluteraldehyde and osmium fixation. Lead citrate staining. Magnification 27,000 ×. From Fawcett, D. W. 1966. *The Cell*. W. B. Saunders Co. Philadelphia. p. 28. Fig. 12.

and are often called ribonucleic protein (RNP) granules. There is no limiting membrane surrounding the nucleolus, but calcium ions have been reported as necessary to keep it intact.

Surrounding the nucleolus are small bodies of associated chromatin varying in size and shape. Cytochemical techniques show that the chromatin of these peripheral nucleoli satellites penetrate the nucleolus as irregular strips or regular lamellae. As a result, this interchromatin can indicate a positive Feulgen reaction. Where the nucleolus lies close to the inner nuclear membrane, but not in direct contact, it is separated by a region of nucleohistones which may be interrupted by canals leading to the pores in the inner membrane.

Vacuoles are frequently found in nucleoli. When they are present, they do not appear to have a limiting membrane. The vacuoles may or may not have visible formed elements, but proteins have been identified in them with varying concentrations.

During ordinary cell division, the nucleolonema fragments and becomes less of an entity as it disperses so that it is difficult to follow. But some evidence shows that the nucleolonema is self-reproducing. In amitosis the nucleolonema reproduces by budding or producing spherules which pass to the resulting cells. This reproducing condition has also been reported for cells where karyokinesis was not accompanied by cytokinesis.

2. Chromatin and Chromosomes

A stained interphase nucleus usually exhibits chromatin as a densely stained region and a region which is more diffuse and unstained. These regions represent chromosomes. The chromatin fibers can be extremely long. Fibers as long as 2.2 cm have been reported from lymphocyte nuclei. Studies on isolated chromatin have identified the presence of DNA, RNA, histones, and non-histone proteins.

In some cells a condition of the chromosomes exists which is called heteropycnosis. This term is used to describe those chromosomes, such as sex chromosomes, which remain condensed in the interphase nucleus. These condensed interphase chromosomes have been called heterochromosomes; the non-condensed ones, euchromosomes. Following this terminology, a chromosome may contain both heteropycnotic and non-heteropycnotic regions. Extending this terminology in common-day usage, these regions of the chromosomes are referred to as heterochromatin and euchromatin. Thus, heterochromatin refers to the deeply stained and condensed regions of chromosomes and euchromatin is the light and diffuse regions of chromosomes in the interphase nucleus. The morphological distinction disappears during cell division when all of the chromosomes become condensed. Photomicrographs from electron microscope studies have confirmed that the condensed chromatin fiber is continuous with the diffuse chromatin fiber. Electron studies have also shown that heteropycnotic chromosomes are thick fibers of chromatin which appear identical to non-heteropycnotic chromosomes; they differ in tight folding. Heterochromatin is associated with tight folding and coiling of the fiber.

The thick region of the native chromatin fiber has a diameter of about 200–300 A. The thin region of the fiber is about 35–60 A in diameter. The diameter of a DNA molecule is about 20 A. A model of the chromatin fiber seems to be that there is a single DNA molecule wrapped in a sheath of protein forming a diffuse and thin part of the chromatin fiber. But as more protein is added to the sheath to form the thicker part of the fiber, coiling of the long DNA molecule is induced and heterochromatin is formed. It has been shown that histones stabilize coiling of long DNA molecules.

Heterochromatin seems to be relatively inert metabolically, while euchromatin appears to be the more active part of the chromatin fiber. With electron microscope autoradiography, it has been demonstrated that tritium-labeled thymidine, a DNA precursor, was incorporated almost totally in diffuse chromatin. Euchromatin has also been shown to be the primary site for incorporating the RNA precursor, tritium-labeled uridine.

3. Nuclear envelope

The nuclear envelope separates the nucleoplasm from the cytoplasm. Although it is not resolved by the light microscope, it is seen as a refractile surface in the living cell. In a stained cell, the envelope is visible because stainable substances adhere to both the nuclear and cytoplasmic surfaces. The electron microscope reveals the envelope to be composed of two membranes, an inner membrane and an outer membrane. Electron micrographs indicate that each membrane is about 75 A thick. The two membranes are separated from each other by a space called the perinuclear cisterna which varies in width from about 400 to 700 A. The outer membrane may be continuous at various points with the membrane system of tubules and cisternae lying in the cytoplasm known as the endoplasmic reticulum. There is some evidence that the endoplasmic reticulum forms vesicles in some cells around the chromosomes during telophase of cell division. Apparently these vesicles coalesce and reconstitute the nuclear envelope (Fig. 2-2).

Around the periphery of the envelope, the inner and outer membranes fuse with each other with no regular pattern. At each of these points where the membranes coalesce, pores or annuli (Fig. 2-3) are formed. It is not clear whether or not the pores permit free diffusion. Electron micrographs at high magnification (Fig. 2-4) indicate that a septum extends across the pore, but in Figure 2-3 there is indicated an area of rarefaction on either side of the envelope at the pore site that may be an area of diffusion substance. Also RNP granules (Fig. 2-5) as

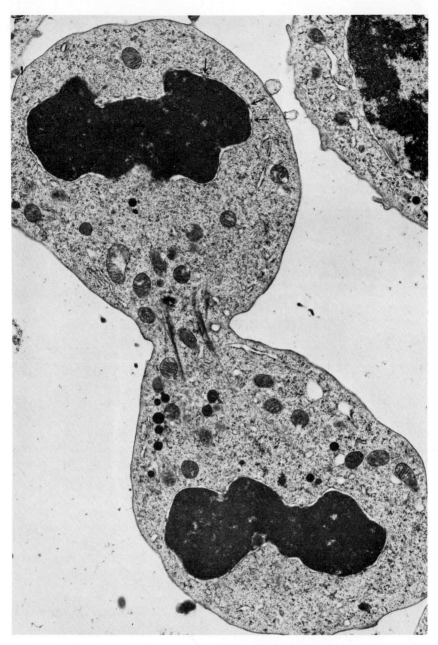

Fig. 2-2. In the telophase of cell division, the formation of the nuclear envelope around the coalescent mass of chromosomes is essentially complete. A number of pores in the nuclear envelope are already evident (at the arrows) around the daughter nucleus above. Formation of the nucleolus has begun in the nucleus below. Dividing heterophil myelocyte from guinea pig bone marrow. Cacodylate buffered gluteraldehyde and osmium fixation. Uranyl acetate and lead citrate staining. Magnification 14,000 ×. From Fawcett, D. W. 1966. *The Cell*. W. B. Saunders Co. Philadelphia. pp. 46 and 47. Fig. 24.

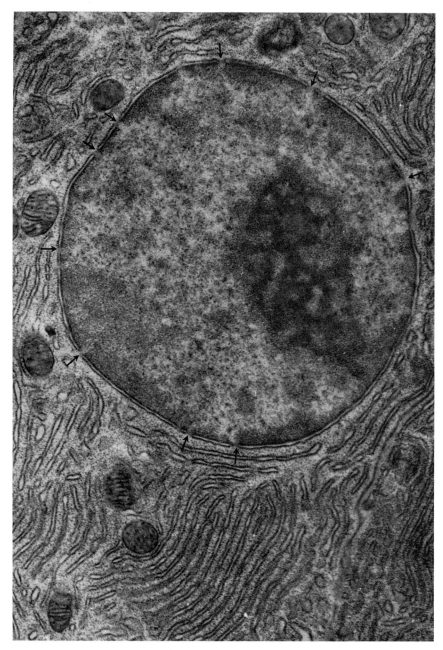

FIG. 2-3. Here the two membranes of the nuclear envelope are clearly visible. The outermost of these is studded with ribosomes just as are the membranes of the endoplasmic reticulum in the surrounding cytoplasm. At several points indicated by the arrows around the circumference of the nucleus, the perinuclear cisterna is traversed by nuclear pores. The sites of pores are betrayed by small areas of lower density that interrupt the otherwise continuous peripheral accumulation of chromatin. The invariable occurrence of such areas of rarefaction in the nucleoplasm opposite the pores has been interpreted as indirect evidence for the passage of material through them. Pancreatic acinar cell from the bat. Collidine buffered osmium fixation. Lead hydroxide staining. Magnification 17,000 ×. From Fawcett, D. W. 1966. *The Cell.* W. B. Saunders Co. Philadelphia. pp. 36 and 37. Fig. 17.

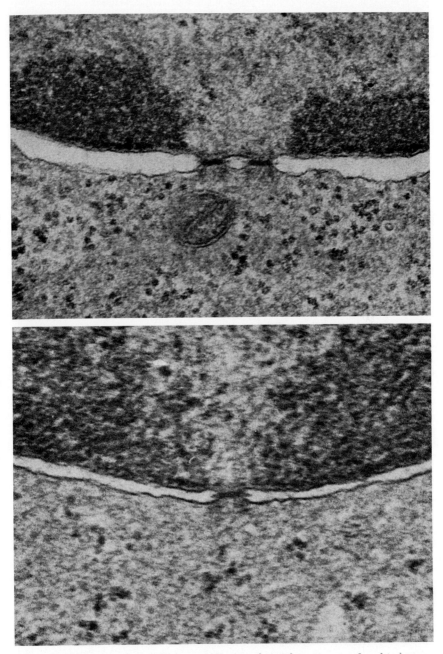

FIG. 2-4. When examined at high magnification, the nuclear pores are found to have a fairly complex structure. Instead of open bridges or channels between the nucleoplasm and cytoplasm, they appear to be closed by a septum that extends across the middle of the pore. In addition there is a thin flange that encircles the pore at the level of the diaphragm, projecting from its limiting membrane into the perinuclear cisterna. This can be seen on the two pores of the upper figure but is more prominent on the single pore in the lower figure. Polychromatophilic erythroblasts from guinea pig bone marrow. Magnifications 55,000 × and 75,000 ×. From Fawcett, D. W. 1966. *The Cell.* W. B. Saunders Co. Philadelphia. pp. 38 and 39. Figs. 18 and 19.

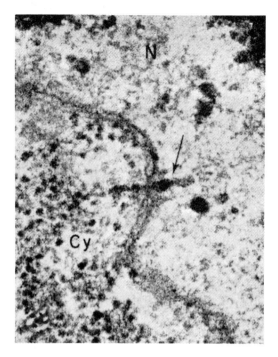

Fig. 2-5. An RNP granule is seen passing through a pore in a nuclear envelope from the nucleus (N) to the cytoplasm (Cy) of a *Chironomus* salivary cell. Osmium fixation. Uranyl and lead stain. Magnification 100,000 ×. From Stevens, B. J. and H. Swift. 1966. RNA transport from nucleus to cytoplasm in Chironomus salivary glands. J. Cell Biol. *31:* 72. Fig. 22.

---

well as gold particles (Fig. 2-6) seem to pass through the pores from the nucleoplasm to the cytoplasm. The first electron micrograph to show the nuclear envelope as two membranes, the outer one with pores, appeared in 1950 by Callan and Tomlin.

A nuclear envelope may form attachment points with either nuclear or cytoplasmic organelles. Long strands extend from the envelope to mitochondria, yolk granules, and vesicles. In Figure 2-7 chromatin fibers are shown attached to the envelope. There also appears in some cells a filamentous layer about 300 A of uniform thickness lying next to the inner membrane which seems to be of support nature (Fig. 2-8). This reinforcement layer has been termed fibrous lamina.

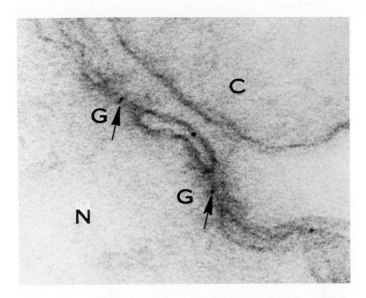

Fig. 2-6. A portion of an ameba fixed 1 minute after injection. In this section, perpendicular to the nuclear envelope, two annuli can be seen, both containing centrally located gold particles (G). C, ground cytoplasm: N, nucleus. Magnification 210,000 ×. From Feldherr, C. M. 1962. The nuclear annuli as pathways for nucleocytoplasmic exchanges. J. Cell Biol. *14:* 71. Fig. 4.

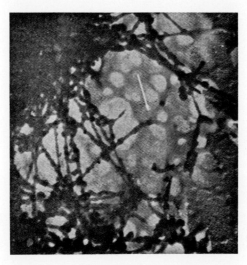

Fig. 2-7. The arrow points to a chromatin fiber attached to a pore in the nuclear envelope. Magnification 47,000 ×. From DuPraw, E. J. 1965. The organization of nuclei and chromosomes in honeybee embryonic cells. Proc. Nat. Acad. Sci. U. S. A. *53:* 164. Fig. 6.

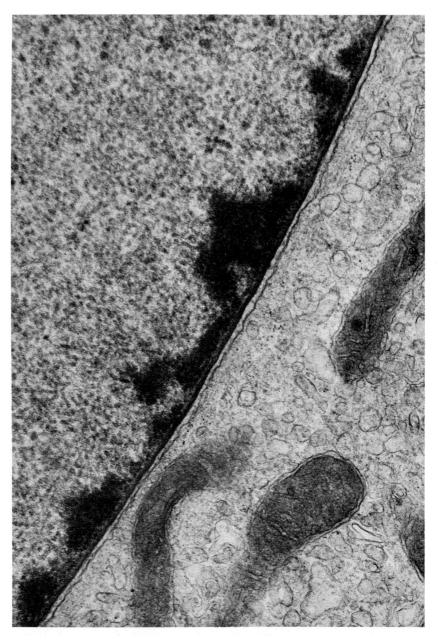

Fig. 2-8. The inner membrane of the nuclear envelope of some cell types is reinforced by a moderately dense, 300 A thick, filamentous layer called the fibrous lamina. The coarse granules and dense nucleolar strands in the nucleoplasm are held away from the inner membrane of the nuclear envelope by the dense fibrous lamina. Intestinal epithelium of *Amphiuma tridactylum*. Collidine buffered gluteraldehyde and osmium fixation. Uranyl acetate and lead citrate staining. Magnification 30,000 ×. From Fawcett, D. W. 1966. *The Cell*. W. B. Saunders Co. Philadelphia. pp. 40 and 41. Fig. 20.

C. Cytoplasm

1. Endoplasmic Reticulum

The outer nuclear membrane is continuous with a system of membranes which ramify throughout the cytoplasm in both animal and plant cells. Although the manner of origin of this membrane system is not known precisely, some evidence indicates that the membranes are budded off from the nuclear envelope. In both animal and plant cells, light microscope studies have demonstrated that a portion of the cytoplasm stains with basic dyes. As a result this cytoplasmic portion of basicity has been termed basophilic bodies, chromophilic substance, Nissl bodies, and ergastoplasm. In 1953 studies by Porter with the electron microscope revealed that the basophilic component of the cytoplasm was a channel system made of membranes. This membranous organelle has been named the endoplasmic reticulum.

The intercommunicating membranous system exists in the form of tubules. These tubules may vary from 400 to 700 A in thickness. A membrane is about 60 A thick. Each membrane appears in electron micrographs as two osmiophilic layers separated by an osmiophobic layer which is characteristic of cytoplasmic membranes. This trilaminar appearance has been referred to as a unit membrane by Robertson. The membranes can appear as flat sac-like expansions of the tubules in which case they are referred to as cisternae. Apart from the interconnecting channel system of tubules and cisternae, there are vesicles isolated in the cytoplasm. These vesicles are considered to be a part of the membranous organelle, and vary from one cell type to another in abundance and size.

As the resolution of the electron microscope was improved and ultrathin tissue sections became possible, a small particular component of the cytoplasm was identified by Palade in 1953. The particle was found to be rich in RNA, and, consequently, it was termed the ribosome. Ribosomes may be found free in the cytoplasm, or studded along the outside margin of the outer nuclear membrane as well as along the outer margin of the endoplasmic reticulum. It is the ribosomes that take basophilic stains for identifying aggregates of the endoplasmic reticulum in light microscope studies which are termed ergastoplasm. When ribosomes appear associated with membranes of the endoplasmic reticulum it is referred to as rough or granular endoplasmic reticulum. Smooth or agranular endoplasmic reticulum does not support ribosomes. Those areas of cytoplasm containing smooth endoplasmic reticulum usually stain with acid dyes. Therefore, these areas are not a part of the ergastoplasm identified by light microscopists on the basis of basic dyes.

Both the granular and agranular endoplasmic reticulum, however, appear to be continuous with each other. But they seem to differ in several ways. For example, only the granular endoplasmic membranes confluence with the outer nuclear membrane, thereby providing communication between perinuclear cisterna and the space in the tubules. It is conceivable that substances in the nucleoplasm could move across the inner nuclear membrane and into the channels of the endoplasmic membrane, and that the nuclear pores could provide a pathway to the continuous phase of the cytoplasmic ground matrix. It is the agranular form that is closely associated with the Golgi complex, and seldomly does it form cisternae as will the granular endoplasmic reticulum. Although Chapter 4 deals more specifically with membranes and membrane function, it may be mentioned that the agranular endoplasmic reticulum has been found to be associated primarily with non-protein synthesis whereas the granular form is primarily associated with protein synthesis. In the liver the agranular type has been related to lipid and cholesterol metabolism, lipid transport into intestinal epithelial cells, and is also related to the synthesis of steroid hormones in the adrenal cortex, the corpus luteum of the ovary, and the interstitial cells of the testis. In striated muscle cells there is an abundance of agranular endoplasmic reticulum called the sarcoplasmic reticulum. The smooth-surfaced tubular system forms an orderly system that is related to the cross striations of the myofibrils in the striated muscle cell. The relation of smooth sarcoplasmic reticulum to impulse transmission in striated muscle cells is discussed in Chapter 4.

Granular endoplasmic reticulum is highly developed in secretory cells. Gland cells elaborate a secretory substance especially rich in proteins. Membrane function indicates that the endoplasmic reticular channels function as a depository and transport region for nascent proteins. The secretory product of protein synthesis may be transported via vesicles to the Golgi complex where it sometimes appears in the form of droplets or as granules. These forms of the secretory product may be stored in the cytoplasm and ultimately passed out of the cell.

2. Golgi complex

The Golgi complex has been a controversial issue in cell structure since 1898 when it was described as an internal reticular apparatus.

Since that time it has been demonstrated in nearly all cells either by osmium or silver impregnation. The point of controversy was whether or not the exhibited network was an artifact resulting from prolonged fixation. Against this view were two facts. One was that a short fixation time demonstrated the organelle in some cells. The other was that it seemed to be related to secretion since highly active cells showed variations in size of the Golgi complex. It has been described in the literature by over 100 names. Among these names are apparatus, body, complex, dictyosome, golgiosome, material, membranes, and vesicles. Of the names, the ones most used at the present time are apparatus and complex; these are used interchangeably. Complex is the name preferred here only in that the functions in the organelle have not been clearly defined. But it is known that the Golgi complex is related to secretion processes.

The Golgi complex is a disc-shaped organelle comprised of unit membranes which typically lie closer together than those of the endoplasmic reticulum. The membranes lie near the nuclear envelope and close to the centrioles forming an aggregation for easy identification although distinctive in various cell types. In 1952 electron micrographs identified the Golgi complex as an entity. A year later the organelle was described as being comprised of three kinds of membrane-bounded parts: flattened, membrane-enclosed cisternae, dense vesicles, and large vacuoles.

The cisternae of the Golgi complex are flat or curved and give the appearance of lamellae stacked on top of each other as shown in Figure 2-9. Intercisternae spacing is about 200–300 A and the intracisterna space is about 150 A. The cisternae are of variable length, and apparently anastomose with each other. A cisternal membrane appears as a unit membrane about 60 A thick. Vesicles of about 400–800 A associate around the ends of cisternae or along their convex outer surface. Distended vacuoles associate with the inner concave cisternae. In fact it has been reported that the Golgi packet exhibits a structural polarity so that two faces are formed: an outer-forming face where new lamellar cisternae of flat, empty sacs are added, and an inner-maturing face in which the distended sacs are lost. The filled vacuoles seem to break away from the terminal expansions of the inner cisternae as the volume and density of their contents increase as shown in Figure 2-10. The Golgi membranes are smooth, and their association with the agranular endoplasmic reticulum may indicate that it differentiates to form the membrane-bound vesicles on the forming face of the cisternae. A mem-

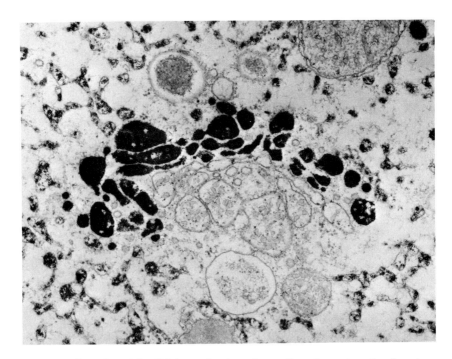

Fig. 2-9. A portion of the Golgi complex from the rat liver. Dense deposits of osmium appear in the outermost cisterna and vesicles extend beyond their membrane limited compartments. Magnification 54,000 ×. Unpublished. Courtesy of Dr. D. S. Friend, University of California, San Francisco Medical Center.

---

brane-flow hypothesis has been proposed since there seemingly is a continuity of the membranes forming the nuclear outer membrane, the endoplasmic reticulum, and the Golgi complex. This flow would account for rapid transport through the endoplasmic reticulum and the Golgi complex spaces. But other evidence indicates that the cavities of the Golgi complex are not continuous with the lumen of the endoplasmic reticulum. Here transport of newly formed protein in the endoplasmic reticulum is effected by vesicles budded off from the cisternae margins, and these smooth surfaced vesicles coalesce with the sac-like membranes of the Golgi complex; the secretory product is then concentrated in the vacuoles of the Golgi complex.

3. Ribosomes

Ribosomes must be ubiquitous in all animal and plant cells since no alternate mechanism for protein synthesis has been established. Ribo-

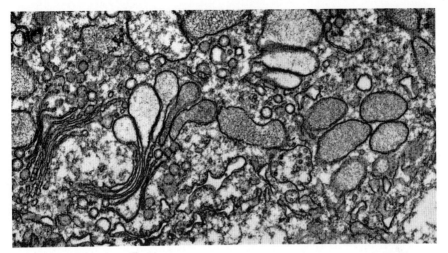

FIG. 2-10. Section of an outer root-cap cell of the primary root of *Zea mays* showing the Golgi complex producing secretory vacuoles. Osmium fixation. Magnification 29,000 ×. From Mollenhauer, H. H. and W. G. Whaley. 1963. An observation on the functioning of the Golgi apparatus. J. Cell Biol. *17:* 223. Fig. 1.

---

somes have been investigated in the bacteria perhaps more than any other group. Here mostly ribosomes have been found to have a sedimentation coefficient (= Svedberg unit) of the 70 S type. It is comprised of two subunits, one 30 S and the other a 50 S. In protein synthesis mRNA and tRNA bind to the 30 S subunit. The growing polypeptide chain with the attached tRNA is bound to the 50 S component. The 50 S has a particle weight of 1.8×10^6; the 30 S has a particle weight of 0.85×10^6. Both the 50 S and the 30 S subunit contain 63–64% RNA and 36–37% protein. Some polyamines are strongly bound.

The 50 S and 30 S subunits have been analyzed further. The former contains a single molecule of 23 S rRNA plus another species of rRNA with a sedimentation coefficient of about 5 S; the latter 30 S contains a single molecule of 16 S rRNA. All three ribosomal RNA's—23 S, 16 S, and 5 S—are distinctly different.

It seems that DNA has segments complementary to rRNA in base sequence, and that rRNA may be synthesized on the DNA template perhaps in the same way as mRNA. Apparently DNA has cistrons for rRNA. For example, each cistron for 23 S and 16 S does not compete in hybrid formation of a DNA-RNA hybrid duplex incorporating rRNA. The molecular weights of 23 S and 16 S rRNA's have been estimated at 7.3×10^5, and 4.5×10^5, respectively.

The appearance of ribosomes are associated with cellular activity in various animal cells. Electron micrographs show that lymphocytes have a great number of free ribosomes scattered at random and a sparse endoplasmic reticulum which is mostly agranular. Granulocytes are sparse in ribosomes. The plasma cell has an extensive rough endoplasmic reticulum which is also characteristic of hepatic parenchyma cells and pancreatic exocrine cells. The Nissl bodies of neurons are constituted by a rough endoplasmic reticulum. Other cells engaged in protein synthesis have a rough endoplasmic reticulum such as serous cells of the submaxillary gland, osteoblasts, chief cells of the glandular stomach, thyroid cells, and mammary gland cells. Cells which synthesize specific protein and retain them tend to have an abundance of free ribosomes. Among these types of retaining cells are erythroblasts, developing muscle cells, skin, and hair. Mature mammalian erythrocytes do not have ribosomes. The general type of rRNA found in animal cells has a sedimentation coefficient of 80 S; its subunits are of the type 60 S and 40 S.

The principal site of nuclear protein synthesis is the nucleolus. One class of proteins made here appears to be histones, but a great bulk of protein synthesized in the nucleolus seems to be ribosomes. The nucleolus synthesizes the RNA of ribosomes, but not the protein component which is generated in the extranucleolar chromatin.

Ultracentrifugation is ordinarily used in isolation of ribosomes. But estimating purity of ribosomes, *i.e.*, free of binding contaminants such as proteins, can be difficult. One criterion that is useful in purity analysis is moving boundary electrophoresis. A single sharp boundary indicates that proteins have been removed from ribosomes. Chemical analysis of their RNA content is another means of determining purity of ribosomes. Although electron micrographs can identify ribosomes so clearly, they cannot be used to differentiate between ribosomes which bind extraneous proteins and those ribosomes which are highly purified. The shape of its ultraviolet absorption curve is also a criterion used for the purity of a ribosome preparation. Maximum absorption of RNA is about 260 mμ, but proteins absorb more strongly at shorter wavelengths. If a ratio of absorbances at 260 mμ and 235 mμ is used, the higher the ratio will be for the lesser protein content. A ratio of about 1.4 indicates a preparation with about 20% protein in it. Although the ratio of absorbances at 260 mμ and 280 mμ is sometimes used, it should be recognized that absorption at 280 mμ measures chiefly aromatic amino acids and low absorption does not necessarily mean absence of extraneous protein.

Ribosomes of plants tend to resemble those in animal cells, rather than those in the bacteria, in relation to size and composition. They contain about 40% RNA, and the sedimentation coefficient for intact particles has been shown for types 80 S, 60 S, and 40 S.

Polyribosomes have been identified in cell-free systems of animals, higher plants, and bacteria. Evidence continues to accumulate that polyribosomes are related to protein synthesis. They can occur either free or loosely bound to the membranes forming the endoplasmic reticulum, or firmly bound even after repeated homogenization. Electron micrographs have confirmed that the ribosomes of the ergastoplasm are attached to the endoplasmic reticular membranes via a short stalk or strand. Isolated polyribosomes consist of a linear array of ribosomes interconnected by a strand of variable length about 10–20 A thick which has been identified as mRNA, and thought to be a single molecule.

In a polyribosome where protein synthesis occurs, it seems that each ribosome attaches itself to the mRNA strand at the same place on the strand. Each ribosome then proceeds sequentially along the mRNA strand reading the message and synthesizing the polypeptide chain. As the chain is completed the ribosome becomes disengaged from the strand. As protein synthesis decreases, the proportion of polyribosomes to single ribosomes decreases. It has been shown that there is one nascent polypeptide chain per ribosome, and that chain length can be influenced by the position of the ribosome on the mRNA strand.

4. Lysosomes

The existence of a cytoplasmic organelle was reported by deDuve in 1955 as a result of biochemical studies made on centrifuged fractions of cells from rat liver. The biochemical evidence indicated that the particle was filled with acid hydrolases surrounded by a single unit membrane. The presence of a membrane was postulated because the saclike structure demonstrated latency. When the homogenates were disrupted by thawing, freezing, or use of detergents or hypotonic media, there was an increase in the activity of the hydrolytic enzymes. The name given to this membrane-bound, sac-like structure containing acid hydrolases was lysosome, meaning lytic (digestive) body. It seemed that lysosomes functioned in the digestive systems of the cell whereas ribosomes serve as centers of protein synthesis, and mitochondria conduct primary energy transfers. The following year morphological confirmation from electron microscopy was given to the biochemical identification of the lysosome as a sac-like, membrane-bound entity.

Apparently all of the enzymes described in association with lysosomes are not found in a single lysosome. And it does not appear that only a single enzyme associates with a single lysosome. But evidence does indicate that various hydrolytic enzymes associate with different lysosomes in a variety of cells. The hydrolytic enzymes have an acid optimum of about pH 5. It should be kept in mind, however, that all of the hydrolytic enzymes in a cell do not appear in lysosomes, and that other specific enzymes—other than acid hydrolases—have been identified in lysosomal particles.

Although a variety of animal cells have been shown to possess lysosomes, little is known about their occurrence in plant cells. But particles have been identified in plant cells which fit the description used for lysosomes in animal cells. Mostly, identification of lysosomes in plant cells has been made by electron microscope studies employing cytohistochemistry.

Lysosomes may be identified histochemically by the Gomori reaction for acid phosphorus, the acid hematein test, the periodic acid-Schiff (PAS) reaction, and with vital dyes. But fixation methods and tissue preparations can cause an inhibition of enzyme activity and make identification extremely difficult. Histochemical staining for acid phosphatase (which led to the discovery of lysosomes) reveals cytoplasmic bodies of different appearances in electron microscope studies. Acid phosphatase is one of the acid hydrolases which has been identified histochemically in lysosomes shown in electron micrographs. Others are β-glucuronidase, arylsulphatases, and the organic phosphorus-resisting esterases. The acid hydrolases which have been identified otherwise from lysosomal fractions are

β-galactosidase
β-glucuronidase
β-N-acetylglucosaminidase
α-glucosidase
α-mannosidase
cathespin A (acid protease)
cathepsin B (acid protease)
arylsulfatase A
arylsulfatase B
acid ribonuclease
acid deoxyribonuclease
acid phosphatase
acid lipase

phospholipase A
phosphotidic acid phosphatase
hyaluronidase
phosphoprotein phosphatase
aminopeptidase A
dextranase
saccharase
lysozyme (muramidase)
Mg^{++}-activated ATPase
indoxylacetate esterase
plasminogen activator

A remarkable property of lysosomes is that they are able to keep their hydrolytic enzymes inside the single membrane, but can be ruptured (Fig. 2-11). Isolated lysosomes behave much like red blood corpuscles in suspension in their susceptibility to membrane-active

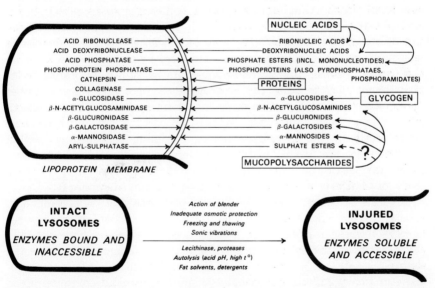

FIG. 2-11. Diagram showing the biochemical concept of the lysosome. This model applies mainly to lysosomes of rat liver. Above, different hydrolytic enzymes and substrates on which they act. Below, indicating the intact lysosomes and the effect of different agents that disrupt the membrane of the lysosomes. From DeRobertis, E. D. P., W. W. Nowinski, and F. A. Saez. 1965. *Cell Biology*. 4th edition. W. B. Saunders Co. Philadelphia. p. 372. Fig. 20-11.

agents. Cholesterol, cortisone, cortisol, and chloroquine are agents which stabilize lysosome membranes. Among the many agents which labilize lysosome membranes are progesterone, endotoxins, testosterone, vitamins A and E, proteases, digitonin, x-irradiation, ultraviolet irradiation, and bile salts.

Based on both morphological and functional criteria a variety of different forms can be distinguished which are collectively referred to as lysosomes that form an intracellular digestive system. The substances undergoing digestion in this system may originate exogenously by heterophagy or endogenously by autophagy. This whole system has been called the vacuolar apparatus. Although the origin of lysosomes is poorly understood, some evidence has accumulated for a developmental concept. First, there seems to be a prelysosome whose enzymes have never engaged in hydrolytic events. Secondly, there are lysosomes which are sites of present digestive activity, or where digestion has occurred. Thirdly, postlysosomes are those which have lost their enzymes. Some cytochemical evidence indicates that prelysosomes may form via the Golgi complex which sometimes stains positively for acid phosphatase. Assuming that the hydrolytic enzymes form at the site of attached ribosomes, they could transfer across the endoplasmic reticulum into the channel system. The acid hydrolases could form an endocytotic vesicle, the prelysosome. But, as shown in Figure 2-12, phagosomes may indicate the best evidence for a prelysosome. Other views offered are that prelysosomes may form from endoplasmic reticulum, or arise from autophagic vacuoles, or perhaps the acid hydrolases occur freely in the cytoplasm before becoming concentrated within lysosomes. In any event, whether by heterophagy or autophagy, lysosomes can discharge their hydrolytic products by exocytosis, or keep them until finally the cell becomes pathologically deranged.

In terms of current concepts one can postulate that, in heterophagy, substances are engulfed by a cell through the action of endocytosis thereby forming a phagosome. After lysosomal enzymes have digested the substances, the entity forms a residual body containing undigested debris. The debris, generally lipid, is ejected from the cell by exocytosis. Autophagy follows functionally a similar degradation route. Even intracellular substances (such as mitochondria) may be hydrolyzed.

Not only is lysosomal activity being studied in relation to normal physiology, but also in relation to pathology. Studies linking lysosomes to pathology include fever, congestive heart failure, hepatitis, pylonephritis, hypertension, mitosis, joint injury, and leukocyte granules and tissue injury.

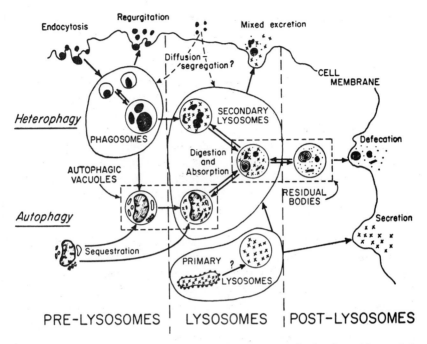

FIG. 2-12. Illustration of the various forms of lysosomes and related particles, and the different types of interactions which they may exhibit with each other and with the cell membrane. Acid hydrolases are represented by crosses. From de Duve, C. and R. Wattiaux. 1966. Functions of lysosomes. Ann. Rev. Physiol. 26: 468. Fig. 6.

Lysosomes also have physiological roles. The concentration of lysosomal enzymes increase in the metamorphosis of the amphibian tail. Isolated acrosomes from spermatozoa disperse the cumulus oophorus and corona radiata of newly ovulated rabbit eggs. Such acrosomal preparations contain hyaluronidase, protease, acid phosphatase, β-glucuronidase, arylsulphatase, β-glucosaminidase, and phospholipase activity. Vital staining reactions of acrosomes are also identical with those of lysosomes. Not only is the evidence strong that the acrosome facilitates entry of the sperm into the egg, but also that the acrosome functions as a specialized lysosome.

5. Mitochondria

Not only was the year 1898 notable for the description of the Golgi complex, but for the mitochondrion (= thread + grain) by Benda.

Many cytologists had observed and reported them as granular and thread-like components of the cytoplasm. In fact, mitochondria had been teased from insect muscle 10 years earlier by Kölliker. He noted then that the granules possessed a membrane, and that they swelled in water. Systematic observations on their occurrence was enhanced by Altmann's study in 1890 when he noted that the granules were somewhat specific for fuchsin staining. And Benda, in his description, had shown mitochondrial staining action with alizarin and crystal violet and by 1900 Michaelis demonstrated the supravital selective staining of mitochondria by Janus green. This development was important in that it demonstrated that oxidation-reduction changes in living cells could be brought about by mitochondrial action. Later, in 1913, Warburg showed that oxygen uptake by liver extracts was associated with mitochondria.

The use of the term mitochondrion was not accepted immediately by cytologists. They continued to use paranyms for mitochondria such as fuchsinophilic granules, parabasal bodies, plasmosomes, plastosomes, fila, vermicules, and chondriosomes. As with the Golgi complex there were dozens of terms put in the literature for mitochondria. Of these many terms, the chondriosome has persisted, but enthusiasm for it has apparently waned in present-day usage.

Besides the early studies concerned with staining and fixation methods for mitochondria, they were also studied in tissue culture. These studies in living cells demonstrated that mitochondria frequently changed their shape, and moved freely.

By 1948, differential centrifugation of cell homogenates allowed separation of mitochondria for chemical analyses. Some of the biochemical studies subsequently showed that mitochondrial fractions catalyze (1) all reactions of the Krebs citric acid cycle (= tricarboxylic acid cycle = TCA cycle), (2) fatty acid oxidation, and (3) coupled phosphorylation.

Mitochondria have been shown to be present in both higher plant and animal cells, but not in bacteria. Like chloroplasts, individual mitochondria move autonomously. Typically, mitochondria are in motion. In many cells their distribution is restricted. For example, in renal distal tubule cells mitochondria lie near and perpendicular to the cell membrane. In striated muscle cells they are found between muscle fibrils and frequently they are found closely aligned to lipid droplets in the cytoplasm in such cells as cardiac muscle and liver.

The number and size of mitochondria can vary in cells. Typically, a cell may have 200–300 mitochondria, but they may vary from a few to 1000 or more. Some algal cells may have only 1 mitochondrion. In filamentous mitochondria, the diameter size may be 0.2–1 μ and the length may vary from 2 to 10 μ. Spherical mitochondria can form from the filament type, and then reform the filamentous appearance. In observing mitochondria in living cells under phase contrast microscopy, one can easily confirm earlier findings that they change shape readily. In cells that are highly differentiated, mitochondria have unique appearances. For example, they appear ring-shaped in pancreatic parenchyma cells when surrounding lipid droplets. They appear round in brown adipose tissue. In the mammalian sperm tail, they are arranged in a helix.

Electron micrographs in 1952 by Palade revealed that a mitochondrion has an outer membrane and an inner membrane. These two membranes are separated by a space containing somewhat of a fluid matrix which is characteristic of the rest of the mitochondrion as shown in Figure 2-13. Each membrane possesses a good tensile strength, stability, and flexibility. The inner membrane forms many enfoldings which Palade named cristae mitochondriales. In animals the cristae usually form parallel arrays, but not in plant cells. Here the cristae frequently form closed loops by connecting with one another. High resolution electron microscopy has shown that each membrane appears as a trilaminar unit membrane with its characteristic outer and inner osmiophilic layers (20–25A) separated by an osmiopholic layer (25 A). This sandwich structure for membranes was postulated in 1935. The space separating the outer and inner membranes is about 80–100 A wide.

Matrix granules about 300–500 A in diameter may be found in the intercristal space. These dense granules have been identified as insoluble inorganic salts.

The inner surface of the inner membrane appears covered by thousands of spherical structures in mitochondria prepared by negative staining, e.g., with phosphotungstic acid. On cristae they are spaced at about 100 A intervals; each sphere, about 75–100 A in diameter, is attached to the membrane by a stalk some 50 A in length. The present concept is a lollipop model. At one time it appeared as if these elementary particles contained all of the enzymes for electron transport and oxidative phosphorylation, and would be appropriately termed as electron transport particles (ETP). But later studies have shown that removal of these knobs does not prevent electron transport. The ETP view has now been largely abandoned in favor of the oxysome hypothe-

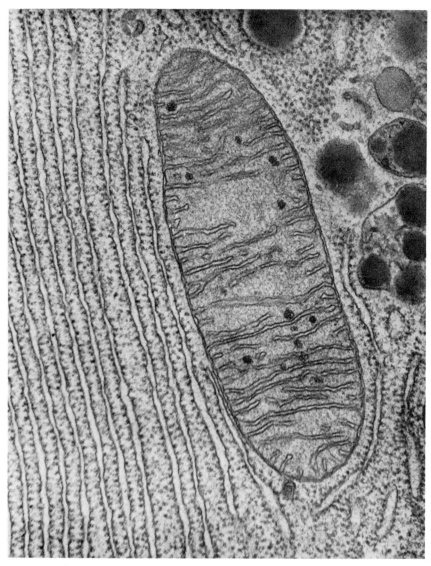

FIG. 2-13. In thin sections of cells, mitochondria exhibit the highly characteristic structural organization illustrated in the accompanying electron micrograph. The mitochondrion extending diagonally across the field is bounded by a smooth-contoured outer membrane about 70 A thick. Within this limiting membrane and separated from it by a clear space about 80 A wide is a second membrane. The inner membrane has numerous infoldings that project into the cavity of the mitochondrion. These thin folds, the *cristae mitochondriales*, are of variable length and form a series of incomplete transverse septa. In most cell types there are dense mitochondrial granules 300–500 A in diameter scattered at random in the matrix between the cristae. (Micrograph of bat pancreas by Dr. K. R. Porter.) Magnification 64,000 ×. From Fawcett, D. W. 1966. *The Cell*. W. B. Saunders Co. Philadelphia. pp. 64 and 65. Fig. 35.

sis. This view has it that highly ordered replicating assemblies carry out oxidative phosphorylation and electron transport distributed in the cristae. These oxysomes would be analogous to the functional molecules, the quantosomes, constituting the lamellae of the chloroplast where photophosphorylation occurs. Fractionation studies indicate that fatty acid and TCA cycle enzymes are in the mitochondrial matrix between the cristae, but the enzymes for electron transport and coupled phosphorylation are firmly bound to the membranes—the cristae.

The gross chemical composition of mitochondria varies in different cells of both plants and animals. Typically, however, by dry weight mitochondria are about 75% protein, 25% lipid, and contain small amounts of DNA and RNA. Cholesterol may account for 5% or less of the lipid and the phospholipid as much as 90%. The remaining lipid consists of free fatty acids and triglycerides. A high phosphatide content is characteristic of membranes, and one would expect as much with mitochondria. The protein includes a structural protein.

Mitochondrial origin has been studied in vivo. Not only can they fragment, but they can fuse. In those algal cells containing a single mitochondrion, its duplication is synchronous with mitotic duplication. But in higher plant and animal cells mitochondria do not divide precisely, but are distributed at random to newly formed cells during mitosis. Cinemicrophotography with the phase contrast microscope gives direct evidence in time lapse studies of an entire life cycle of cell that the randomly distributed mitochondria gradually lengthen, and then fragment or divide to form new mitochondria in the newly formed cells. It has been suggested, based on electron micrograph studies, that mitochondrial formation is an important activity of the interphase nucleus. In this case the inner nuclear membrane invaginates at specialized regions to form intranuclear mitochondria which are later extruded into the cytoplasm.

It was shown in 1959 and confirmed by 1963 in electron micrographs that mitochondria contain DNA, and by 1965 that there was a general occurrence of mitochondrial DNA in representatives of both invertebrates and vertebrates as well as in some plants. It is now established that mitochondria have DNA, and it is present as a closed ring. The double stranded circular DNA of mammalian mitochondria has a molecular weight of 1.1×10^7. And there is evidence that mitochondria contain ribosomes, tRNA, and activating enzymes. The general conclusion is that mitochondria do possess some capacity for some protein synthesis in intact cells. But to code for any specific mitochondrial proteins, it has been estimated that mitochondrial DNA would have to be five times its molecular weight.

When mitochondria are isolated from a cellular environment and disrupted, some of the enzymes associated with matrix are released as soluble proteins while other enzymes remain firmly bound to the membranes. There have been about 70 enzymes at present identified with mitochondria and about a dozen coenzymes. Soluble enzymes from mitochondria include all enzymes of the TCA cycle except some dehydrogenases, those that catalyze β-oxidation of fatty acids, and others that catalyze transamination of amino acids and synthesis of protein. Membrane bound enzymes of mitochondria are those for electron transport chain and oxidative phosphorylation. Mitochondria do not contain the enzymes of anaerobic glycolysis which occur in solution in extra-mitochondrial cytoplasm.

In the mitochondria are located the processes for controlled oxidation of respiratory substrates. It is here that energy, mostly from the combustion of carbohydrates and fatty acids, is conserved as bond energy in adenosine triphosphate (ATP) which is formed from ADP and inorganic phosphate concomitant with oxidation processes. This energy is then released in forms usable by those functions in the cell which require energy at specific sites such as cell membranes, ribosomes, and contracting myofibrils.

Mitochondria are essential to a cell in that they are the organelles in which the TCA cycle functions along with the flavoprotein-cytochrome system often called the respiratory chain. This is a chain of enzymes which transfer hydrogen (removed from various stages in the TCA cycle) to oxygen, forming water. The chain of electron carriers consists of flavoproteins, ubiquinone, and cytochromes b, c_1, c, a, and a_3. The TCA cycle functions as a device for producing carbon dioxide and hydrogen by oxidation reactions. In the general scheme, acetyl coenzyme A is produced primarily by pyruvic acid from carbohydrate precursors, and by fatty acids and certain amino acids as a result of lipid and protein degradation. The 2-carbon acetyl CoA formed condenses with the 4-carbon oxaloacetic acid to form the 6-carbon citric acid which enters the TCA cycle to initiate the cycle.

As the cycle completes a revolution, two molecules of CO_2 are evolved, and four dehydrogenation reactions occur. Those hydrogens reduce coenzymes. Coenzymes, in contradistinction to enzymes, are organic, non-protein substances which may catalyze a variety of reactions, and act as hydrogen and electron carriers for reaction products.

Simple ions may also act as carriers. In a redox reaction, one substance becomes oxidized as the other reaction is reduced. Oxidation may be described as a substance combining with oxygen, the loss of hydrogen, or the loss of electrons, but reduction is the reversed process. Many coenzymes can be reduced, *i.e.*, accept and carry hydrogen. For example, when a substrate (R) is oxidized, a coenzyme (CoE) is reduced and acts as a hydrogen carrier according to

$$R{-}OH + CoE \rightarrow R{=}O + CoEH$$

When the four dehydrogenation reactions mentioned above occur in the TCA cycle, a coenzyme called nicotinamide adenine dinucleotide = NAD = (DPN = diphosphopyridine nucleotide = coenzyme I) accepts the hydrogen thereby becoming reduced as NADH, and the hydrogen is carried by the flavoproteins to ubiquinone (coenzyme Q) in the electron transport chain. Succinic acid dehydrogenase also feeds hydrogen into this flavoprotein-cytochrome system, presumably at the site of ubiquinone which is abundant in mitochondria. Although NADH occurs both in cytoplasm and mitochondria, only the NADH reduced in the mitochondria is oxidized by the respiratory chain since NADH does not pass through the mitochondrial membrane. The coenzymes, then, could be viewed as connecting the TCA cycle to the electron transport chain.

During the transfer of protons or electrons along the oxidative pathways in the flavoprotein-cytochrome system to oxygen to ultimately form water, ATP is formed. Its formulation occurs at specific points coupled to oxidation, and the process is often called oxidative phosphorylation as ADP is phosphorylated to ATP.

6. Chloroplast

In higher plants the chloroplast is a well-defined plastid containing chlorophyll. In light microscopy (LM) studies, the chloroplast appears as a saucer-shaped body about 1–2 μ thick and about 4–10 μ in diameter. The chlorophyll appears mostly in bodies about 0.4 μ in diameter called grana. Each granum contains so much chlorophyll it is totally absorbing and appears black to the eye. About 50–60 grana are present in the matrix of the chloroplast. Two unit membranes, each about 50 A thick, surround the chloroplast.

Each granum is composed of some 5–25 discs which appear in electron microscopy (EM) as a stack of coins. Each of the discs appears as a membranous compartment, enclosing a space, termed a thylakoid. The membranes forming the disc separate its contents from the stroma of chloroplast. The stack of discs appears as dark (osmiophilic) membranes about 65 A thick. These alternate with light (osmiophobic) layers of about the same thickness as shown in Figure 2-14. In some cases it appears as if the grana are interconnected by intragranal membranes running through the stroma giving the appearance of stromal lamellae, but current evidence of swelling in hypotonic solution indicates the granum is a structural unit (Fig. 2-15). The best evidence now shows that chlorophyll is evenly distributed in both grana and stromal lamellae.

It is in the chloroplast where photosynthesis occurs. As Engelmann demonstrated in 1884, light absorption and oxygen evolution occurred within the chloroplast. By using *Spirogyra* with its large chloroplast and *Pseudomonas*, the oxyphilic bacteria, he ingeniously proved that upon illumination, the bacteria moved to the site of oxygen formation

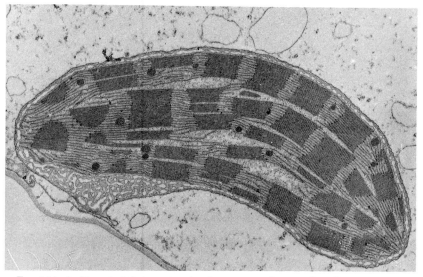

FIG. 2-14. A chloroplast from maize showing grana and stroma lamellae. Magnification 17,000 ×. From Shumway, L. K. and T. E. Weirer. 1967. The chloroplast structure of iojap maize. Amer. J. Bot. 54: 774. Fig. 1.

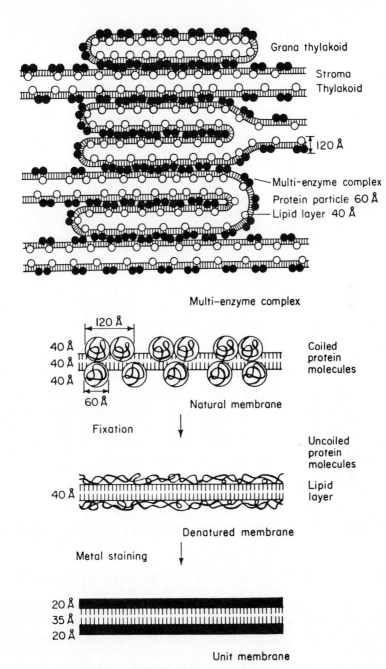

FIG. 2-15. Model of chloroplast granum (upper) and membrane (lower). From Muhlethaler, K. 1966. The ultrastructure of the plastid lamellae. In *Biochemistry of Chloroplasts*. Vol. 1. T. W. Goodwin (editor). Academic Press, New York. pp. 61 and 62. Figs. 15 and 16.

in the chloroplast. By working with isolated chloroplasts to study photosynthesis, Hill in 1937 found that he could produce oxygen photochemically if a suitable electron acceptor was present. The reaction became known as the Hill reaction

$$4Fe^{+++} + 2H_2O \xrightarrow[\text{chloroplasts}]{\text{radiant energy (hv)}} 4Fe^{++} + 4H^+ + O_2$$

and later it was shown by Reuben and Kamen, with ^{18}O experiments, that the oxygen evolved in photosynthesis is derived from water, not carbon dioxide, as in

$$CO_2 + 2H_2O^{18} \xrightarrow{hv} C(H_2O) + H_2O + O_2^{18}$$

By 1954 Arnon and his coworkers, with ^{14}C experiments, reported that enzymes for carbon dioxide fixation reside in the chloroplasts. This is the basis for the current belief that both the light and dark reactions of photosynthesis occur within the grana of the chloroplast.

During the assimilation of carbon dioxide in green plants, light quanta (radiant energy) is trapped by chlorophyll and converted to chemical energy. In the light reaction, the chlorophyll-containing lamellae absorbs light quanta. When a photon of light moves an electron of an atom into a higher orbit, or outer orbit, the atom is said to be excited and unstable. As the electron returns to its lower energy level by stages, its return is accompanied by release of some of the energy acquired in excitation. After absorbing light, electrons in the chlorophyll molecule reach an excited state, and this energy of excitation is utilized to make ATP through an electron transport chain. All subsequent reactions of carbon dioxide fixation and reduction for carbohydrate synthesis are dark reactions, and are catalyzed by enzymes in the stroma. Photosynthesis may be considered as occurring in three steps as:

1. photophosphorylation in which ATP is formed;
2. photolysis of water;
3. carbon dioxide fixation and reduction to carbohydrate.

Finding that dark and light reactions in photosynthesis could be separated in time led to the establishment of the chlorophyll unit, later known as the photosynthetic unit. This concept was based on experimental evidence and resulting calculation of the number of chlorophyll molecules required to reduce one molecule of carbon dioxide, or yield one oxygen molecule. This number is usually given as 200–300 chlorophylls per photosynthetic unit. In terms of function, the photosynthetic unit can be considered a physiological unit. Searching for an expression, in terms of morphology for the physiological unit, chlorophyll lamellae were investigated. The search was for the smallest fragment of lamella that could perform the Hill reaction. EM studies on lamellae fragment by the shadow-cast technique revealed repeating structures on the inner surface of the lamellar membrane. It has been suggested that these structures are the morphological unit of the quantosomes which are about 160 A by 180 A and about 100 A thick. Their lamellar packing can be from the least common crystalline type to the most common linear type. They can also appear in a random array. From the chemical composition of a quantosome, it appears that one quantosome contains about the same number of chlorophyll molecules as contained in a photosynthetic unit.

There appear to be two general methods for chloroplast reproduction. Division of mature chloroplasts are most often reported for lower plants such as algae and ferns. In higher plants, chloroplasts arise from proplastids. At first these have no chlorophyll lamellae present. As chlorophyll synthesis occurs normally, the proplastid develops and undergoes division. The inner membrane usually invaginates to form the internal lamellae, but the inner membrane can form vesicles which then fuse to form lamellae.

Specific DNA and RNA have been found present in chloroplasts. Although the DNA content is only about 0.5%, its presence has been confirmed by the Feulgen reaction, cesium chloride density gradient studies, autoradiographic techniques which showed tritiated thymidine incorporation into chloroplasts, and EM micrographs. The content of RNA in chloroplasts is about 5%. Its presence is directly supported by EM micrographs of chloroplasts containing ribosomes. Chloroplast ribosomes are smaller than cytoplasmic ribosomes as confirmed by sedimentation and electron microscopy; also the base composition of the former has a higher $A + U/G + C$ ratio than the latter. The presence of rRNA implies the existence of mRNA and tRNA. Evidence exists that DNA codes for proplastid differentiation into the chloroplast from labeling studies. This evidence supports the hypothesis that DNA codes for mRNA which is translated to protein synthesis necessary for chloroplast development from the proplastid.

The presence of DNA in cytoplasmic organelles—mitochondria, chloroplasts, and centrioles and basal bodies—indicates that the inheritance of an organism may be constituted by cytoplasmic gene pools as well as a nuclear gene pool.

7. Cilia and Flagella

Cilia and flagella are extensions of the cell surface. They are so similar that no definite morphological or physiological characteristic can be used to separate one from the other. Often the two terms are used interchangeably. Generally, a description of a cilium implies one that is shorter than the size of the cell whereas a flagellum implies a longer dimension. Cilia tend to beat with a coordinated rhythm, but flagella usually beat independently. Cilia are more numerous on a cell surface than flagella. There is a great range in movement in both cilia and flagella. In cilia, there may be a stiff and active thrust with a limp recovery, or the ciliary action may be rotary. The beat frequency may vary from one cell type to another, and may be up to 1500 beats per minute. A flagellum usually exhibits an undulatory movement. In this case a wave of contraction passes from the base of the flagellum to the tip.

Cilia and flagella are characteristic of many different plants and animals. In plants, these range from gamete types in algae and aquatic fungi to mosses and ferns and some seed plants. In animals, both Protozoa and Metazoa contain ciliated, or flagellated cell types. In fact, two classes of Protozoa are characteristically named Ciliata and Flagellata. In the Metazoa, every phylum has representatives with one or more types of ciliated epithelium.

In both plant and animals, and in both cilia and flagella, there is a remarkable pattern of fibril arrangement. From light microscopy, various cytologists in the 19th century reported a single dark line running through the central axis of each cilium. This was called the axial filament. In the latter part of the century, some evidence indicated that fibrils could be seen when sperm flagella were subjected to pressure. In the early part of the 20th century, longitudinal fibrils were reported for both cilia and flagella with stained material. Then interest in cilia and flagella was later enhanced with the advent of the electron microscope.

Electron micrographs of plant and animal cilia and flagella demonstrate a remarkable constantancy of 11 longitudinal fibrils. These are arranged in the common "9 + 2" arrangement of microtubules. The axial filament of the light microscopists form the two central fibrils.

These are in turn surrounded by nine fibrils located peripherally. Each of these is composed of two subfibrils. One of these is designated as subfibril A and the other as subfibril B. There can be minute differences between the two, but generally the microtubular wall of B projects on that of A. Sometimes A may possess walls with arms about 50 by 150 A and a common wall of 50 A. High resolution electron microscopy shows that the walls of each fibril is in turn made-up of 10–12 parallel subfibrils with a center-to-center spacing of about 60 A. This arrangement has been reported for both cilia and flagella.

The diameter of the shaft of a cilium is about 0.2 μ and can extend to about 10 μ above the cell surface. Flagella varies in length from about 10 μ up to about 150 μ. The cell membrane, here about 90 A thick, extends over the cilium or flagellum. The nine doublet fibrils are spaced circumferentially toward the periphery about 1.5 μ in diameter. A peripheral doublet fibril is about 200 A wide and 350 A long. The two single central fibrils do not share a common wall, each is about 250 A in diameter, each wall is about 60 A thick, and the center-to-center spacing approaches 350 A. Figure 2-16 shows a diagram of a cilium and its related part, the basal body, embedded in the cytoplasm of the cell.

8. Centrioles and Basal Bodies

The precise origin of cilia and flagella has just begun to be formulated. It has been known that development of cilia and flagella is definitely associated with the basal body. This structure is found at the base of each cilium and flagellum. Other terms synonymously used for basal body in the literature have been basal corpuscle, basal granule, kinetosome in protozoan ciliates, and blepharoplast in protozoan flagellates.

A basal body is a hollow cylinder with the identical structure of a centriole. Here the internal structure varies somewhat from the typical "9 + 2" fibrillar arrangement of the fibrillar microtubules in cilia and flagella. In the basal body and centriole the two central fibrils are lacking, but the nine peripheral fibrils are present. But these nine fibrils appear as triplets, not as doublets as in cilia and flagella. The three tubules of each triplet appears fused together so that they appear as three circles in a row as shown in Figure 2-16F. The innermost tubule of the triplet is known as subfibril A, the middle-placed one is subfibril B, and the outermost subfibril has been designated C. Each triplet forms a circumferential pattern about 150 mμ

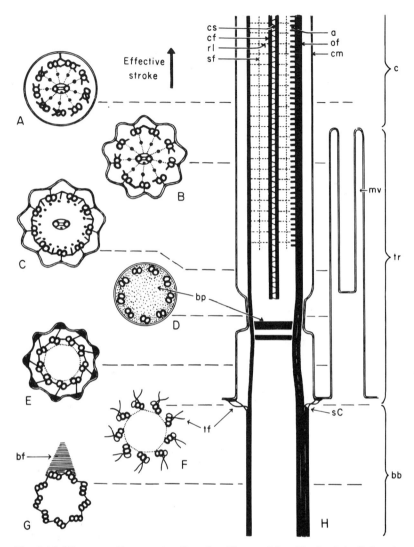

FIG. 2-16. Diagrammatic reconstruction of a cilium and basal body. A to G show transverse sections at the levels indicated; they are all oriented with the plane of beat up and down in the figure, with the effective stroke toward the top. H shows a median longitudinal section in the plane of the central axial fibrils; the outer fibril on the right is turned sideways to show it more clearly. a, arm; bb, basal body; bf, basal foot; bp, basal plate; c, cilium proper; cf, central axial fibril; cm, ciliary membrane; cs, central sheath; mv, microvillus of brush border; of, outer fibril; rl, radial link; sC, distal end of subfibril C; sf, secondary fibril; tf, transitional fibril; tr, transitional region of cilium. From Gibbons, I. R. 1961. The relationship between the fine structure and direction of beat in gill cilia of a lamellibranch mollusc. J. Biophys. Biochem. Cytol. *11:* 184. Fig. 4.

in diameter. The centriole varies from 300 to 500 mμ in length, closed at one end and open at the other.

The formation of centrioles, basal bodies, and cilia have been poorly understood until recently. Work on the fetal rat lung by Sorokin now allows a scheme for centriole formation and ciliogenesis. New centrioles designed to become a part of the achromatic figure form during interphase. These centrioles also form primary cilia which are transitional. The procentriole arises directly off the wall of the pre-existing centriole; they lengthen into cylinders and triplet fibers develop. At about half of its maximum length, the procentriole moves away to the cytoplasm to complete its maturation. The new centriole may become part of the achromatic figure, or give rise to transitional cilia as indicated in Figure 2-17.

Permanent ciliogenesis occurs somewhat differently. Precursor fibrogranular material lies in the cytoplasm close to the Golgi complex. Granules form, increase in size, and aggregate to form spheroid bodies which organize the development of procentrioles about them as shown in Figure 2-18. The newly formed, mature centrioles align themselves under the cell membrane in rows. From each of these basal bodies, a cilium is produced apically.

Typically in animal cells the new and smaller centriole of the achromatic figure appears oriented at right angles to the mature centriole which measures about 5000 A long and 1500 A in diameter. The newly formed centriole is structurally similar to the mature one, but it is shorter (Fig. 2-19). Both of these appear in a specialized zone of cytoplasm called the centrosome. The centrosome usually lies close to the nucleus. Typically, no centrioles appear in plant cells. DNA has been localized in both basal body and centriole.

9. Microtubules

Besides the fibrils of the cilia and flagella, the cytoplasm of cells contains other filamentous microtubules. During mitotic cell division, different staining reactions reveal the presence of these cytoplasmic fibrils. They can also be seen in living cells in phase contrast microscopy, or even studied with a micromanipulator. Electron micrographs have revealed the abundance of these fibrils in both plant and animal mitosing cells as well as their scarcity during interphase in most cell types. Microtubules are several microns in length. Their diameter varies in different cells from about 150 to 250 A with a wall thickness of about 60 A. Each wall is constituted by about a dozen subunits

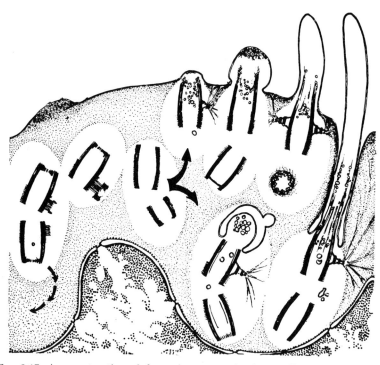

Fig. 2-17. A reconstruction of the major events of primary ciliogenesis presented sequentially from left to right. At the left two centrioles are sketched in a diplosomal configuration, much as they occur in an interphase cell. Near its basal end, each centriole begins to form a daughter centriole by extending fibrillar material out from its wall. Before mitosis occurs the centrioles separate (dashed arrows) and move to opposite poles of the cell. The subsequent fate of only one of these centrioles is depicted. Its daughter centriole, or procentriole, begins to take on form near the tips of the fibrils that extend for a distance of 50–70 mμ from the centriolar wall. Osmiophilic matter accumulates about the fibrils, and the procentriole first becomes recognizable as an annular structure with a diameter of about 100 mμ. Subsequently the procentriole lengthens into a cylinder whose long axis parallels the fibrils. At the same time it widens to approach the diameter of a mature centriole (150 mμ). When half-grown the procentriole loses its right-angle orientation to the centriolar wall and may slip down, to adopt a new position facing away from the base of the parent centriole. When fully grown the procentriole resembles its parent, and the two constitute the new diplosome. A primary cilium is generated from the distal end of the more mature centriole, which becomes modified into a basal body by producing satellite arms from its wall. These appear to aid in fixing the position of the organelle. If the diplosome normally resides in the apical cytoplasm, as in many epithelial cells (upper arrow), the basal body becomes closely attached to the apical plasmalemma, and the ciliary bud grows out into the epithelium-lined cavity. If the diplosome normally rests in the neighbourhood of the nucleus (lower arrow), then the ciliary shaft lengthens within the confines of a ciliary sheath. In most cases the sheath appears to originate from cytoplasmic vesicles, which eventually fuse with the apical plasmalemma. Ciliary fibres develop in a centrifugal direction with the lengthening shaft and may reach the tip. The second centriole of the pair meanwhile moves between the various preferred positions indicated. From Sorokin, S. P. 1968. Reconstructions of centriole formation and ciliogenesis in mammalian lungs. J. Cell Sci. *3:* 207. Fig. 1.

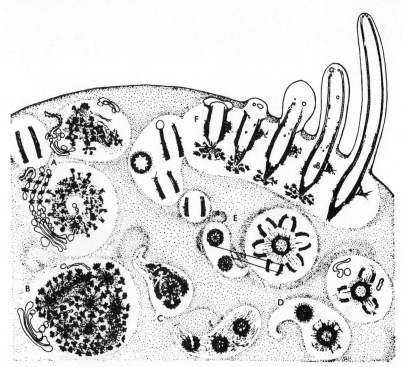

Fig. 2-18. A pictorial summary of developmental stages encountered in the epithelium of the bronchial tree during the formation of basal bodies and cilia. The stages have been arranged into a plausible sequence (A–F). The objects illustrated are all drawn roughly to scale. The onset of basal-body formation in immature, non-ciliated cells is marked by the appearance of fibrogranular aggregates in the apical cytoplasm near the site of the cell's previously existing pair of centrioles (A). At first the filamentous matrix appears to predominate, but later highly osmiophilic, granular elements come into prominence. In some sections these components seem thrown together in disarray (A, above); in others they appear to be fashioned into a cored structure (A, below). Annulate lamellae and Golgi elements frequently border on the aggregates. At a more advanced stage of development the aggregates become rearranged into a bundle (B) of strands composed of parallel coursing matrical filaments surrounded by the osmiophilic granules. In cross-sections 8–12 granules rim the individual strands. Tubular extensions from the Golgi apparatus freely ramify in the space between strands. Eventually the bundle becomes subdivided into smaller units, and individual strands are released to the cytoplasm. Through continued deposition of osmiophilic material at the periphery of the strands, the ring of granules becomes consolidated into fewer elements (C, above). These round up into radial structures whose axis of symmetry lies perpendicular to the long axis of the strand (C, below). The structures are released to the cytoplasm and develop into deuterosomes (D), which are represented in surface view and in section. Fine filaments extend radially outward to form a corona about the deuterosome. Elaboration of procentrioles (E) occurs at the outer limits of the corona. As many as 14 evenly spaced procentrioles may develop simultaneously about a deuterosome; for this reason, sections of procentriolar clusters frequently pass through symmetrical arrays of 2, 3 (illustrated), 4, 5, or 6 (illustrated) procentrioles, whose long axes point to the centre of the deuterosome. Procentrioles develop similarly, whether about a deuterosome or against a centriole (see Fig. 2-17). From Sorokin, S. P. 1968. Reconstructions of centriole formation and ciliogenesis in mammalian lungs. J. Cell. Sci. *3:* 207. Fig. 2.

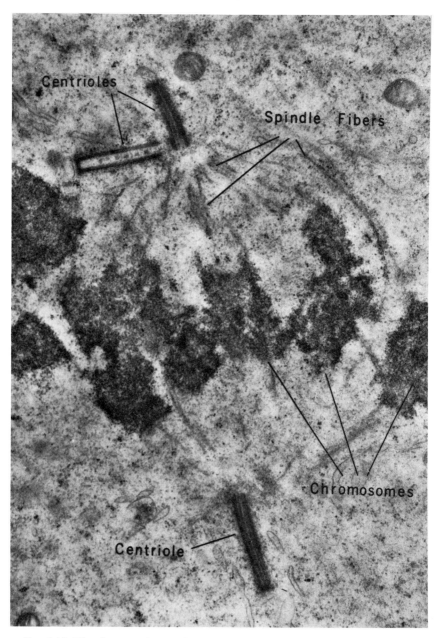

FIG. 2-19. The electron micrograph shows two pairs of centrioles after replication positioned at the poles of the nucleus. In this metaphase view can also be seen the spindle fibers and chromosomes in a spermatocyte from cock testis. Collidine buffered osmium fixation. Lead hydroxide staining. (Micrograph by Dr. T. Nagano.) Magnification 20,000 ×. From Fawcett, D. W. 1966. *The Cell.* W. B. Saunders Co. Philadelphia. p. 55. Fig. 27.

which are also tubular. During mitosis, the microtubules extend from the chromosomes to the centriole pair at each pole in animal cells. They are the fibers of the spindle apparatus. Their functional significance is obscure, but they may be related to chromosomal movement to the poles since electron micrographs show that spindle fibers are attached to the kinetochore of the chromosome. It has been reported that in the hydra nematocyst, they confluence with small spherical vesicles of the Golgi complex, and hence, may be functional in the transport of water, ions, or small molecules.

10. Tonofibrils

These fibrils appear in the cytoplasm of most cells, but in some cells they are most abundant, *e.g.*, deep epidermal cells. Individual fibrils measure about 30 A–60 A and vary in length. When they occur in bundles they are resolved in light microscopy. Tonofibrils often terminate in dense plaques, the desmosomes, which are present at points of cell contact. Fibrils may also appear in random assortment in the cytoplasm. Whether or not they are of the same composition as tonofibrils is unknown.

11. Cell Membrane

From observations on cells of animals, plants, and bacteria with light microscopy, it is obvious that a cell is separate from its environment. But it is not so obvious what provides the separation. Investigators as early as 1855 believed a cell boundary existed as a permeability barrier and they studied it with physiological methods. It was soon recognized that a cell responded to different solutes. A cell swells in a hypotonic solution and shrinks in a hypertonic solution, and some solutes can pass readily into a cell; others not so readily. Some substances can concentrate in a cell, but others cannot. Results such as these lead to an assumption that the boundary existed as a membrane which was selectively permeable. Among the early studies on permeability was an indication that solute permeation was generally related to its lipid solubility. These results implied that the barrier was mainly lipidal. Other workers found that molecular size influenced permeability. Later, micromanipulators were used to stretch a cell membrane, or even puncture it as evidenced by cytoplasm flowing out to the environment and in transplantation of nuclei by micromanipulation from one cell to another; a membrane could be even observed to repair itself.

In 1935 the classical "sandwich" model for membranes was proposed by Danielli and Davson as a result of indirect experiments on the properties of membranes including surface tension and thickness estimates. This view was that the membrane consisted of a trilaminar structure in which a lipid layer was bounded on either side by an adsorbed protein layer. This hypothesis gave rise to the 1938 Harvey-Danielli membrane model with bimolecular lipid leaflet structure. Credulence for this view came later from electron microscopy studies. Ultrastructure micrographs of the cell membrane portrayed it to consist of an outer and inner dark (osmiophilic) layer. These two layers were separated by a light (osmiophobic) layer at high magnification, and depending on the kind of preparation, the clear area measures about 30 A thick, and each dark layer measures about 25 A in thickness.

This trilaminar appearance corresponds in a general way to the D-D membrane model. Extensive investigations on membranes of cells showed that the trilaminar pattern seemed universal among cell types. This remarkable universality led Robertson in 1959 to propose the concept of a "unit" membrane with a trilaminar appearance. Because chemical and physical preparation of membranes for electron microscopy may bring about transformation of the cell membrane into a stable configuration, a uniform layering of the membrane constituents may resolve into a unit membrane in ultrastructure micrographs. In Figure 2-20 the Harvey-Danielli membrane model, based on physico-chemical evidence, may be compared to the Robertson model for a unit membrane as shown in Figure 2-21.

The unit membrane has been found to be characteristic of both the inner and outer mitochondrial membranes, both the inner and outer membranes of the nuclear envelope, the endoplasmic reticulum, the Golgi complex, lysosomes, and the chloroplast as well as the cell membrane.

Although the membranes of the cell resolve into a trilaminar appearance, their thickness can vary. In the same cell as well as in different cells, this thickness may vary from 75 to 100 A, although there seems to be a rather constant pore size of 7–8 A and some cells may have an external coat on its cell membrane.

This coating has long been known as the basement membrane and is not a restrictive term as many histologists use it. It has been found to be rich in polysaccharides, and to vary in thickness. Examples are plant cell walls, cuticles on epithelia of insects, the zona pellucida of ova, and the basement membrane of epithelial cells, the intercellular substance of connective tissue cells, and the sarcolemma of striated

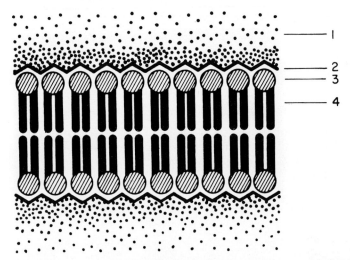

FIG. 2-20. The classical model of plasma membrane structure: 1, globular protein; 2, protein in extended form; 3, hydrophilic ends of lipid molecules; 4, hydrocarbon chains of lipid molecules. From Maddy, A. H. 1966. The chemical organization of the plasma membrane of animal cells. Int. Rev. Cytol. 20: 4. Fig. 1.

---

muscle cells. Other terms used synonymously for the basement membrane are glycocalyx, basal lamina, boundary layer, glycoprotein mantle, and external lamina. Perhaps externa lamina is the most apt description for a layer that completely envelops a cell. Firm evidence indicates that the layer is secreted by the underlying cells.

The prepared cell membranes of erythrocytes, called ghosts, have primarily been studied for information regarding chemical composition of cell membranes. Direct chemical analysis is a problem in membrane study since it deals with isolation of the membrane in a pure state. The purity of the membrane must be such that there can be no membrane contaminents. And the non-membrane contaminents must be recognized in the membrane preparation. An example of this difficulty is indicated with the ambiguous relation between hemoglobin and the ghost membrane. When the ghost is isolated as much of it as 50% in dry weight is constituted by hemoglobin. And this hemoglobin content may change during ghost isolation by varying the pH. Removing the hemoglobin can partially disintegrate the membrane. Electrophoretic properties of the membrane-bound hemoglobin are

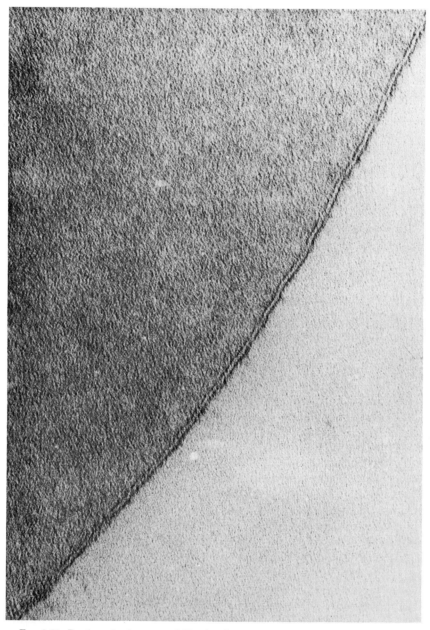

Fig. 2-21. Portion of a human red blood cell, fixed with permanganate and sectioned, showing the unit membrane structure bounding the cell. Magnification 275,000 ×. From Robertson, J. D. 1964. Unit Membranes. In *Cellular Membranes in Development*. M. Locke (editor). Academic Press. New York. p. 3. Fig. 2.

different from non-bound hemoglobin in the erythrocyte. Generally, membrane ghosts have a gross chemical composition of 20–40% lipid, 60–80% protein, and about 5% carbohydrate distributed between both the lipid and protein.

In the lipid fraction, phospholipids constitute the largest class. Glycerol and phosphate make up a phospholipid molecule along with a fatty acid tail; the former is water soluble, the latter is highly insoluble. Neutral lipids follow, and are mostly free (unesterified) cholesterol, but also present is esterified cholesterol, triglycerides, and some which are unspecified. The chemical composition of erythrocyte membranes may, or may not, be typical of cell membrane in general. In the membranes of mammalian erythrocytes, the major phosphatide components are lecithin, cephalins, and sphingomyelin. Major fatty acids identified from several mammalian species of erythrocyte membrane ghosts are palmitic, stearic, oleic, linoleic, linolenic, and arachidonic acids. Glycolipids have also been identified from several mammalian species. In membranes of erythrocytes in humans, the glycolipid is a globoside containing hexosamine, but no sialic acid which is also characteristic of pig, rabbit, sheep, and goat. Little hexosamine is found in cat, dog, and horse preparations although sialic acid is present. Glycolipid from ox ghosts contains both hexosamine and sialic acid. The red cell membrane is not static, but appears to be labile. Its dynamic aspect is indicated by a turnover of membrane components. The rise and fall of fatty acid levels in diet-controlled experiments indicate that there is an exchange between the serum and membrane of the phospholipid fraction. An exchange between cell and serum free cholesterol has also been confirmed with ^{14}C-labeled cholesterol. Lipid turnover, however, in membranes of other cell types needs further investigation.

There is a paucity of confirming evidence for proteins in the cell membrane as well as for enzymes and the components of the active transport mechanism. The basic problem relates to lipid-protein separation. Extracting the membrane lipid by an alcohol-ether solvent results in a denatured protein residue. But butanol can be used to solubilize the membrane protein and separation can be done on a column of sephadex. When prepared in this manner protein can be separated in various components. Little also is known about the binding between lipid and protein. Several types of interactions may exist between protein and lipid in membranes since the lipid can be separated into loose, weak, and firmly bound fractions. Undoubtedly, the array of components of proteins, lipids, lipoproteins, glycoproteins, enzymes, and those involved in actual transport mechanisms interact so as to make the cell membrane one of great complexity exhibiting many models of structure.

Actually, another hypothetical membrane model was proposed by Lucy in 1964. This was a micellar model, based on the concept that a micelle is a colloidal aggregation of molecules. In this micellar model for natural membranes, the globular, lipid micelles can be in dynamic equilibrium with the bimolecular leaflet structure. Figure 2-22 shows a view of a membrane cross-section in which globular micelles of lipid are in dynamic equilibrium with a bimolecular leaflet or lipid. It may be that in an area of a micellar membrane, lipid micelles maybe replaced by globular protein molecules, and in this way enzymes could become a part of the membrane. In support of the micelle model fracture planes revealed by freeze-etching preparations of cells have shown in micrographs globular subunits of about 90–130 A in diameter on the surfaces. Spherical particles on, or in, membranes about 40 A have been reported by combined electron microscopy and x-ray diffraction studies.

Before any one hypothetical membrane model is accepted, more knowledge must be gained. Perhaps an understanding of mechanisms of synthesis of new membranes, now obscure, will aid directly.

FIG. 2-22. A cross-sectional view of a membrane in which globular micelles of lipid are in dynamic equilibrium with a bimolecular leaflet of lipid. A layer of protein or glycoprotein or both is shown on each side of the lipid layer. The structure of each lipid molecule is illustrated in a diagrammatic fashion: only a polar group (○) and a non-polar moiety (wavy line) are shown, and the lipid may be phospholipid or non-phospholipid. One globular micelle of lipid has been replaced by a globular protein molecule (Pr) which may be a functional enzyme. From Lucy, J. A. 1968. Ultrastructure of membranes: Micellar organization. Brit. Med. Bull. 24: 128. Fig. 1.

12. Modifications of the Cell Surface

In 1949 micrographs from electron microscopy demonstrated that the striated border of intestinal epithelium was composed of microvilli. Studies in 1953 revealed the microvilli, each limited by the cell membrane, to be peculiar to absorptive cells, *e.g.*, the so-called brush border of the proximal convoluted tubules of the kidney. Later EM studies have also shown microvilli to be present in distal convoluted tubules of the kidney, but are so scanty they usually are not visible with light microscope techniques. Cilia and flagella have already been mentioned.

Sometimes contact between cells result in a modification of the cell membrane. One type of contact is exhibited by some epithelial cells. This has been called the junctional complex shown in Figure 2-23.

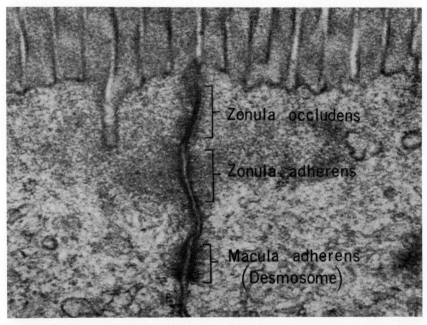

FIG. 2-23. The junctional complex from the intestinal epithelium from the hamster showing the tight junction (zonula occludens), the loose junction (zonula adherens), and the desmosome (macula adherens). Phosphate-buffered osmium fixation. Lead citrate staining. (Micrograph by Dr. Elliott Strauss.) Magnification 53,000 ×. From Fawcett, D. W. 1966. *The Cell*. W. B. Saunders Co. Philadelphia. p. 367. Fig. 193.

Toward the lumen there is an area called the tight junction (zonula occludens) with no intercellular space. Below the tight junction is the zonula adherens with an intercellular space of about 200 A with a dense filamentous area in the adjacent cytoplasm.* Below the zonula adherens is the desomosome (macula adherens) with an intercellular space about 250 A wide. Two dense plaques on opposing cell surfaces form half desmosomes. In each may be found fine cytoplasmic filaments converging on the more dense filamentous plaques. Desmosomes are not confined to junction complexes but may appear separately in various epithelial tissues. Another type of cell-to-cell contact is represented by the intercalated disc of light microscopy found in cardiac muscle. Although opposed membranes may vary in specialization, the junction complex membrane modification may exhibit as the tight junction, loose junction and desmosome.

13. Cell Wall

There are three general regions of the cell wall. These are the primary wall, secondary wall, and middle lamella. The middle lamella lies between two new cell walls which form at the cell plate in adjacent cells. Mostly pectin and lignins concentrate in the middle lamella. The primary wall forms first, and is thinner than the secondary wall. There are few layers of cellulose fibers seen in the primary wall by electron microscopy. Since the primary wall forms first, it is the most peripheral part of the cell wall. Pits are seen in the primary wall. The pits bear plasmodesmata or filaments of cytoplasm running from one cell to another. The thick secondary wall lies adjacent to the cell membrane.

Plant cells—called protoplasts—have cell walls which are best studied in cells undergoing cell division when a cell plate forms. The formation of a cell plate begins in the metaphase stage. It seems that proteolytic enzymes first decompose the protein structure of the cytoplasm and spindle-fibers. Lipid globules then accumulate on the equatorial plate and are hydrolyzed by lipase. Esterase is the first enzyme identified in the cell plate followed by lipase; acid phosphatase has also been found in the cell plate. During the anaphase and telophase stages the Golgi complex appears to produce a large number of vesicles. At first these vesicles are small and appear between the spindle fibers.

---

* From LM studies, early microscopists recognized the terminal bar as a heavy, darkstaining line encircling cells at the cell surface; if used in that context, terminal bar also includes the tight junction and zonula adherens, the loose junction.

The vesicles condense and merge, grow larger, and move to the equator of the cell where some fuse to add to the forming cell plate. Golgi vesicles then fuse to form new cell membranes. The formation of the cell plate is also characteristic for the formation of the walls of pollen tubes, root hairs, root-tip cells, and green algae. The endoplasmic reticulum associates closely with the formation of the cell plate. The endoplasmic reticulum bridges the newly formed wall and becomes a component of the plasmodesmata.

By the freeze-etching technique for electron microscopy, the outer surface of the newly formed cell membrane adjacent to the newly formed cell wall has been found to be covered by particles of about 150 A in diameter. They do not appear on the inner membrane surface. The particles in some way are involved in the synthesis of cellulose fibrils as indicated in Figure 2-24. The Golgi complex produces the substances of the non-cellulose matrix which are released to the cell surface of the protoplast. The particles on the outer membrane become

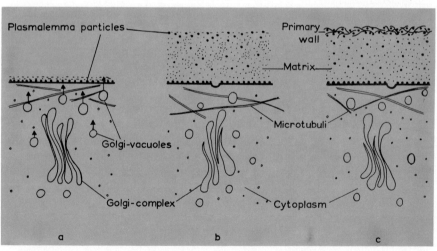

Fig. 2-24. Scheme of the subsequent steps in cell wall formation of the green alga *Chlorella* sp: a, a cortical region of the cell with the plasmalemma (cell membrane), covered with particles and a subadjacent Golgi complex. b, accumulation of matrix material, carried to the cell surface by Golgi vacuoles. The plasmalemma particles become detached and move to the outer periphery of the matrix. c, primary wall formation in the region where the particles are concentrated. From Muhlethaler, K. 1967. Ultrastructure and formation of plant cell walls. Ann. Rev. Plant Physiol. *18:* 14. Fig. 4.

detached and move to the outer surface of the non-cellulose matrix. At this time the first cellulose fibrils become apparent.

The synthesis of the cellulose is maintained by the formation of more elementary fibrils. These have been identified from electron micrographs in cell walls and bacterial cellulose as being about 35 A in diameter. They can aggregate into coarse bundles of cellulose fibrils. An elementary fibril model proposes that the fibril represents one molecule which is folded into a ribbon that is subsequently wound as a tight helix.

Cellulose is the world's most abundant organic compound, and is a polysaccharide. On complete hydrolysis, it yields glucose. The fact that hydrolysis of fully methylated cellulose yields 2,3,6—tri—O—methylglucose indicates no branching, and supports the elementary fibril model of one fibril—one molecule. When cellulose is partially hydrolyzed, the disaccharide cellobiose is formed. Its repeating units, joined by $\beta,1,4$ linkages are indicated in a Haworth hexagon perspective as

As cellulose fibers are laid down in the primary cell wall matrix, other substances such as hemicellulose, pectin, and lignins are deposited, and are added to the thickening secondary cell wall. And proteins are always present. The most abundant amino acid present is hydroxyproline which is the common amino acid found in collagen; it may be significantly related to the growth of the cell wall through the action of auxins.

Suggested Reading—Chapter 2

Allison, A. C. Lysosomes. In *The Biological Basis of Medicine*. Bittar, E. E. and Bittar, N. (editors). Academic Press. New York. 1968.

Arnon, D. I., Allen, M. B., and Whatley, F. R. Photosynthesis by isolated chloroplasts. Nature (London) *174:* 394, 1954.

Baker, J. R. New developments in the Golgi controversy. J. Roy. Micr. Soc. *82:* 145, 1963.

Beams, H. W. and Kessel, R. G. The Golgi apparatus: structure and function. Int. Rev. Cytol. *23:* 209, 1968.

Behnke, O. A preliminary report of "microtubules" in undifferentiated and differentiated vertebrate cells. J. Ultrastruct. Res. *11:* 139, 1964.
Benedetti, E. L., Bont, W. S., and Bloemendal, H. Electron microscopic observation on polyribosomes and endoplasmic reticulum fragments isolated from rat liver. Lab. Invest. *15:* 196, 1966.
Bonner, J. and Varner, J. E. (editors). *Plant Biochemistry.* Academic Press. New York. 1965.
Bensley, R. R. and Hoerr, N. The preparation and properties of mitochondria. Anat. Rec. *60:* 449, 1934.
Byers, B. and Porter, K. R. Oriented microtubules in elongating cells of the developing lens rudiment after induction. Proc. Nat. Acad. Sci. USA *52:* 1090, 1964.
Callan, H. G. and Tomlin, S. G. Experimental studies on amphibian oocyte nuclei. I. Investigation of the structure of the nuclear membrane by means of the electron microscope. Proc. Roy. Soc. Biol. *137:* 367, 1950.
Chèvremont, M., Chèvremont—Comhaire, S., and Baeckeland, E. Action of neural and acid desoxyribonucleases on living somatic cells cultured in vitro. Arch. Biol. (Liege). *70:* 811, 1959.
Christensen, A. K. and Fawcett, D. W. The normal fine structure of opossum testicular interstitial cells. J. Biophys. Biochem. Cytol. *9:* 653, 1961.
Claude, A. "Microbodies" et lysosomes: une étude au microscope électronique. Arch. Int. Physiol. *68:* 672, 1960.
Copenhaver, W. M. *Bailey's Textbook of Histology.* 15th edition. The Williams & Wilkins Co. Baltimore. 1964.
Dalcq, A. M. Le centrisome. Bull. Acad. Roy. Sci. Belg. *50:* 1408, 1964.
Dalton, A. J. and Felix, M. D. Studies on the Golgi substance of the epithelial cells of the epididymus and duodenum of the mouse. Amer. J. Anat. *92:* 277, 1953.
Dalton, A. J. Golgi apparatus and secretion granules. In *The Cell.* Vol. 2. Academic Press. New York. 1961.
Dalton, A. J. and Haguenau F. (editors). *The Nucleus.* Academic Press. New York. 1968.
Davies, H. G. Electron-microscope observations on the organization of heterochromatin in certain cells. J. Cell Sci. *3:* 129, 1968.
deDuve, C. The lysosome. Sci. Amer. *208(5):* 64, 1963.
deDuve, C. and Wattiaux, R. Function of lysosomes. Ann. Rev. Physiol. *28:* 435, 1966.
DuPraw, E. J. *Cell and Molecular Biology.* Academic Press. New York. 1968.
DuPraw, E. J. The organization of nuclei and chromosomes in honeybee embryonic cells. Proc. Nat. Acad. Sci. USA *53:* 161, 1965.
deReuck, A. V. S. and Cameron, M. P. (editors). *Lysosomes. Ciba Foundation Symposium.* London. Churchill. 1963.
Farquar, M. G. and Palade, G. E. Junctional complexes in various epithelia. J. Cell Biol. *17:* 375, 1963.
Fawcett, D. W. *The Cell.* W. B. Saunders Co. Philadelphia. 1966.
Fawcett, D. W. On the occurrence of a fibrous lamina on the inner aspect of the nuclear envelope in certain cells of vertebrates. Amer. J. Anat. *119:* 129, 1966.
Fishman, A. P. (editor). Symposium on the plasma membrane. Circulation *26:* 983, 1962.
Feldherr, C. M. The nuclear annuli as pathways for nucleocytoplasmic exchanges. J. Cell Biol. *14:* 65, 1962.
Fernández-Morán, H. Electron microscopy of nervous tissue. In *Metabolism of the Nervous System.* Richtec, D. (editor). Second Inter. Neurochem. Sym. Aarhus, Denmark. Pergamon Press. New York. 1957.
Gahan, P. B. Histochemistry of lysosomes. Int. Rev. Cytol. *21:* 1, 1967.
Gorgiev, G. P. The nature and biosynthesis of nuclear ribonucleic acids. Progr. Nucl. Acid Res. *6:* 259, 1967.
Gibbons, I. R. The relationship between the fine structure and direction of beat in gill cilia of a lamellibranch mollusc. J. Biophys. Biochem. Cytol. *11:* 179, 1961.

Glauert, A. M. The fine structure of bacteria. Brit. Med. Bull. *18:* 245, 1962.
Goodwin, T. W. (editor). *Biochemistry of Chloroplasts.* Academic Press. New York. 1966.
Granger, B. and Baker, R. F. Electron microscope investigation of the striated border of intestinal epithelium. Anat. Rec. *103:* 459, 1949.
Granick, S. The chloroplasts: inheritance, structure, and function. In *The Cell.* Vol. 2. Brachet, J., and Mirsky, A. E. (editors). Academic Press. New York. 1961.
Granick, S. and Gibor, A. The DNA of chloroplasts, mitochondria, and centrioles. Progr. Nucl. Acid Res. *6:* 143, 1967.
Hay, E. D. and Revel, J. P. The fine structure of the DNP component of the nucleus. J. Cell Biol. *16:* 29, 1963.
Heilbrunn, L. V. *An Outline of General Physiology.* W. B. Saunders Co. Philadelphia. 1952.
Heitz, E. Das heterochromatin der Moose. Jahrb. Wiss. Botan. *69:* 762, 1928.
Hicks, R. M. The function of the Golgi complex in transitional epithelium. Synthesis of the thick cell membrane. J. Cell Biol. *30:* 623, 1966.
Hirsch, J. G. Digestive and autolytic functions of lysosomes in phagocytic cells. Fed. Proc. *23:* 1023, 1964.
Hoffman, H. J. and Grigg, G. W. An electron microscopic study of mitochondria formation. Exp. Cell Res. *15:* 118, 1958.
Hokin, L. E. and Hokin, M. R. The chemistry of cell membranes. Sci. Amer. *213(4):* 78, 1965.
Holtfreter, J. Observations on the physico-chemical properties of isolated nuclei. Exp. Cell Res. *7:* 95, 1954.
Iterson, W. V. Symposium on the fine structure and replication of bacteria and their parts. II. Bacterial cytoplasm. Bact. Rev. *29:* 299, 1965.
Kellenberger, E., Schwab, W., and Ryter, A. L'utilisation d'un copolymere du groupe des polyesters comme matérial d'inclusion en ultramicrotomie. Experientia *12:* 421, 1956.
Kirkman, H. and Severinghaus, A. E. A review of the Golgi apparatus. Parts 1, 2, and 3. Anat. Rec. *70:* 413, 557; *71:* 79, 1938.
Korn, E. D. Cell membranes: structure and synthesis. Ann. Rev. Biochem. *38:* 263, 1969.
Kurosumi, K. Electron microscopic analysis of the secretion mechanism. Int. Rev. Cytol. *11:* 1, 1961.
Ledbetter, M. C. and Porter, K. R. Morphology of microtubules of plant cells. Science *144:* 872, 1964.
Littau, V. C., Allfrey, V. G., Frenster, J. H., and Mirsky, A. E. Active and inactive regions of nuclear chromatin as revealed by electron microscopy autoradiography. Proc. Nat. Acad. Sci. U. S. A. *52:* 93, 1964.
Luck, D. J. L. Genesis of mitochondria in *Neurospora crassa.* Proc. Nat. Acad. Sci. USA *49:* 233, 1963.
Lucy, J. A. Globular lipid micelles and cell membranes. J. Theor. Biol. *7:* 360, 1964.
Lyttleton, J. W. Isolation of ribosomes from spinach chloroplasts. Exp. Cell Res. *26:* 312, 1962.
Maaloe, O. and Birch-Anderson, A. On the organization of the nuclear material in Salmonella typhimurium. Soc. Gen. Microbiol. *6:* 261, 1956.
Maddy, A. H. The chemical organization of the plasma membrane. Int. Rev. Cytol. *20:* 1, 1966.
Malhotra, S. K. A comparative histochemical study of the Golgi apparatus. Quart. J. Micr. Sci. *102:* 83, 1961.
Mollenhauer, H. H. and Whaley, W. G. An observation on the functioning of the Golgi apparatus. J. Cell Biol. *17:* 222, 1963.
Montgomery, T. H. Some observations and considerations upon the maturation phenomena of the germ cells. Biol. Bull. *6:* 137, 1904.

Morowitz, H. J. and Tourtellotte, M. E. The smallest living cells. Sci. Amer. 206(3): 117, 1962.
Muhlethaler, K. Ultrastructure and formation of plant cell walls. Ann. Rev. Plant Physiol. 18: 1, 1967.
Nass, M. M. K. and Nass, S. New intramitochondrial fibers with DNA characteristics. I. Fixation and electron staining reactions. J. Cell Biol. 19: 593, 1963.
Nass, M. M. K. and Nass, S. Intramitochondrial fibers with DNA characteristics. II. Enzymatic and other hydrolytic treatments. J. Cell Biol. 19: 613, 1963.
Northcote, D. H. The biology and chemistry of the cell walls of higher plants, algae, and fungi. Int. Rev. Cytol. 14: 223, 1963.
Northcote, D. H. (editor). Structure and function of plant cell membranes. Brit. Med. Bull. 24: 99, 1968.
Novikoff, A. B. Mitochondria (chondriosomes). In *The Cell*. Vol. 2. Brachet, J., and Mirsky, A. E. (editors). Academic Press. New York. 1961.
Novikoff, A. B. Lysosomes and related particles. In *The Cell*. Vol. 2. Brachet, J., and Mirsky, A. E. (editors). Academic Press. New York. 1961.
Novikoff, A. B., Esner, E., and Quintana, N. Golgi apparatus and lysosomes. Fed. Proc. 23: 1010, 1964.
Osawa, S. Ribosome formation and structure. Ann. Rev. Biochem. 37: 109, 1968.
Palade, G. E. The fine structure of mitochondria. Anat. Rec. 114: 427, 1952.
Palade, G. E. An electron microscopic study of the mitochondrion structure. J. Histochem. Cytochem. 1: 188, 1953.
Palade, G. E. A small particulate component of the cytoplasm. J. Appl. Physics. 24: 1419, 1953.
Palade, G. E. and Porter, K. R. Studies on the endoplasmic reticulum. I. Its identification in cells in situ. J. Exp. Med. 100: 641, 1954.
Palade, G. E. The endoplasmic reticulum. J. Biophys. Biochem. Cytol. 2: suppl., 85, 1956.
Palay, S. L. On the appearance of absorbed fat droplets in the nuclear envelope. J. Biophys. Biochem. Cytol. 7: 391, 1960.
Perry, R. The nucleolus and the synthesis of ribosomes. Progr. Nucl. Acid Res. 6: 219, 1967.
Petermann, M. *The Physical and Chemical Properties of Ribosomes*. Elsevier Publ. Co. Amsterdam. 1964.
Pierce, G. B., Jr., Beals, T. F., Sri Ram, J., and Midgley, A. B. Basement membranes. IV. Epithelial origin and immunologic cross reactions. Amer. J. Path. 45: 929, 1964.
Ponder, E. The cell membrane and its properties. In *The Cell*. Vol. 2. Brachet, J., and Mirsky, A. E. (editors). Academic Press. New York. 1961.
Porter, K. R. Observations on a submicroscopic basophilic component of the cytoplasm. J. Exp. Med. 97: 727. 1953.
Porter, K. R. The ground substance; observations from electron microscopy. In *The Cell*. Vol. 2. Brachet, J., and Mirsky, A. E. (editors). Academic Press. New York. 1961.
Porter, K. R., and Bonneville, M. A. *Fine Structure of Cells and Tissues*. 3rd edition. Lea & Febiger. Philadelphia. 1968.
Porter, K. R., Claude, A., and Fullam, E. F. A study of tissue culture cells by electron microscopy. J. Exp. Med. 81: 233, 1945.
Racker, E. The membrane of the mitochondrion. Sci. Amer. 218(2): 32, 1968.
Rhodin, J. A. G. *An Atlas of Ultrastructure*. W. B. Saunders Co. Philadelphia. 1963.
Rich, A. Polyribosomes. Sci. Amer. 209(6): 44, 1963.
Robertson, J. D. Unit membranes. In *Cell Membranes in Development*. Ronald Press. New York. 1964.
Roodyn, D. B. The classification and partial tabulation of enzyme studies on subcellular fractions isolated by differential centrifuging. Int. Rev. Cytol. 18: 99, 1965.
Sanborn, E., Koen, P. F., McNabb, J. D., and Moore, G. Cytoplasmic microtubules in mammalian cells. J. Ultrastruct. Res. 11: 123, 1964.

Shumway, L. K. and Weirer, T. E. The chloroplast structure of iojap maize. Amer. J. Bot. *54:* 774, 1967.
Siekevitz, D. Protoplasm: endoplasmic reticulum and microsomes and their properties. Ann. Rev. Physiol. *25:* 15, 1963.
Sjöstrand, F. S. Electron microscopy of mitochondria and cytoplasmic double membranes. Nature (London) *171:* 30, 1953.
Sjöstrand, F. S. A comparison of plasma membrane, cytomembranes, and mitochondrial membrane elements with respect to ultrastructural features. J. Ultrastruct. Res. *9:* 561, 1963.
Sjöstrand, F. S. and Rhodin, J. The ultrastructure of the proximal convoluted tubules of the mouse kidney as revealed by high resolution electron microscopy. Exp. Cell Res. *4:* 426, 1953.
Slautterback, D. B. Cytoplasmic microtubules. I. Hydra. J. Cell Biol. *18:* 367, 1963.
Smith, H. W. The plasma membrane, with notes on the history of botany. Circulation *26:* 987, 1962.
Solari, A. J. The ultrastructure of chromatin fibers. Exp. Cell. Res. *53:* 553. 1968.
Sorokin, S. P. Reconstructions of centriole formation and ciliogenesis in mammalian lungs. J. Cell Sci. *3:* 207, 1968.
Stasny, J. T. and Crane, F. L. Separation of elementary particles from mitochondrial christae. Exp. Cell Res. *34:* 423, 1964.
Stevens, B. J. and Swift, H. RNA transport from nucleus to cytoplasm in *Chironomus* salivary glands. J. Cell Biol. *31:* 55, 1966.
Toner, P. G. and Carr, K. E. *Cell Structure.* Williams & Wilkins Co. Baltimore. 1968.
Turner, F. R. W. and Whaley, W. G. Intercisternal elements of the Golgi apparatus. Science, *147:* 1303, 1965.
Vincent, W. S. Structure and chemistry of nucleoli. Int. Rev. Cytol. *4:* 269, 1965.
Vincent, W. S. and Miller, O. I., Jr. (editors). *International Symposium on the Nucleolus, Its Structure and Function.* Nat. Can. Inst. Monograph no. 23. 1967.
Warner, J. R. and Rich, A. The number of growing polypeptide chains on reticulocyte polyribosomes. J. Molec. Biol. *10:* 202, 1964.
Weissmann, G. Lysosomes. Blood *24:* 594, 1964.
Weissmann, G. Lysosomes. New Eng. J. Med. *273:* 1084, 1143, 1965.
Weissmann, G. The many-faceted lysosome. Hosp. Proct. Feb., 1968.
Whaley, W. G. The Golgi apparatus. In *The Biological Basis of Medicine.* Bittar, E. E., and Bittar, N. (editors). Academic Press. New York. 1968.
Wilson, K. The growth of plant cell walls. Int. Rev. Cytol. *17:* 1, 1964.
Wolff, S. Strandness of chromosomes. Int. Rev. Cytol. *25:* 279, 1969.
Work, T. S., Coote, J. L., and Ashwell, M. Biogenesis of mitochondria. Fed. Proc. *27:* 1174, 1968.
Yamamoto, T. On the thickness of the unit membrane. J. Cell Biol. *17:* 413, 1963.

Chapter Three

Matrix: Physical and Chemical Characteristics

A. PROPERTIES OF AQUEOUS SOLUTIONS
  1. SOLVENT
  2. SOLUTES
  3. MOLECULAR AND IONIC SOLUTIONS
  4. COLLOIDS

B. MACROMOLECULES
  1. CARBOHYDRATES
    a. Classification
    b. Monosaccharides
    c. Derived Monosaccharides
    d. Oligosaccharides
    e. Polysaccharides
  2. PROTEINS
    a. Classification
    b. Properties
    c. Amino Acids and Peptides
  3. NUCLEOPROTEINS AND NUCLEIC ACIDS

C. PORPHYRINS

D. LIPIDS
  1. SIMPLE LIPIDS
  2. COMPOUND LIPIDS

E. VITAMINS
  1. FAT-SOLUBLE VITAMINS
  2. WATER-SOLUBLE VITAMINS

F. MINERALS

G. ACIDS, BASES, AND SALTS

H. pH AND BUFFERS

In order to have a better understanding of cell structure and function, it is necessary to keep in mind the relation of the cell organelles to the cell matrix. The organelles are suspended in, or provide a support structure for, the matrix. And how the organelle functions depends on the nature of the matrix. It has certain physiochemical properties peculiar to living matter. These properties of protoplasm include irritability, conductivity, movement, excretion and secretion, absorption, assimilation, growth, reproduction, and metabolism. In addition, there are six atoms selected by nature in forming molecules peculiar to the living system. These are carbon, hydrogen, nitrogen, oxygen, phosphorus, and sulfur. A dozen others are known to be essential for living systems: calcium, chlorine, cobalt, copper, iodine, iron, magnesium, manganese, molybdenum, potassium, sodium, and zinc. Often boron, vanadium, or silicon is required. The essentiality of fluorine is still open to question. The term matrix in this sense is used synonymously with the classical term protoplasm.

Protoplasm exists inside the cell membrane as a colloid. The particles of a colloid are suspended in water. This infers that the chemical reactions occurring inside a cell do so in an aqueous solution. However, substances do enter and leave the cell through the cell membrane. The substances can pass through the membrane because they possess certain characteristics. Size and charge are the most important characteristics. The particles are either molecules or ions dissolved in an aqueous solution. Once inside the cell the molecules or ions can take part in chemical reactions *per se*, or they can, by aggregation, form colloid particles. These are in a dynamic state and can be degradated to molecular and ionic forms.

It is seen then that the organelles of a cell lie in a colloid solution composed of solute particles in a water solvent. Substances involved in chemical reactions pass in and out of a cell in a molecular or ionic solution. Comprising the colloids are macromolecules of carbohydrates, proteins, and nucleic acids. Foodstuffs also present include lipids, vitamins, minerals, enzymes, and hormones. All of these take part in chemical reactions mediated by acids, bases, and salts as modified by buffers and pH.

A. Properties of Aqueous Solutions

1. Solvent

Water is the biological solvent for all living cells. Its ubiquitous distribution is essential for functions in cells. It is a well known fact that water exists as a liquid over the physiological ranges of temperature for normal cell function, and as such it has certain properties which aid in the interaction of molecules. The hydrogen and oxygen atoms of water, H—O—H, do not lie in a straight line. The two covalent bonds holding the two hydrogens to oxygen form an angle of 105°. Each bond is 0.99 A long. The hydrogens are positively charged and the oxygen is negatively charged, and the configuration of the bond angle makes the electric charge on the molecule unevenly distributed. The water molecule therefore tends to orient itself as a dipole in an electric field which implies a dipole moment.

Because of the unequal charge distribution the water molecule can form hydrogen bonds by electrostatic attractions. These are weak bonds as compared to the O—H covalent bond in water which has an energy of

110 kcal/mole; the O · · · · H hydrogen bond has an energy of 4.5 kcal/mole. The hydrogen bonds formed between two neighboring water molecules may be represented as

$$\begin{array}{c} H \\ \diagdown \\ _{d_-}O^{d_-}\!\cdots\!H \\ \diagup \\ H \end{array} \quad \begin{array}{c} O \\ \diagup \quad \diagdown \\ H \qquad H \end{array}$$

where d_+ represents the partial positive charge of the hydrogen atoms and d_- represents the partial negative charge of the oxygen atom. The electrostatic attraction between the oxygen and hydrogen forms a hydrogen bond.

Water can also act as an electrolytic solvent in the following way. Since water has a strong permanent dipole moment, it also has a high dielectric constant. This term is used as a measure of the ability of water to reduce the intensity of an external electric field. In an external field, each single water molecule (acting as a dipole) orients itself as if it were in a magnetic field. The minute fields of each separate charge of each water molecule add together. A resulting field is formed, by these minute fields, in such a direction so as to oppose the original external electric field and thereby reduce its original intensity. So in an electrolyte solution, the pairs of oppositely charged ions have the electrostatic attraction between them greatly reduced by the high dielectric constant of water. After electrolytic dissociation, each ion can behave as a single molecule.

A property of water that aids in the thermal stabilization of a cell is the high specific heat of water. As the temperature rises the hydrogen bonds gradually break due to the absorption of energy. The higher the specific heat of a substance the less the change in temperature when heat is absorbed by it.

Water also has a high heat of vaporization. This property aids a living cell, or organism, to maintain a constant temperature. A large amount of heat can be dissipated by water vaporization by sweating.

There are other properties of water which make it essential for a cell. It is immiscible with nonpolar liquids. In the section on cell membranes, the lipid nature of the membrane was described. Since water does not form linkages with nonpolar solvents to release enough energy to break the hydrogen bonds in water, water molecules remain stable and unmixed with the cell membrane. Therefore the lipid nature of the membrane limits the passage of water in and out of a cell. Water may mask active groups on proteins from substrates by forming ice-like structures about the group. This would tend to stabilize proteins in aqueous solutions.

2. Solutes

Substances which dissolve in a solvent are termed solutes. There exists a relation between water and solutes, and adding a solute to water has an effect on its behavior. One important effect is that the addition of a solute to water lowers the chemical potential of the water by increasing its entropy. Entropy is a term used to indicate the probable state of orderliness or arrangement of molecules of a substance. An increase in entropy describes that a highly disordered state is more probable. A decrease in entropy describes that a highly ordered state is more probable. If 1 gm molecular weight is added to a liter of water, the freezing point will be decreased by 1.86° C. Another important relation between solutes and water is osmotic pressure. Osmotic pressure develops when a semipermeable membrane separates water from a solution whose solute cannot pass through the membrane. Under these conditions the membrane is permeable only to the water. The pressure necessary to counterbalance the flow of water through the membrane to attain a state of equilibrium is called osmotic pressure. Osmotic pressure and its relation to membrane function are discussed in Chapter 4.

3. Molecular and Ionic Solutions

Solutes dispersed in water may, or may not, pass through a cell membrane. The limiting factors of permeability relate to both the characteristics of the solute and the nature of the membrane. In any event, substances which do pass through the membrane of a cell are usually dissolved into molecules or ions whose particle size is less than 1 mμ in diameter. By definition as employed arbitrarily, solute particles larger than 1000 mμ tend to settle out of solution, on standing, and may be termed suspensions. In and around a cell, solutes exist in a molecular or ionic form. Thus, solutes such as glucose, amino acids, fatty acids, electrolytes, minerals, vitamins, hormones, and enzymes have the potential of passing in and out of a cell.

4. Colloids

The physical characteristic of the complex substance making up the cell exists inside the cell membrane as a colloid; it is this polyphasic substance that has retained the descriptive term of protoplasm. Although the percentage of water may vary among cells, and even within cells, water largely constitutes protoplasm. This percentage may be less than 1% in a dormant seed to over 95% in cells such as in the hydra.

On an arbitrary basis of particle size, a colloid exists when the particles in solution vary from about 1 to 1000 mμ. Colloid particles do not settle out on solution, but tend to remain suspended. They can be observed to have an erratic, dancing movement. This is a result of water molecules, which are in constant motion, striking the colloid particle. The phenomenon is called Brownian movement. Colloid particles are visible with the light microscope. Colloids also exhibit the Tyndall effect. A beam of light passing through a colloid solution at right angles in a darkened room can be seen reflected from the colloidal particles. So a colloid differs from a molecular or ionic solution whose solute size is smaller, not visible in light microscopy, does not exhibit the Tyndall effect or Brownian movement, and can pass through a cell membrane.

Colloids may be prepared in different ways. Molecules may be aggregated, or coarse particles may be dispersed. Colloids may be separated from noncolloids, such as salts, by dialysis since colloid particles do not pass through a membrane. Other separation methods include electrophoresis, ultrafiltration, and ultracentrifugation.

Colloids may have a high affinity for water in which case they are termed hydrophilic colloids. They usually require a large amount of electrolytes for precipitation. The charge on a particle may be changed by varying the pH of the solution. Charge and hydration of the particle prevent the flocculation and aggregation of the colloid. Salts in excess act as discharging and dehydrating agents.

Colloid particles present a large surface area at interfaces which relates to surface tension. Substances such as proteins which decrease surface tension tend to accumulate at surfaces while substances such as sodium chloride which increase surface tension tend to be less concentrated at surfaces. An increase in the concentration of a substance at a surface is known as adsorption. Although the adsorption mechanism is obscure, it may relate to the formation of protein boundaries of unit membranes in cells.

B. Macromolecules

The chemical composition of protoplasm varies from one cell type to another. The variations include both the nature and quantity of chemical substances present, and the reactions involved. An example of one variation showing a cause and effect relation is the insistent demand for salt by those on a vegetarian diet. The selective storage of potassium by plants with little or no storage of sodium may explain the salt hunger of vegetarians. In nonvegetarians, *i.e.*, on a meat diet, the disproportion between potassium and sodium is not as great as in a vegetarian diet, and a salt hunger is not created.

Numerous analyses have been made for various tissues, organs, and whole animals. In Table 3-1 is shown a comparison of classes of compounds for a sea urchin egg and a newborn human. The percentages for lipids and ash appear high for the newborn because the whole body contains nonprotoplasmic material.

The inorganic composition of protoplasm includes various elements. Present are cations such as potassium, sodium, calcium, magnesium and anions as chlorides, sulfates, and bicarbonates, and phosphates. Trace elements with essential physiological functions are also present such as copper, cobalt, zinc, iodine, manganese, and molybdenum.

There are some constituents of protoplasm which exist as monomers but can link together by covalent bonds to form repeating units. These substances are carbohydrates, proteins, and nucleic acids. Because of their ability to form repetitive units they are sometimes grouped under the heading of macromolecules. A small number of monomers bonded together forms a molecule called an oligomer, but a molecule containing a large number of monomers is called a polymer. A polymer made up of all the same polymers is a homopolymer, but one made with different polymers is a heteropolymer. Polymers are formed by methods of polymerization which may result in branched or unbranched arrays. Sugar monomers form the polysaccharides of carbohydrates. Amino acids are

TABLE 3-1
A Comparison of Residues of a Newborn Human with a Sea Urchin Egg

Residue	Newborn Human	Sea Urchin Egg
	%	%
Water	66.0	77.3
Protein	16.0	15.18
Lipids	13.0	4.81
Carbohydrates	0.61	1.36
Ash	5.0	0.34

From Heilbrunn, L. V. 1952. *An Outline of General Physiology.* W. B. Saunders Co. Philadelphia. p. 22.

linked together by peptide bonds to form polypeptides and proteins. Polymers of polynucleotides and nucleic acids are formed from nucleotides. Recognizing that monomers are organic building blocks of three large classes of organic compounds, it is not difficult to understand the importance of macromolecules as constituents of living matter.

1. Carbohydrates

a. Classification. Carbohydrates are compounds containing hydrogen and oxygen usually, but not always, in a 2:1 ratio since many sugars have an empirical formula of $C_nH_{2n}O_n$. For this reason the French applied the name "hydrate de carbone," and the name was retained although it is not descriptive of the nature of the substances. Carbohydrates are compounds which are polyhydroxy aldehydes and ketones (monosaccharides), dimers (disaccharides), or polymers (oligosaccharides and polysaccharides) formed by the combination of these substances. The term is extended to also include oxidation and reduction products of carbohydrates and their simple derivatives such as amino and phosphorylated sugars. Carbohydrate compounds, therefore, are usually classified as

        Monosaccharides
        Derived monosaccharides
        Oligosaccharides (2–10 carbon units)
        Polysaccharides (high molecular weight).

A general formula for an aldehyde is

$$\begin{array}{c} H \\ | \\ C=O \\ | \\ R \end{array}$$

and for a ketone it is

$$\begin{array}{c} R \\ | \\ C=O \\ | \\ R' \end{array}$$

where R represents a hydrocarbon residue. If the R in an aldehyde is replaced by hydrogen the compound is formaldehyde, a gas, but not considered to be a carbohydrate although conforming to the empirical formula. The 2-carbon compound, hydroxyacetaldehyde, is not considered a carbohydrate.

b. Monosaccharides. The smallest molecules termed carbohydrates are monosaccharides called trioses. These 3-carbon sugars are

$$\begin{array}{cc} \text{H} & \text{CH}_2\text{OH} \\ | & | \\ \text{C}=\text{O} & \text{C} \\ | & \| \\ \text{CHOH} & \text{O} \\ | & | \\ \text{CH}_2\text{OH} & \text{CH}_2\text{OH} \\ \text{Glyceraldehyde} & \text{Dihydroxyacetone} \end{array}$$

Other monosaccharides are named also on the basis of the length of the carbon chain. For example,
  tetrose —4-carbons (erythrulose)
  pentose—5-carbons (ribose, deoxyribose, arabinose, xylulose)
  hexose —6-carbons (glucose, fructose, galactose)
  heptose—7-carbons (sedoheptulose)
Ketoses are named in a similar manner as aldoses but the ending is -ulose, *e.g.*, hexulose.

Monosaccharides are carbohydrates which cannot be hydrolyzed, or split further into simpler compounds. Most of the monosaccharides occurring in nature are the pentoses and hexoses. Ribose, a pentose, is important as a component of RNA. Deoxyribose, likewise a pentose, is a component of DNA. There are some coenzymes such as nicotinamide adenine dinucleotide (NAD) and NAD phosphate (NADP), ATP, and coenzyme A (CoA) that contain ribose. Fructose phosphate esters are important intermediates of glucose utilization. Glucose is the principle carbohydrate utilized by animal cells. All other carbohydrates utilized by animal cell may be converted to glucose.

The general formula for a monosaccharide may be represented as a Fittig-Baeyer model

$$\begin{array}{c} \text{CHO} \\ | \\ (\text{CHOH})_n \\ | \\ \text{CH}_2\text{OH} \end{array}$$

Many sugars of this type having one or more asymmetrical carbon atoms exhibit optical isomerism. Optical isomerism is the ability of a compound to bend polarized light. A carbon atom is asymmetrical in a compound if it has four different groups attached to it. If an object can be bisected into two equal (identical and superimposable) parts, it is symmetrical; if not, it is asymmetrical. The carbon atom has four valences which seem to be distributed in three dimensions so that the bonds point to the angles of a regular tetrahedron. If any two of the groups bonded to the carbon atom are cut in a plane to make the halves superimposable the carbon atom is symmetrical, but if all the groups are different the molecule is asymmetrical. Carbon atoms with double or triple bonds do not fit in this category. Molecules possessing an asymmetrical carbon atom exist in two forms which are mirror images of each other. These antipodes are usually represented in plane projections as

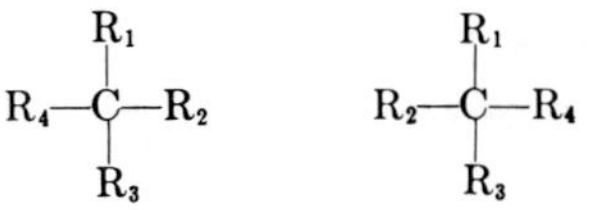

$$R_4-\underset{R_3}{\overset{R_1}{C}}-R_2 \qquad R_2-\underset{R_3}{\overset{R_1}{C}}-R_4$$

because of the difficulty in drawing numerous tetrahedrons. It is seen that the two above structures are not the same, but are mirror images of each other.

If the number of asymmetrical carbon atoms is represented by n in a molecule, then 2^n represents the number of possible isomers. The pairs forming mirror images have the same name, but are differentiated according to their structural formula. Glucose may be used as an example.

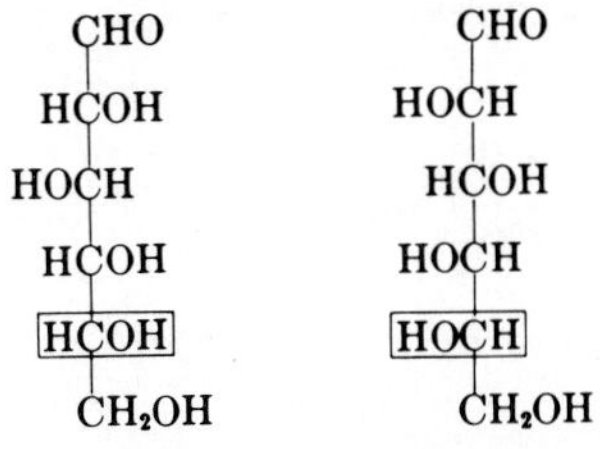

```
      CHO              CHO
      HCOH             HOCH
      HOCH             HCOH
      HCOH             HOCH
     |HCOH|           |HOCH|
      CH₂OH            CH₂OH
    D-Glucose        L-Glucose
```

The sugar is called D if it is structurally related to the parent substance (D-glyceraldehyde with the OH of its asymmetrical carbon on the right), and L if it relates spatially to L-glyceraldehyde. The assignment of D or L to a sugar depends on the configuration of the asymmetrical carbon atom adjacent to the primary alcohol group. D and L do not refer to the sugar rotating polarized light. The actual rotation of polarized light by the sugar to the right is said to be dextrorotatory and is designated by a plus (+) sign. Rotation of the beam to the left by a sugar is levorotatory and a minus (−) sign designates this characteristic. Geometric isomerism is an independent property from optical isomerism.

There is conclusive evidence that sugars exist in their nature state as hemiacetals. This compound is formed when an aldehyde or ketone group condenses with an alcoholic hydroxyl group in the same molecule. A cyclic structure results, with D-glucose, for example, when the aldohexose in an aqueous solution exists in equilibrium with two isometric ring compounds shown by Fischer formulation:

$$\begin{array}{ccc} \text{HO} \quad \text{H} & \text{H} \quad \text{O} & \text{H} \quad \text{OH} \\ \text{C} & \text{C} & \text{C} \\ | & | & | \\ \text{HCOH} & \text{H}^2\text{COH} & \text{HCOH} \\ | & | & | \\ \text{HOCH} \quad \text{O} & \rightleftharpoons \quad \text{HO}^3\text{CH} \quad \rightleftharpoons & \text{HOCH} \quad \text{O} \\ | & | & | \\ \text{HCOH} & \text{H}^4\text{COH} & \text{HCOH} \\ | & | & | \\ \text{HC} & \text{H}^5\text{COH} & \text{HC} \\ | & | & | \\ \text{CH}_2\text{OH} & {}^6\text{CH}_2\text{OH} & \text{CH}_2\text{OH} \\ \beta\text{-D-Glucose} & \text{D-Glucose} & \alpha\text{-D-Glucose} \end{array}$$

The ring compounds predominate and they exist because an additional asymmetrical carbon is introduced into the molecule. Carbon 1 is the locus for this asymmetry. The α-form of D-glucose has the hydroxyl group on the right of the carbon 1, in the β-form it is on the left. In L-glucose, the α-form has the hydroxyl group on the left, the β-form on the right. Therefore, α-D-glucose is the mirror image of α-L-glucose. There is a different melting point for the α- and β-forms, and each has a different optical rotation. The two ring compounds above may be drawn by the Haworth hexagonal structure showing numbered carbon atoms:

β-D-Glucose   α-D-Glucose

The Haworth perspective is a more accurate representation of the molecule than the Fischer formulation, and permits an easier visualization of the oligosaccharides and polysaccharides. A compound is called a glucoside when it is formed from a cyclic form of glucose by combination of the carbon 1-hydroxyl group with an alcohol. If the compound is related to α- or β-glucose, it is specifically called α- or β-glucosides. Most other sugars are similar to the properties and nomenclature of the cyclic forms of glucose. Depending on the sugar from which they originate they are called galactosides, fructosides, etc. These compounds analogous to glucosides are generically called glycosides. Glycosides are compounds containing a carbohydrate and a noncarbohydrate in the same molecule. The noncarbohydrate is attached by an acetyl linkage to carbon atom 1 of the carbohydrate residue.

Care must be used in visualizing a planar representation of a three dimensional object such as an aldohexose by the Haworth formulation. The molecule does not lie in a plane, but exists as a chair or boat form because of the angles of the bonds. A concept of conformational formulation is helpful in understanding the reactivity of hydroxyl groups in aldohexoses. One should keep in mind that shapes of molecules, especially polymers, are difficult to visualize from two dimensional drawings.

c. Derived Monosaccharides. The derived monosaccharides are similar to the monosaccharides, but differ in regard to the aldoses and ketoses. Some of these are sugar acids, sugar alcohols, sugar phosphates, deoxy-sugars, and amino sugars. A sugar acid of biological importance in both plant and animals is vitamin C. Of the sugar alcohols, glycerol is most important in lipid chemistry. Phosphorylated sugars are intermediates in glycolysis and are components of nucleic acids, nucleotides, and polysaccharides. A deoxy-sugar has its hydroxyl group replaced by hydrogen; of importance is 2-deoxyribose which shares with ribose the sugar function in nucleic acids and related compounds. Replacement of a hydroxyl group by an amine group yields an amino sugar. Two amino sugars of importance are glucosamine and galactosamine. The former occurs in polysaccharides and is the hydrolytic product of chitin; the latter occurs in chondroitin sulfate of cartilage.

d. Oligosaccharides. Oligosaccharides are composed of monosaccharide units or their derivatives held together by glycosidic linkages. On hydrolysis, the oligosaccharides yield from 2 to 10 monosaccharide residues. The definition is an arbitrary one, and merely refers to those polysaccharides of low molecular weight. Although many oligosaccharides have been described, most of them do not occur naturally. The most important of the oligosaccharides are the disaccharides since most of them do occur naturally as constituents of both plant and animal cells.

Disaccharides are formed by the union of two monosaccharides with the elimination of a water molecule. Conversely, disaccharides may be hydrolyzed as

$$C_{12}H_{22}O_{11} + H_2O \rightarrow C_6H_{12}O_6 + C_6H_{12}O_6$$

Although the disaccharides have in common the molecular formula $C_{12}H_{22}O_{11}$, their properties differ. Those properties depend not only on the component monosaccharides, but also on the points of linkage and the manner of linkage. For example, if the two potential aldehyde or ketone groups are involved in the linkage, the sugar will have no reducing properties, but if one of them is not bound the sugar will be a reducing one. The three most common disaccharides are sucrose, maltose, and lactose.

Sucrose, common table sugar, is composed of D-glucose and D-fructose linked together by the aldehyde and ketone carbon; hence, it is not a reducing sugar. However, both maltose and lactose are reducing sugars since both have an unlinked potential aldehyde. Maltose is composed of two units of D-glucose, and lactose of D-glucose and D-galactose. Although the relative sweetness of sugar is difficult to determine, it is usually done so from 1925 data. Because the relative sweetness of sugar is difficult to determine, it is usually done by taking sucrose as a standard, and then measuring the greatest dilution of another sugar at which the sweetness is still detected. On this basis, sucrose is 100, maltose 32, lactose 16, and fructose 173. Honey is sweet because of its fructose content.

Maltose is the major product in the enzymatic hydrolysis of starch, a polysaccharide. Maltose can be further hydrolyzed to glucose which is readily absorbed and utilized in the body by cells in glycolysis. Maltose, like sucrose, is readily soluble in water, but unlike sucrose, does not occur abundantly in nature. It is called malt sugar.

Lactose is milk sugar. It constitutes 5% of the volume of milk. Neither is it very sweet, nor very soluble. Souring of milk can be caused by bacteria found in milk converting lactose into lactic acid.

Enzymes with known specificities are used to determine the α- or β-configuration of the glycosidic bond. The enzymes which hydrolyze glycosides are called glycosidases. A specific example is β-glycosidase. This enzyme hydrolyzes β-glycoside linkages. Lactose is hydrolyzed by β-glycosidase proving that the sugar is a glucose-β-galactoside.

There are other oligosaccharides. Some examples of these are:
    trisaccharides   —raffinose, mannotriose, rabinose, rhamminose, gentianose, melezitose

tetrasaccharides —stachyose, scorodose
pentasaccharides—verbascose

e. Polysaccharides. Like oligosaccharides, polysaccharides are polymerized from a variety of monosaccharides and their derivatives. They differ, however, from oligosaccharides in that the number of their monosaccharide monomers vary from 10 to several thousand units. Thus, the polysaccharides have a high molecular weight. Their molecules are of colloidal size. As condensation polymers, the polysaccharides are joined together by glycosidic linkages. They form patterns of molecular configuration which may be branched, linear, or cyclic. Some are helical. Polysaccharides may be homosaccharides made up of a single kind of monosaccharide as in starch, glycogen, or cellulose. Others are heterosaccharides when they contain two or more kinds of monosaccharides or their derivatives: examples are mucopolysaccharides.

Polysaccharides may serve as food storage in plants (starch) and animals (glycogen). Other functions include structural patterns as cellulose in plants, chitin in insects and crustaceans, and chondroitin sulfate in cartilage. They are also found in synovial fluids, vitreous body of the eye, in mucus, heparin, and blood.

Starch is the principal carbohydrate of the usual human diet. In the USA, up to 60% of the food intake is starch. Starch occurs in many plants as storage food. The molecular weight of starch may vary from 5000 to 10^6, and does not refer to any specific compound. Starch is a polymer of glucose units, linked together in an α-1,4 glycosidic linkage which enzymes can break. Thus, starch is digestible. Cellulose is nondigestible because it has a β-1,4 glycosidic linkage of repeating maltose units. Bond linkage is the main difference between starch and cellulose.

A starch granule contains amylose and amylopectin. Both of these substances are polysaccharides, and each is a polymer of glucose. Amylose ratio to amylopectin is about 1:3. About 300 glucose units are linked together by α-1,4 glycosidic bonds which tend to cause a spiraling into a helix-like structure. Amylopectin likewise is composed of glucose units in an α-1,4 glycosidic linkage, but occasionally there appears α-1,6 glycosidic bonds which cause branching. The amylopectin structure formed is a branch-to-branch arrangement of amylopectin with each branch consisting of about 25 glycose units whereas amylose is an unbranched chain. Starch can be hydrolyzed to lower weight polysaccharides, and finally splits off maltose or glucose. The enzymes hydrolyzing starch are called amylases, and are either of plant or animal origin. They may be α-amylases, or β-amylases occurring in germinating seeds and potatoes. Partial hydrolysis of starch results in substances known as dextrins. These carbohydrates are often mixed with maltose and used in feeding infants. Dextrins are easily digested, and their physical properties are such that they prevent the formation of heavy curds in the infant stomach when mixed with milk.

Like starch, glycogen does not refer to any specific compound. It is the counterpart of starch in animals. It is particularly abundant in mollusks, liver, and muscle. Structurally, glycogen resembles amylopectin of starch except the branching chains average about 12 glucose units, not 25 or so. There is an α-1,4 glycosidic linkage and chains are formed at α-1,6 glycosidic bonds as in amylopectin. The glycogen units are collections of polymers. Molecular weights have been reported ranging from about 3×10^5 to 10^8.

Cellulose is a term representing a group of high molecular weight substances which yields glucose on final hydrolysis. In this respect, it resembles starch and glycogen. In the discussion on cell walls, it was mentioned that the repeating units of cellulose was cellobiose joined by β-1,4 glycosidic bonds. Like starch, cellulose has been synthesized in the laboratory with plant enzymes. It is of interest to note that although the human intestine mucosa does not secrete a cellulose-splitting enzyme, cellulose does supply bulk which aids the peristaltic movement. Inulin, agar-agar, hemicelluloses, and pectins are other polysaccharides. Scattered widely in the body, particularly in the connective tissues, are heteropolysaccharides called mucopolysaccharides. These are hyaluronic acid, heparin, and the chondroitin sulfates. Hyaluronic acid acts as a cementing substance in connective tissue. It is present in synovial fluid of joints where its viscosity contributes to the lubricating quality of these fluids. Chondroitin sulfates serve as a matrix for bone formation. Heparin is a blood anticoagulant present in liver, lung, thymus, spleen, and blood. There are also various blood-group polysaccharides. They are present in red corpuscles, saliva, gastric mucin, and other body secretions. They are generally polymers of a simple sugar and D-glucosamine or D-galactosamine. Specific polysaccharides have been isolated from different species of bacteria. Some of these carbohydrates have been found to be responsible for the immunological specificity of bacteria.

2. Proteins

Proteins are polymers of high molecular weight which consist of chains of primary α-amino acids joined together by peptide linkages.

Their molecular weights vary from about 13,000 to millions, and like polypeptides and lipids they are of colloidal dimensions. In general, they contain some 20 different amino acids which, unlike the monomers forming polysaccharides, are repeating units of heteropolymers. As a result of the polymerization of different amino acids, innumerable and various proteins can exist. Proteins can be denatured and coagulated by agents causing intramolecular rearrangement.

a. Classification. No satisfactory system for classifying proteins exists because of the differences and similarities of thousands of proteins. Present-day classifications are generally made on a 1907 scheme based primarily on solubility properties. This scheme recognized primary proteins and conjugated proteins where each group contained subclasses with varying solubility properties. Although more is known today about structure of proteins than in 1907 it would be an ideal basis for classification, but present knowledge is still inadequate concerning protein structure.

However, it is recognized that some proteins exist as elongated molecules consisting of coiled polypeptide chains tightly linked together and insoluble. These are fibrous proteins. They constitute the collagens of connective tissue, elastins of blood vessels and tendons, and keratins, the proteins of skin, hair, horns, hoofs, and nails. There are other insoluble proteins which in solution appear as globules; hence they are called globular proteins. Included in this category are enzymes and protein hormones.

As mentioned above, proteins can also be classed according to their solubility properties. These are called simple proteins since they yield mostly α-amino acids on complete hydrolysis. Among these are albumins (egg albumin, serum albumin, milk albumin), globulins (serum, muscle, plant seeds), globins (hemoglobins), histones (nucleoproteins), prolamins (wheat gliadin, corn zein), and protamines (fish sperm). It has been mentioned that no system of protein classification is complete. An example of this is classifying the aforementioned fibrous proteins as the albuminoids, found only in animals and grouping them with the simple proteins. The simple proteins have the following solubility characteristics:

albumins —soluble in water and dilute salt solutions
globulins —insoluble in water, but soluble in dilute neutral salt solutions
globins —soluble in ammonium hydroxide

histones —soluble in water, and insoluble in ammonium hydroxide
prolamins —insoluble in water, and soluble in 70% ethyl alcohol
protamines —soluble in ammonium hydroxide
albuminoids—insoluble in water and all neutral solvents

Conjugated proteins are those combined with a nonprotein moiety called a prosthetic group. Examples of these are nucleoproteins, lipoproteins, mucopolysaccharides, mucoproteins, phosphoproteins, chromoproteins, and metalloproteins. Nucleoproteins are proteins combined with nucleic acids. Lipoproteins are proteins combined with lipids and found particularly associated with serum, brain, and other nerve tissue. Mucopolysaccharides are proteins combined with the predominating carbohydrates. Mucoproteins are proteins in combination with polysaccharides behaving more like proteins. Phosphoproteins are proteins joined to phosphoric acid. Chromoproteins contain different prosthetic groups such as hemoglobins, flavoproteins, and cytochromes. Metalloproteins are proteins joined to metals which are not a part of the prosthetic group such as carbonic anhydrase containing zinc.

Derived proteins are those compounds produced by partial hydrolysis of proteins. These products are formed when proteins are denatured or coagulated.

b. Properties. Proteins have varied properties aiding in cell structure and function. Proteins are amphoteric compounds, *i.e.*, they can act as both acids and bases. Being electrolytes, they can migrate in an electric field as shown by electrophoresis. The solubility of proteins depends on the pH of the solution. As the solution becomes more acid, or more alkaline, solubility increases. Conversely, solubility decreases as the isoelectric point of the protein is reached, and is minimal at that point. They can bind salts by forming ionic bonds. Proteins are precipitated by salts of heavy metals which is applicable in antidotes for metallic poisons. The fact that certain acids precipitate proteins is important in determining proteins in urine and body fluids. Proteins may be sedimented in the ultracentrifuge. Most proteins, as some polysaccharides and lipids, have antigenic properties. Species specificity is also characteristic of proteins.

c. Amino Acids and Peptides. Any compound possessing one or more amino groups and one or more carboxyl groups is an amino acid. The amino acids of biological importance to the cell ordinarily contain one amino group in the α-position to the carboxyl group. The general formula for a primary α-amino acid showing the amino and carboxyl groups attached to the same carbon atom is

$$R-\underset{H}{\overset{NH_2}{C}}-COOH$$

All of the amino acids are optically active except one, glycine. Here R is replaced by hydrogen, but the other amino acids have an asymmetry about the α-carbon atom. Configuration about the α-carbon atom is indicated by D and L, and nomenclature is similar to the carbohydrate series. Optical rotation is designated also by plus (+) and minus (−) signs. Specific values of optical rotation for any amino acid vary with temperature, pH, and concentration.

There are about 20-odd naturally occurring amino acids which are present in animal and plant protein molecules. With the exception of glycine, all are of the L-form in which case the —NH₂ group should be placed above the α-carbon atom and the —COOH group to the right if the carbon chain is written horizontally. D-Amino acids are known to occur in nature although only L-amino acids have been isolated from proteins.

Although different R groups are attached to the α-amino acids, they are neutral, acidic or basic depending on the number of amino acid carboxyl groups present. Two amino acids are imino acids in which the nitrogen of the imino group is in a ring, but can still form peptides. In Table 3-2 are listed the commonly occurring amino acids in proteins. Besides the formula, other characteristics are noted. These are pK_a values, isoelectric points, and formula weight.

There are other amino acids and derivatives which play an important role in metabolism. One of these is β-alanine, a component of the vitamin pantothenic acid, and coenzyme A. Thyroxine occurs only in thyroglobulin elaborated by the thyroid gland. The vasodepressor, histamine, is a decarboxylation product of histidine. Ornithine and citrulline are important intermediates in the biosynthesis of urea. And γ-amino butyric acid (GABA) is a constituent of brain and plant tissues.

Amino acids can exist as dipolar ions in aqueous solutions since at least one carboxyl group and one amino group is present. In this fashion they can react as either acids or bases. The form in which the carboxyl group is dissociated and the amino group is associated is called a zwitterion and is represented as a dipolar ion as

$$R-\underset{H}{\overset{NH_3^+}{C}}-COO^-$$

Thus, all amino acids have in common a series of reactions, besides specific side-chain reactions, which include alkylation, acylation, ester formation, and amide formation because they exist as dipolar ions.

The essentiality of amino acids have been established for different species. The essential amino acids for man are isoleucine, leucine, methionine, phenylalanine, threonine, tryptophan, and valine; arginine and histidine are probably essential.

Where two amino acids combine, a dipeptide is formed. Fischer and Hofmeister, working independently, provided the first satisfactory hypothesis of how this might be done. The postulation is that the carboxyl group of one amino acid combines with the amine group of the other. In the process, the hydroxyl group in the carboxyl group combines with one of the hydrogens in the amine group to form a water molecule which is split off and lost in the reaction. The —C—N— junction is termed a peptide bond or peptide linkage. In the same way as the dipeptide is polymerized, a third amino acid can be added to form a tripeptide. Therefore, a peptide chain can be built from peptide bonding with the ultimate formation of polypeptide chains. Not only can the chain be straight, but it can take the form of a helix. The helical arrangement results from hydrogen bonding between CO and NH groups as shown by Pauling and Corey in 1949. This gives a spatial or three-dimensional configuration to the polypeptide.

The peptide linkage may be broken by hydrolysis. In the intestine, proteolytic enzymes quickly and completely add water across the bond at specific linkage points in the polypeptide to aid absorption. Once the amino acid passes through the cell membrane, enzymatic reactions can form polymers into colloid particles which become a part of the protoplasmic millieu.

There are also hormones consisting of peptides. Oxytocin and vasopressin are elaborated by the neurohypophysis; it also liberates melanocyte-stimulating hormones. Oxytocin is released immediately before parturition; it causes contraction of the uterus and initiates the secre-

TABLE 3-2
Amino Acids Commonly Occurring in Proteins

Name	pK_a Values	pI	Formula Weight	Formula
Aliphatic amino acids				
1. Glycine	2.34; 9.60	5.97	75.07	H–C(H)(NH$_2$)–COOH
2. Alanine	2.35; 9.69	6.02	89.10	CH$_3$–C(NH$_2$)(H)–COOH
3. Valine	2.32; 9.62	5.97	117.15	(CH$_3$)$_2$CH–C(NH$_2$)(H)–COOH
4. Leucine	2.36; 9.60	5.98	131.18	(CH$_3$)$_2$CH–CH$_2$–C(NH$_2$)(H)–COOH
5. Isoleucine	2.36; 9.68	6.02	131.18	CH$_3$–CH(C$_2$H$_5$)–C(NH$_2$)(H)–COOH
6. Serine	2.21; 9.15	5.68	105.10	HO–CH(H)–C(NH$_2$)(H)–COOH

TABLE 3-2—Continued

Name	pK$_a$ Values	pI	Formula Weight	Formula
7. Threonine	2.63; 10.43	6.53	119.12	$CH_3-\underset{HO}{\underset{|}{C}}-\underset{H}{\overset{NH_2}{\underset{|}{C}}}-COOH$
Aromatic amino acids				
8. Phenylalanine	1.83; 9.13	5.98	165.20	$C_6H_5-CH_2-\overset{NH_2}{\underset{H}{\underset{|}{C}}}-COOH$
9. Tyrosine	2.20; 9.11 (α-amino); 10.07 (phenolic hydroxyl)	5.65	181.20	$HO-C_6H_4-CH_2-\overset{NH_2}{\underset{H}{\underset{|}{C}}}-COOH$
10. Tryptophan	2.38; 9.39	5.88	204.23	indole-$CH_2-\overset{NH_2}{\underset{H}{\underset{|}{C}}}-COOH$
Sulfur-containing amino acids				
11. Cysteine	1.71; 8.33 (sulfhydryl); 10.78 (α-amino)	5.02	121.16	$HS-CH_2-\overset{NH_2}{\underset{H}{\underset{|}{C}}}-COOH$

TABLE 3-2—*Continued*

Name	pK$_a$ Values	pI	Formula Weight	Formula
12. Cystine	1.65; 2.26 (carboxyls); 7.85; 9.85 (aminos)	5.06	240.31	
13. Methionine	2.28; 9.21	5.75	149.22	
Acidic amino acids				
14. Aspartic acid (occurs mostly as asparagine)	2.09 (α-carboxyl); 3.86 (β-carboxyl); 9.82	2.87	133.11	
14a. Asparagine	2.02; 8.8	5.41	132.12	
15. Glutamic acid (occurs mostly as glutamine)	2.19 (α-carboxyl); 4.25 (γ-carboxyl); 9.67	3.22	147.14	
15a. Glutamine	2.17; 9.13	5.65	146.15	

TABLE 3-2—Continued

Name	pK, Values	pI	Formula Weight	Formula
Basic amino acids				
16. Lysine	2.18; 8.95 (α-amino); 10.53 (ε-amino)	9.74	146.19	$CH_2-CH_2-CH_2-CH_2-\underset{NH_2}{\overset{NH_2}{C}}-COOH$
17. Hydroxylysine (occurs only in collagen and gelatin)	2.13; 8.62 (α-amino); 9.67 (ε-amino)	9.15	162.19	$CH_2-CH-CH_2-CH_2-\underset{H}{\overset{NH_2}{C}}-COOH$ with NH_2, OH
18. Arginine	2.17; 9.04 (α-amino); 12.48 (guanidino)	10.76	174.21	$H-N-CH_2-CH_2-CH_2-\underset{H}{\overset{NH_2}{C}}-COOH$ with $C=NH$, NH_2
19. Histidine	1.82; 6.0 (imidazole); 9.17	7.58	155.16	imidazole-$CH_2-\underset{H}{\overset{NH_2}{C}}-COOH$
Imino acids				
20. Proline	1.99; 10.60	6.10	115.14	pyrrolidine-$CH-COOH$

TABLE 3-2—Continued

Name	pK$_a$ Values	pI	Formula Weight	Formula
21. Hydroxyproline (occurs only in collagen and gelatin)	1.92; 9.73	5.83	131.14	HO–CH–CH$_2$ H$_2$C–CH–COOH N H

pK$_a$ = ionization constant of the acid
pI = isoelectric point; example: pI of glycine

$$pI = \frac{pK_1(COOH) + pK_2(NH_3^+)}{2}$$

$$= \frac{2.34 + 9.60}{2}$$

$$= 5.97$$

tion of milk. Vasopressin is an antidiuretic hormone important in the regulation of water balance in mammals. In its absence, urine cannot be concentrated. The adenohypophysis secretes, among other hormones, adrenocorticotropic hormone (ACTH) which stimulates the secretion and growth of the adrenal cortex. Some peptides isolated from bacteria are antibacterial agents. Among these are tyrocidin and gramicidin produced by *Bacillus brevis*, and bacitracin by *Bacillus subtilis*.

The helix configuration for certain polypeptide molecules led Linderstrøm-Lang in 1953 to suggest how peptide chains could be arranged to indicate protein structure. A protein could be formed by a primary structure in which a sequence of amino acids could be joined together by peptide linkages. The secondary structure of a protein indicated a helical form for the polypeptide chains stabilized by hydrogen bonds. The tertiary structure was the configuration in which the chains were held rigidly by —S—S— bonds which folded the helix into ellipsoid or other forms. A quaternary structure was later postulated for such molecules as hemoglobin in which the primary, secondary, and tertiary structures were joined to form clusters and layers of subunits.

Conformation in proteins is related to the kind and formation of bonds present. Primary structure is stabilized by the peptide bond. Secondary and tertiary structures are maintained principally by bond types as shown in Figure 3-1. There are hydrogen bonds which form between the CO and NH groups of adjacent coils in the α-helix. If a molecule contains many opportunities for hydrogen-bond formation, they will reinforce each other in stabilizing the molecule. Hydrogen bonds may also form between side chains. Ionic bonds or hydrophobic (apolar) bonds may form. Ionic bonds can form either strong or weak interactions. The importance of the hydrophobic bond of proteins is being vigorously pursued.

The sequence of amino acids bestows remarkable characteristics to a protein. An example of this is hemoglobin. As first shown by Ingram in hemoglobin A (normal) there exists 574 amino acids in the molecule. The conformation of the molecule is such that four polypeptides exist; two α-chains and two β-chains, each pair of chains having the same amino acid sequence, but different sequences in the α- and β-chains. The α-chains each have 141 amino acids, and the β-chains each have 146 amino acids. Hemoglobin S (sickle-cell anemia) was discovered by a combination of chromatography and electrophoresis called fingerprinting. Subsequent amino acid analyses of the S fingerprints, or peptide residues, showed it to be similar to hemoglobin A with one exception. There is a substitution of glutamic acid by valine at position 6

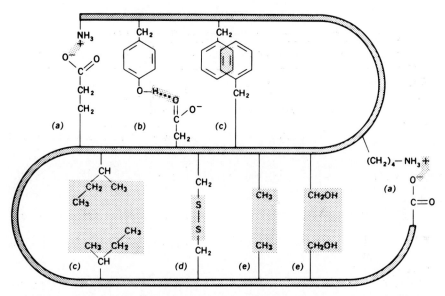

FIG. 3-1. Bonds which stabilize 2° and 3° structural features of proteins: (*a*), salt linkages, electrostatic interactions; (*b*), hydrogen bonding; (*c*), interaction of nonpolar side chains; (*d*), disulfide linkage between cysteine residues; (*e*), van der Waal's interactions. From Bennett, T. P. and E. Frieden. 1966. *Modern Topics in Biochemistry*. The Macmillan Co. New York. p. 19. Fig. 3-2.

---

from the N-terminal in each of the β-chains of hemoglobin S. As a result, there is a change in the shape of the hemoglobin S molecule as well as a change in the oxyhemoglobin affinity and electrophoretic patterns. Differences in some human hemoglobins are shown in Table 3-3.

3. Nucleoproteins and Nucleic Acids

The nucleoproteins constitute a third group of macromolecules, along with carbohydrates and proteins, which are characteristic of all living plant and animal cells. Even viruses seem to be composed entirely of nucleoprotein. The conjugated proteins are formed by a nonprotein prosthetic group, the nucleic acid, in association with a simple protein. The nucleoproteins are associated with the cell nucleus and with the cytoplasm. Chromatin is mostly made up of nucleoprotein which suggests for it a role in transmission of hereditary factors. Complete hydrolysis of a nucleoprotein by acid or enzymes result in several

TABLE 3-3
Differences in Composition of Human Hemoglobins

Hemoglobin Type	Probable Chain Composition	Differences from Hemoglobin A (A_1)*
$A_{(1)}$	$\alpha_2^A \beta_2^A$	—
A_2	$\alpha_2^A \delta_2^{A_2}$	δ replaces β
F	$\alpha_2^A \gamma_2^F$	γ replaces β
Barts	γ_4^F	Four γ^F chains
C	$\alpha_2^A \beta_2^C$	Lys replaces Glu in β residue No. 6 (same locus as S)
D_α St. Louis ($G_{Philadelphia}$, $G_{Bristol}$)	$\alpha_2^{DSt.\ Louis} \beta_2^A$	Lys replaces AspN in α residue No. 68
D_{Punjab} (Chicago, Cyprus)	$\alpha_2^A \beta_2^{DPunjab}$	GluN replaces Glu in β residue No. 121 (same locus as O_{Arabia})
E	$\alpha_2^A \beta_2^E$	Lys replaces Glu in β residue No. 26
$G_{Honolulu}$ (Singapore, Hong Kong)	$\alpha_2^{GHonolulu} \beta_2^A$	GluN replaces Glu in α residue No. 30
$G_{San\ Jose}$	$\alpha_2^A \beta_2^{GSan\ Jose}$	Gly replaces Glu in β residue No. 7
H	β_4^A	Four β^A chains (may be accompanied by $\delta_4^{A_2}$)
I	$\alpha_2^I \beta_2^A$	Asp replaces Lys in α residue No. 16
M_{Boston}	$\alpha_2^{MBoston} \beta_2^A$	Tyr replaces His in α residue No. 58
$M_{Saskatoon}$ (Emory)	$\alpha_2^A \beta_2^{MSaskatoon}$	Tyr replaces His in β residue No. 63 (same as locus Zürich)
$M_{Milwaukee}$	$\alpha_2^A \beta_2^{MMilwaukee}$	Glu replaces Val in β residue No. 67
Norfolk	$\alpha_2^{Norfolk} \beta_2^A$	Asp replaces Gly in α residue No. 57
$O_{Indonesia}$	$\alpha_2^{OIndonesia} \beta_2^A$	Lys replaces Glu in α residue No. 116
O_{Arabia}	$\alpha_2^A \beta_2^{OArabia}$	Lys replaces Glu in β residue No. 121 (same as locus D_{Punjab})
S	$\alpha_2^A \beta_2^S$	Val replaces Glu in β residue No. 6 (same locus as C)
Zürich	$\alpha_2^A \beta_2^{Zürich}$	Arg replaces His in β residue No. 63 (same locus as $M_{Saskatoon}$)

* Residues numbered from N-terminus.

From Cantarow, A. and B. Schepartz. 1967. *Biochemistry*. W. B. Saunders Co., Philadelphia. p. 127. Table 6-1.

components. The first of these are simple protein, usually basic proteins as histones and protamines, and nucleic acid. The nucleic acid is a mixture of nucleotides—a polynucleotide. Further hydrolysis of the nucleotides derives a mixture of nucleosides and phosphoric acid. The nucleosides are further degraded to a mixture of purines and pyramidines, and ribose or deoxyribose. In other words, a nucleotide is the structural unit of nucleic acid, and consists of a purine or pyrimidine + sugar + phosphoric acid. The two general types of nucleotides found in nucleic acids are ribonucleic acid (RNA) and deoxyribonucleic (DNA) acid named as such because the former contains ribose and the latter 2-deoxyribose. The formulae for these two pentoses are

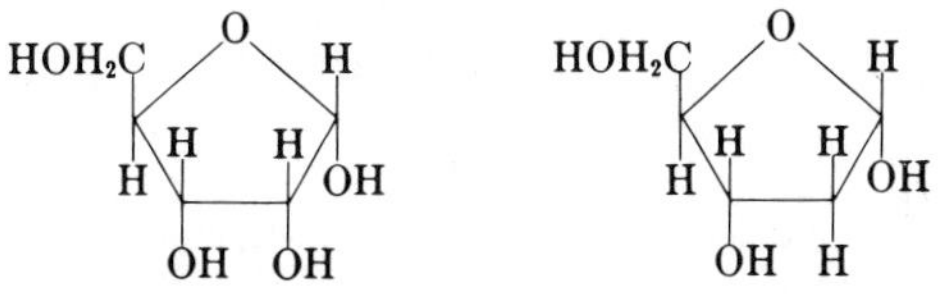

D-Ribose   D-2-Deoxyribose

In Figure 3-2 is shown the scheme of linkage of nucleotide pairs in a segment of the DNA molecule. Phosphoric acid links two sugars, and hydrogen bonding stabilizes the base pairs. When a nucleotidase acts enzymatically on a nucleotide, it splits off phosphoric acid. The resulting structure is a nucleoside which is a purine or pyrimidine base combined with the pentose sugar.

The number system for the pyrimidine and purine rings according to the International System is shown below

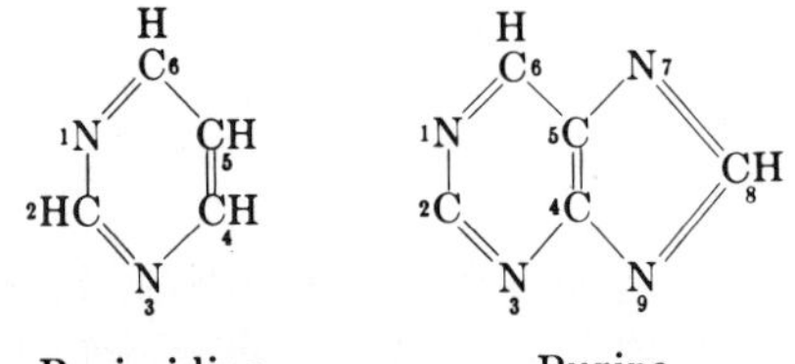

Pyrimidine   Purine

There are three prevalent pyrimidines isolated from nucleic acids. These bases are cytosine, thymine, and uracil. Generally, nucleic acids contain both cytosine and thymine. Uracil, however, is restricted to the ribonucleic acids. Pyrimidine nucleotides are also important in the metabolism of lipids and carbohydrates, and one of the vitamins,

102   CELL ANATOMY AND PHYSIOLOGY

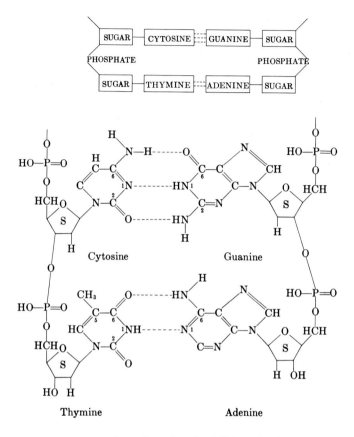

Fig. 3-2. A segment of a deoxyribonucleotide (DNA) molecule showing pairs of nucleotides.

---

thiamine or B_1, is a pyrimidine derivative. There are two purines, adenine and guanine, found in all nucleic acids. The ring structure for purine above shows it to be a pyrimidine ring fused to an imidazole ring.

The nucleotides are the monomers of a nucleic acid containing base (purine or pyrimidine), sugar (ribose or deoxyribose), and phosphoric acid. In a polynucleotide formed by polymerization of nucleotides, the bases may be arranged in any sequence as shown for DNA. In DNA two polynucleotides are such that thymine of one strand is bonded by hydrogen to adenine of the complementary strand. Similarly, cytosine is bonded to guanine. RNA exhibits the same type of base sequence as DNA, but uracil replaces the thymine base. In Figure 3-2 also is shown how base sequence could be arranged in a part of the DNA molecule, how base pairing could occur through hydrogen bonding of nucleotide pairs, and the linkages formed between the sugar and phosphate and the sugar acid base. Stabilization of the DNA helix can be aided by the three hydrogen bonds formed between cytosine and guanine and two between thymine and adenine. Note that the pyrimidine or purine nitrogen is bonded to carbon 1 of the sugar, but a phosphate diester linkage unites individual nucleotides through carbon 3 of one sugar to carbon 5 of the next sugar. It seems that this internucleotide linkage between carbon 3 and carbon 5 is not only true for DNA, but also for RNA.

There are three types of RNA. One is messenger RNA (mRNA). It is a single strand molecule formed in the nucleus as a complementary strand to DNA. Apparently it serves as a means of transcribing genetic information uncoded in DNA to the cytoplasm. Transfer RNA (tRNA) functions as a mediator in protein synthesis by transporting an amino acid to ribosomes. There is at least one tRNA for each amino acid. Ribosomal RNA (rRNA) is associated with ribosomes, but its function is still not settled, although the ribosomes are the site of protein synthesis.

C. Porphyrins

Porphyrins form complexes with metal ions as metalloporphyrins. When these are conjugated to proteins, compounds of extreme importance to biological processes in the cell are formed. Among these are hemoproteins including hemoglobins, myoglobins, cytochromes, and catalyses. Chlorophyll is a magnesium-containing porphyrin. Porphyrins are cyclic compounds formed by four pyrrole rings linked by methylene bridges. As shown below, rings are numbered I, II, III, and IV; hydrogen substituted positions on rings are labeled 1, 2, 3, 4, 5, 6, 7, 8; methylene bridges are labeled α, β, γ, δ.

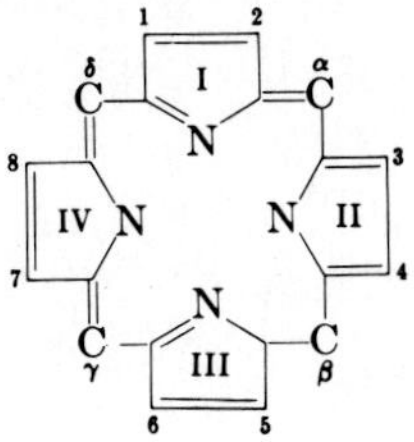

For convenience in showing the substituted positions of the eight hydrogen atoms as numbered above, the Fischer shorthand formula is usually used since the porphyrins found in nature are all compounds with substituted side chains.

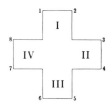

A porphyrin is classed as a type I porphyrin if there is complete arrangement of the substituents.

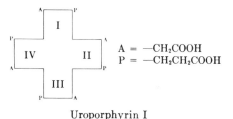

A = —CH₂COOH
P = —CH₂CH₂COOH

Uroporphyrin I

If, however, an asymmetrical substitution occurs as

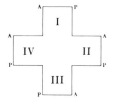

Uroporphyrin III

in ring IV, the porphyrin is classed as a type III porphyrin (or nine since they were designated ninth in a series of isomers isolated by Fischer). Only type I and type III porphyrins occur in nature and the type III series are more abundant.

When metalloporphyrins are formed the metal replaces the two dissociable hydrogen atoms from the nitrogen of pyrrole rings II and IV, and is simultaneously bound by coordinate valences to the nitrogen atoms of pyrrole rings I and III to form a resonating structure which is usually represented (*e.g.*, iron) as

$$\begin{array}{c} N \\ | \\ N-Fe-N \\ | \\ N \end{array}$$

The iron is attached to a histidine residue in both hemoglobin and cytochrome as shown in Figure 3-3. In hemoglobin the iron of heme, protoporphyrin III (9), is conjugated to imidazole nitrogens in two histidine residues of the globin. Iron is closely coordinated to an imidazole group histidine 87 in the α-chain and 92 in the β-chain.

A distinguishing characteristic of all porphyrins regardless of side chains present is a sharp absorption bond near 400 mμ. This band is termed the Soret band and is found in all hemoproteins. Hemoproteins absorb also at other spectral wave lengths.

Of particular importance to the animal body is the catabolism of heme in the formation of bile pigments. Apparently the porphyrin portion of hemoglobin (heme) is broken down by the reticuloendothelial cells of the liver, spleen, and bone marrow. Metabolic degradation of heme is shown in Figure 3-4.

D. Lipids

Lipids are organic substances which are extracted from plant and animal cells by solvents such as ether, chloroform, benzene, hot alcohol, and petroleum ether. This insolubility in water and high solubility in organic solvents indicate the hydrophobic and hydrocarbon nature of the lipids. Since a definition for lipids is not based on structure, classification is difficult but operationable for groups. On this basis lipids may be grouped as simple lipids and compound lipids.

1. Simple Lipids

Simple lipids include the neutral fats and waxes. These neutral fats serve as a diet lipid source, and it is mostly in this form that lipids are stored in animals. As might be expected, neutral fats form the most abundant group of lipids in nature. Neutral fats are esters of fatty acid with glycerol, a trihydroxyl alcohol. When all three of the hydroxyl groups of glycerol are esterified with a fatty acid, the molecule is a

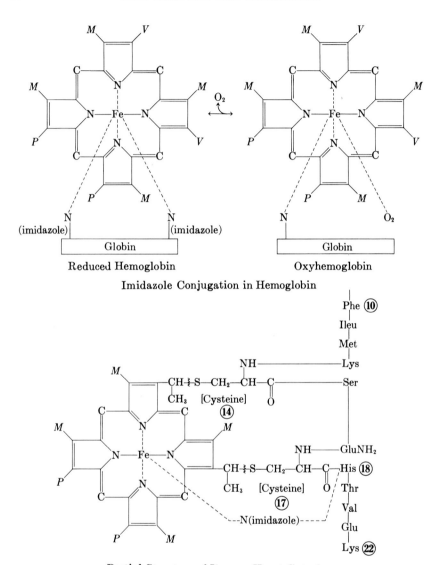

Fig. 3-3. Imidazole conjugation in hemoglobin (above), and cytochrome *c* from human heart (below). From Harper, H. A. 1967. *A Review of Physiological Chemistry*. Lange Medical Publishers, Los Altos, California. p. 76. Figs. 6-6 and 6-7.

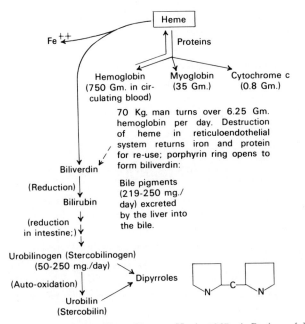

FIG. 3-4. Catabolism of heme. From Harper, H. A. 1967. *A Review of Physiological Chemistry*. Lange Medical Publishers, Los Altos, California. p. 80. Fig. 6-8.

---

triglyceride. Monoglycerides and diglycerides occur in nature, but the triglycerides are the most abundant and important. The general formula for a triglyceride is shown in hydrolysis as

$$\begin{array}{c} R_1{-}COO{-}CH_2 \\ R_2{-}COO{-}CH \\ R_3{-}COO{-}CH_2 \end{array} + 3H_2O \xrightleftharpoons{H^+} \begin{array}{c} R_1COOH \\ + \\ R_2COOH \\ + \\ R_3COOH \end{array} + \begin{array}{c} HOCH_2 \\ HOCH \\ HOCH_2 \end{array}$$

Triglyceride   Fatty Acids   Glycerol

The R groups are indicated as the same or different fatty acids. Triglyceride names are based on the names of the R hydrocarbon residues, *i.e.*, the fatty acids.

When neutral fats are hydrolyzed, they yield three molecules of fatty acids and one molecule of glycerol. In plants and animals, hydrolysis of a lipid is accomplished by lipases. Hydrolysis of fats usually form straight chain derivatives containing an even number of carbon atoms, but some odd-numbered carbon atom fatty acids are found in nature. The chain of a fatty acid may be saturated, or unsaturated with one or more double bonds.

The general formula for a saturated fatty acid is $C_nH_{2n-1}COOH$. Acetic acid is the simplest acid of this series. Palmitic, $CH_3(CH_2)_{14} \cdot COOH$, and stearic, $CH_3(CH_2)_{16}COOH$, are the most abundant saturated fatty acids found in animal lipids. Fatty acids dissociate in water as

$$RCOOH \rightleftharpoons RCOO^- + H^+$$

The unsaturated fatty acids may have more than one double bond. Examples of these fatty acids are oleic with one double bond; linoleic, two; linolenic, three; arachedonic, four; clupanodonic, five. The most abundant of the unsaturated fatty acids in animal lipids are palmitoleic, oleic, linoleic, and linolenic acids. Most of the animal depot fat consists of palmitic, oleic, and steraric acids. The polyunsaturated fatty acids, linoleic and linolenic, apparently are not synthesized by animals. Since they are considered dietary, they have been called essential fatty acids. Of all the fatty acids, oleic acid is the most abundant and widely distributed.

Double bonds can give rise to stereoisomerism. If all of radicals of the molecule appear on the same side of the bond, it is called the "cis" form, and a "trans" form if on opposite sides. Oleic acid is almost symmetrical since its one double bond occurs exactly in the middle of the chain. The "cis" configuration is the one appearing most in nature.

As ester is formed by the condensation of an alcohol and a carboxylic acid, and reconverted to them by hydrolysis. Waxes are esters of high molecular weight fatty acids with alcohols other than glycerol. Cholesterol, a steroid, is one of the most important of these alcohols.

2. Compound Lipids

These lipids, formed in association with other compounds, consist of phospholipids, sphingolipids, glycolipids, terpenes, steroids, and lipoproteins. Because of the complexity and nature of lipids in these complex forms, there is difficulty in following a uniform classification. Therefore, an arbitrary grouping of compound lipids may be made.

The phospholipids, also called the phosphatides, are those lipids containing glycerol, fatty acids and phosphoric acid. Phospholipids are often found associated with cell membranes, and they are particularly abundant in nerve tissue. There are several kinds of these phosphatides. Among them are those lipids containing one of the following nitrogeneous compounds: choline; ethanolamine; or serine. The phosphatidyl cholines, or lecithins, are widely distributed in cells and are important in fat metabolism in the liver. Phosphatidyl ethanolamines and phosphatidyl serines are similar to the lecithins in structure, and are called the cephalins. A diphosphatidyl glycerol, called cardiolipin, has been useful in serological diagnosis of syphillis. Uniquely, it is the only phosphatide with immunological properties. Monophosphatidyl glycerols have been identified in plants. Phosphoinositides contain inositol, and are nitrogen-free compounds. They have been isolated from liver, brain, muscle, and soybean. Plasmogens are similar to lecithins and cephalins, but give a positive reaction to aldehydes with Schiff's test.

The sphingolipids occur as a large number of complex lipids especially in the brain. Instead of glycerol, they contain an amine alcohol, sphingol or sphingosine. The sphingosine phosphatides are known as sphingomyelins since they appear in myelin sheaths of nerves. On hydrolysis, the sphingomyelins yield sphingosine, a fatty acid, phosphoric acid, and choline.

The glycolipids include two major groups, the cerebrosides and the gangliosides. When hydrolyzed, the glycolipid yields sphingosine, a fatty acid, and a monosaccharide, usually galactose or sometimes glucose. The molecule, however, does not contain phosphoric acid or glycerol. The cerebrosides are abundant in the white matter of the brain, and the myelin sheath of nerves. Kerasin, cerebron, nervon, and oxynervon have been isolated and identified as cerebrosides. Gangliosides occur in the gray matter of brain, erythrocyte membranes, and spleen. They are structurally more complex than cerebrosides in that they contain a ceramide linked to the carbohydrate (a ceramide is the amide of sphingosine and a fatty acid). In addition to sphingosine, fatty acids, and carbohydrates contained in cerebrosides, the gangliosides also have neuraminic acid. The complexity of gangliosides arises from the fact that several molecules of carbohydrates may be present. Ceramide oligosaccharides have also been reported. A distinguishing difference between the phospholipids and the glycolipids (glycerosphingolipids) is that only one phospholipid, cardiolipin, mentioned above, is known for immunological activity, but immunological activity is a general property of the glycerosphingolipids.

The terpenes constitute a large and important group of lipids. They are related to each other in that the carbon skeletons contain multiples of five carbon atoms arranged in simple repeating units of isoprene.

Although isoprene is not a naturally occurring compound, it is biologically active in the form of isopentenyl phorphosphate. Condensation of isoprenoid units form such compounds as rubber, carotenes, phytol, squalene, and various carotenoids and steroids. Phytol is the alcoholic fragment obtained from chlorophyll upon hydrolysis. Squalene is an intermediate in the biosynthesis of cholesterol, and is found in shark liver oil; hence, its name from *Squalus*. Terpenoid side chains have been identified in vitamin E, vitamin K, and coenzyme Q.

There have been over 70 carotenoids isolated, besides the carotenes, from plants. It is the β-carotene molecule that is cleaved in the liver to form two molecules of vitamin A as

$$C_{40}H_{56} + 2H_2O \rightarrow 2C_{20}H_{29}OH$$

Other carotenoids besides the carotenes include the xanthophylls. In autumnal coloration, leaves appear golden yellow because of the presence of carotenes and xanthophylls.

The steroids are often found associated with the unsaponifiable residue of fats after saponification. There is a great diversity of physiological activity among the steroid compounds. As a result they constitute a group which highly interests investigators. One reason for attracting the experimentalists is that all known hormones are either steroid compounds, or nitrogeneous compounds such as amino acids, polypeptides, proteins, or derivatives of proteins. There is a cyclic nucleus common to all of the steroids. Shown below is the labeled and numbered carbon ring skeleton of steroids:

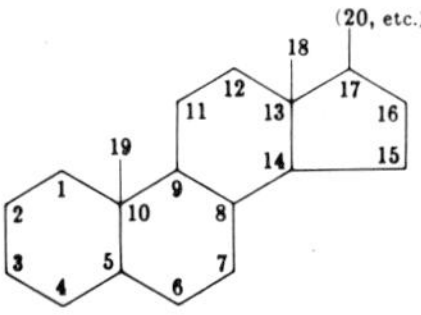

With the exception of relaxin, all of the hormones secreted by the adrenal cortex, testis, and ovary are steroids. All of the hormones secreted by the adenohypophysis and pancreas are large polypeptides and proteins. The neurohypophysis, thyroid, and adrenal medulla secrete hormones all of which are constituted by polypeptides or amino acids. The relations of the steroid hormones are shown in Figure 3-5. Other important steroids include the bile acids, vitamin D, and the sitosterols of plants.

The ring skeleton is a useful aid in depicting steroids related to each

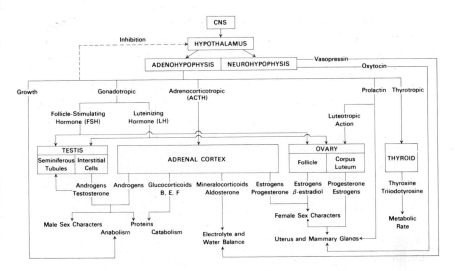

FIG. 3-5. Some interrelations among hormones. From Oser, B. L. (editor). 1965. *Hawk's Physiological Chemistry*. McGraw-Hill Book Co. New York, p. 277. Fig. 11-2.

other. Frequently, there is a side chain at position 17 as in cholesterol. Methyl groups are usually attached at positions 10 and 13; these would form carbon atoms 19 and 18, respectively. If the steroid compound has a hydroxyl group at carbon 3, and about 10 carbon atoms in the carbon 17 side chain, it is a sterol. Cholesterol is widely distributed in all cells of the animal body as well as in blood and bile. It is not a component of plant fat. Another sterol, ergasterol, is an important precursor of vitamin D.

Lipoproteins appear in the blood of mammals as lipids in association with plasma proteins. Lipoprotein complexes of high molecular weight are formed by blood cholesterol in combination mostly with α- and β-globulins.

E. Vitamins

A vitamin is a dietary essential necessary for one or more species. The term comes from vitamine a word used by Funk in 1912 to describe an accessory food factor which he thought to have properties of amines. The final e was dropped when it was later shown that the substances were not related to amines. Not only are vitamins essential in nutrition, but some also function as cofactors. The relations of vitamins and coenzymes in metabolic processes will be discussed later. There are two groups of vitamins identified according to their solubility properties—fat-soluble and water-soluble. The fat-soluble vitamins are A, D, E, and K. The water-soluble vitamins include ascorbic acid (vitamin C), thiamine (B_1), riboflavin (B_2), pantothenic acid (B_3), niacin (nicotinamide) (B_5), pyridoxine (B_6), biotin (B_7), folic acid (B_9), p-aminobenzoic acid, and cobalamin (B_{12}). Inositol and choline are also considered as vitamins by some nutritionists, as is α-lipoic acid.

1. Fat-Soluble Vitamins

Vitamin A exists as A_1, retinol, and A_2 as 3-dehydroretinol. The livers of saltwater fishes contain vitamin A_1 and those of freshwater fishes contain vitamin A_2. Both A_1 and A_2 are usually referred to as vitamin A. The primary precursors, or provitamins, are the carotenes. These provitamin carotenoids are the primary dietary sources of vitamin A. A β-carotene molecule yields two molecules of vitamin A, but the α- and γ-carotenes yield only one. The major location of the conversion of β-carotene to vitamin A in animals is the intestinal mucosa rather than the liver. All vertebrates require a dietary source of vitamin A, or provitamin, for vision, maintenance of epithelium, and normal bone development. Sources of provitamin A include carrots, kale, and broccoli. Milk, cheese, butter, eggs, and mostly liver are the chief dietary sources for vitamin A.

Vitamin D has been called the antirachitic vitamin since it is related to rickets. Rickets is a childhood disease identified by faulty bone formation. Normal bone formation and tooth development depends on normal calcium and phosphorus metabolism which in turn is affected by vitamin D. It has also been called the sunshine vitamin. As long ago as 1822, Sneadecki in Poland recognized that direct sunshine had a preventative and a curative effect on rickets. This effect has been explained as ultraviolet irradiation causing changes in molecular structure of inactive sterols to form activated sterols which then form vitamin D. The provitamins of vitamin D are the inactive sterols. There are about 10 provitamins which yield compounds with antirachitic properties after irradiation. These differ only in the C_{17} side chains, and have been designated D_1, D_2–D_{10}. Of these compounds, two are of particular importance. One is of plant origin and is ergocalciferol (D_2). The other is of animal origin and is cholecalciferol (D_3). Provitamins are widely distributed in plants and in animals. The liver of cod and other fishes is the most abundant source of the common vitamin D. It occurs as cholecalciferol. The flesh and viscera of oily fishes such as herring and sardines are also food sources of D_3. Bile salts are essential for absorption from the intestinal tract of vitamin D.

Vitamin E is known as the "antisterility" vitamin. A chemical group named the tocopherols were found to possess vitamin E activity, *i.e.*, when not included in the diet of males (rats and chickens) they became sterile; in females (rats and birds) this dietary deficiency prevented them from producing normal progeny. There have been seven tocopherols found naturally occurring and of these α-tocopherol is most widely distributed and biologically active. Vitamin E is found in all green plants, and particularly abundant in wheat-germ oil from which it was first isolated. Bile acids are essential for its absorption in the small intestine. They are extremely susceptible to oxidation, and have a sparing action on vitamin A and carotene which are, in the presence of unsaturated fats, sensitive to oxidative destruction. A vitamin E deficiency is also related to muscular dystrophy in rabbits and guinea pigs as well as fetal death in certain rats.

Vitamin K is the antihemorrhagic vitamin. The compounds possessing vitamin K activity are known as the naphthoquinones. There are two naturally occurring vitamins K. These are phylloquinone (K_1) and farnoquinone. Both of these have the same general activity. Vitamin K_1 occurs primarily in plants. Excellent sources are cabbage, cauliflower, and kale and other green vegetables. Other sources are cheese, eggs, and liver. Vitamin K_2 is a metabolic product of bacteria. Most of the higher animals have intestinal bacteria, such as *Escherichia coli*, producing vitamin K_2. Vitamin K is absorbed mostly in the jejunum, and bile salts appear to be essential for optimum absorption. It is unknown by just what process vitamin K participates in blood coagulation, but a deficiency of it influences the concentration of proconvertin and thromboplastin. It stimulates the production of prothrombin by liver cells.

2. Water-Soluble Vitamins

The water-soluble vitamins are termed as such because of their solubility in water. They are no more chemically related to each other than are the fat-soluble vitamins. In the early years of vitamin research, vitamin A was recognized as fat-soluble, and the water-soluble one was vitamin B. In later research, the antiscorbutic factor was termed vitamin C, and the antirachitic factor as vitamin D. Then vitamin B, with its original antiberiberi activity, was shown to have multiple activities. It was consequently designated vitamin B_1. Once isolated, successive factors identified were known as B_2, B_3, etc. The vitamin B-complex is useful in that these factors usually are found in association in nature.

Vitamin C is the antiscorbutic vitamin, and the compound is designated as ascorbic acid. Its activity prevents scurvy. This disease was common in Europe in the 15th century, and was long a dread of ocean voyagers. In 1747, Lind, a British naval surgeon discovered that limes added to sailor rations cured and prevented scurvy. By 1795, official rations included limes and the British sailor became known as a "limey." Ascorbic acid deficiency results in anemia, defects in skeletal calcification, and hemorrhages from the mucous membranes of the mouth and gingiva as well as the gastrointestinal tract. These symptoms result from a weakening of the walls of endothelial cells when the amount of intercellular substance is reduced, and collagen not formed. The best dietary sources of vitamin C are citrus fruits and tomatoes.

The vitamin B complex consists of water-soluble vitamins. Organisms deficient in any one of these factors exhibit one or more species-specific avitaminosis. For example, a thiamine deficiency causes beriberi in man, and polyneuritis in birds. Rats develop alopecia, pediculosis, vascularizations of the cornea, and arrested growth when deprived of riboflavin. Pantothenic acid has a pellagra-preventive activity in the chick, and is a growth-promoting factor in the pigeon. A deficiency of niacin, or nicotinic acid, causes pellagra in man, and blacktongue in dogs. The deficiency of pyrodoxine results in growth failure and acrodynia in rats, and causes convulsions in infants fed on pyroxidine-deficient baby foods. A biotin deficiency can be produced by an egg white protein called avidin in the rat. It is also a microbial growth factor. Folic acid appears widely distributed in the folia of plants. It is an effective factor against sprue, and essential for the growth of microorganisms. Another essential microbial growth factor is p-aminobenzoic acid. It forms a part of the folic acid molecule and has an antagonistic action to the sulfonamide drugs. Vitamin B_{12}, cobalamin, is the antipernicious anemia factor. Pernicious anemia results from defective gastric secretion of mucoproteins (intrinsic factor) essential for normal absorption of B_{12} (extrinsic factor). The source for B_{12} is soil microorganisms since neither plants nor animals synthesize this vitamin. A unique feature of vitamin B_{12} is the presence of cobalt. There has been no other cobalt-containing organic compound found in nature.

Inositol, choline, and α-lipoic acid are not usually considered vitamins although they have some nutritional importance. Inositol deficiency causes alopecia and retarded growth in mice. Choline prevents fatty livers in depancreatized dogs. Anorexia develops in puppies on a low choline intake. An acetate-replacing factor for a specific bacterial growth has been identified as α-lipoic acid.

Vitamin requirements have been established for different species, and in some cases with difficulty. Those vitamins usually listed as essential for man are ascorbic acid, choline, folic acid, niacin, pyridoxine, riboflavin, thiamine, vitamin B_{12}, vitamin A, vitamin D, vitamin E, and vitamin K. Probably biotin and pantothenic acid are also essential.

Cells are apparently unable to handle an excess of some vitamins. In hypervitaminosis A there are gastrointestinal disturbances, scaly dermatitis, alopecia, and joint pains. Hypervitaminosis D finally results in renal failure. Anemia and gastrointestinal disturbances are associated with hypervitaminosis K.

F. Minerals

A mineral is any inorganic homogeneous material found in the earth's crust. There is no uniform distribution of minerals in cells, but some minerals tend to occur in specific regions. Minerals are usually found in the cell in form of salts or combined with carbohydrates, proteins, and lipids. When salts ionize, cations and anions are formed. These ions are important in maintaining osmotic pressure and acid-base balance in body fluids as well as cells.

As in the case of vitamins, different cells in different species may have different requirements for normal metabolic functions requiring minerals. In man, the elements required are calcium, chlorine, magnesium, phosphorus, potassium, sodium, and the so-called trace elements as copper, iodine, iron, manganese, and probably fluorine, molybdenum, selenium, and zinc; cobalt is essential in that it is a part of the vitamin B_{12} molecule. Anemia is caused by iron and copper deficiencies, and a lack of adequate iodine results in goiter. Bone abnormalities result from a calcium deficiency. A calcium and magnesium balance is essential in that a deficiency results in vasodilation and tetany. Magnesium is an activator for some enzyme systems in phosphorylation of compounds such as glucose. Calcium and phosphorus are regulated by the parathyroids. Sodium, potassium, and chlorine are electrolytes controlled by the kidney. Copper, manganese, molybdenum, and zinc are needed by enzymes as cofactors for certain reactions. Iodine is part of the thyroid metabolism. Fluorine apparently prevents dental caries, but, in high concentrations, it causes mottled enamel.

The various elements and inorganic substances may appear in an ionic form in intracellular and extracellular fluids. The ionic concentrations of these fluids inside and outside the cell are different. In the cell K^+ and Mg^{++} can be high, and Na^+ and Cl^- high outside the cell. In muscle and nerve cells a high order of difference exists between intracellular K^+ and extracellular Na^+. Free $Ca^=$ may occur in cells or circulating blood. Phosphates are in tissue fluids and blood as free ions, but may also be bound. Phosphates act as a buffering system tending to stabilize pH of blood and tissue fluids. Tissue ions also include $SO_4^=$, CO_3^-, Mg^{++}, and amino acids. A proper ionic balance is required by cells for normal function.

G. Acids, Bases, and Salts

Acids, bases, and salts have a marked effect on the pH of cells, and on body fluids. A body maintains its acid-base balance by regulating its hydrogen ion concentration. The rates of chemical reaction can be altered greatly, either accelerated or depressed, by just a slight change in hydrogen ion concentration. Although direct measurements of intracellular fluids are questionable, indirect measurements indicate a pH of about 7.0–7.2. In any cell where a rapid rate of metabolism occurs, carbon dioxide concentration increases and the pH is lowered. To prevent either acidosis or alkalosis there is an acid-base buffering system present that prevents immediate excessive changes in hydrogen ion concentration by combining with an excess of either acid or alkali.

An acid or base may be defined in different ways. One way is to define an acid as any substance that is a proton donor, and a base as a proton acceptor. And still another way is to define an acid as any compound which yields positively charged hydrogen ions in solution. A base likewise may be designated as any compound which yields negatively charged hydroxyl ions in solution.

Two types of acids can be recognized. One type is a strong acid which is almost completely dissociated in aqueous solution, and its hydrogen ion concentration is nearly equal to the concentration of the acid present. For example, HCl does not exist as such in aqueous solution, but as H^+ and Cl^-. The other acid type is a weak acid. It ionizes partially in aqueous solution, and has a small hydrogen ion concentration in relation to the total acid concentration present. In an example of acetic acid in solution there exists three forms of the acid: acetic acid, acetate ions, and hydrogen ions; the ions are derived from dissociation of part of the acid. If a weak acid is designated as HA and the anion derived from it as A^- then the equilibrium reaction for the dissociation may be written as

$$HA \rightleftharpoons H^+ + A^-$$

Dissociation of a base may also be expressed in a similar manner as

$$BOH \rightleftharpoons OH^- + B^+$$

Furthermore, acids can react with bases to form salts in a double decomposition reaction indicated as

$$\begin{array}{c} (\text{Acid}) \; HA \; \rightleftharpoons H^+ + A^- \\ + \quad\quad + \\ (\text{Base}) \; BOH \rightleftharpoons OH^- + B^+ \\ \updownarrow \quad\quad \updownarrow \\ HOH + BA \; (\text{Salt}) \end{array}$$

H. pH and Buffers

Recognizing that weak electrolytes ionize makes the understanding of buffer systems more simple. A buffer solution is a solution which resists marked changes in hydrogen ion concentration on the addition of an acid or alkali. The commonest buffer is a mixture of a weak acid and its salt. Any buffer pair may be used to prepare buffers of different hydrogen ion concentrations.

Hydrogen ion concentration is measured in terms of pH. The term pH was introduced in 1909 by Sörensen who defined it as "Für die Zahl p schlage ich den Namen Wasserstoffiionenexponent und die Schreibweise pH vor. Unter dem Wasserstoffionexponenten (pH) einer Lösung wird dann der Briggsche Logarithmus des reziproken Wertes des auf Wasserstoffionen bezagenen Normalitätsfaktors de Lösung verstanden."*

If a solution contains only a weak acid (HA) and its salt (BA), the hydrogen ions can only come from the acid molecules by dissociation as

$$\frac{[H^+][A^-]}{[HA]} = K \qquad (3\text{-}1)$$

---

* The translation into English is: "For the sign p I propose the name 'hydrogen ion exponent' and the symbol pH. Then, for the hydrogen ion exponent (pH) of a solution, the negative value of the Briggsian logarithm of the related hydrogen ion normality factor is to be understood."

CELL ANATOMY AND PHYSIOLOGY

From equation 3-1 it follows that

$$H^+ = K_a \frac{[HA]}{[A^-]} \tag{3-2}$$

In 1916, Hasselbalch rewrote the Henderson equation 3-2 in terms of logarithms as

$$\text{Log } H^+ = \text{Log } K_a + \text{Log } \frac{[HA]}{[A^-]} \tag{3-3}$$

Multiplying both sides of equation 3-3 by -1, it becomes

$$-\text{Log } H^+ = -\text{Log } K_a + \text{Log } \frac{[A^-]}{[HA]} \tag{3-4}$$

Since pH = $-\log H^+$, and $pK_a = -\log K_a$, then a rewriting would be

$$pH = pK_a + \text{Log } \frac{[A^-]}{[HA]} \tag{3-5}$$

Equation 3-5 is called the Henderson-Hasselbalch equation. When a solution contains a weak acid and its salt it may be expressed as

$$pH = pK_a + \text{Log } \frac{\text{Salt}}{\text{Acid}} \tag{3-6}$$

and is most useful in calculating the pH of buffer solutions. For a buffer solution containing acetic acid and sodium acetate, the pH can be calculated by finding the value of pK_a (1.8×10^{-5}) which is equal to 4.7. When sodium acetate = acetic acid then the pH of the buffer solution is

$$pH = 4.7 + \text{Log } \frac{1}{1} \tag{3-7}$$

$$= 4.7$$

By varying the proportion of salt to acid in a solution, the pH range will be about 1.5 units higher or lower than the pK_a value. Other pH ranges can be covered by using other weak acids having suitable pK_a values and their salts.

The importance of a buffer solution is its capacity to resist marked changes in pH. When a base or acid is added to the buffer it may be represented as

Buffer Solution { Acid: $HA \rightleftharpoons H^+ + A^-$; add $OH^- \rightarrow HOH$
Salt: $BA \rightleftharpoons B^+ + A^-$; add $H^+ \rightarrow HA$

Of course, if a greater amount of acid is added to the buffer solution than the equivalent of salt present, the buffer capacity of the solution will disappear. The important property of the buffer is that its capacity depends on the concentration ratio rather than concentration itself.

In an organism's body and interstitial fluids as well as the fluid in its cells, more than one buffer system is present. The main buffer of cells is the carbonate-bicarbonate system. This is so because carbon dioxide is one of the main waste products of cell metabolism, and is constantly being formed. At 25° C the pK_{a_1} of carbonic acid is 6.34 and the pK_{a_2} is 10.25. The phosphate buffer system is also important in the cell. Figure 3-6 shows the three dissociation constants for phosphoric acid.

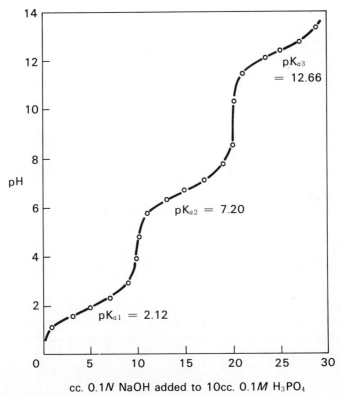

FIG. 3-6. The three dissociation constants are shown when phosphoric acid, a weak acid, is titrated with a base (25° C). From A. C. Giese. 1968. *Cell Physiology*. 3rd edition. W. B. Saunders Co. Philadelphia. p. 189. Fig. 8-3.

Although the carbonate and phosphate buffer systems are most important, there are other buffer systems persent in cell, tissue, and body fluids. Proteins, amino acids, fatty acids, and organic acids can form buffer systems. Hemoglobin is a buffer because the protein molecule contains a large number of acidic and basic groups. In the physiological range of about 7.0–7.8 most of the buffering action by hemoglobin is done by the imidazole groups of histidine.

Like other weak acids, amino acids form buffers in each of the pH regions close to their pK_a values. How the hydrogen ion concentration changes as pH increases may be represented by the various possible ionic forms of amino acids, shown below, which are all considered—to begin with—as in highly acid solutions in cationic form:

$$R\!\!\begin{array}{c}COOH\\NH_3^+\end{array} \underset{}{\overset{K_1}{\rightleftharpoons}} R\!\!\begin{array}{c}COO^-\\NH_3^+\end{array} \underset{}{\overset{K_2}{\rightleftharpoons}} R\!\!\begin{array}{c}COO^-\\NH_2\end{array}$$

Neutral amino acid system

$$R\!\!\begin{array}{c}COOH\\COOH\\NH_3^+\end{array} \underset{}{\overset{K_1}{\rightleftharpoons}} R\!\!\begin{array}{c}COOH\\COO^-\\NH_3^+\end{array} \underset{}{\overset{K_2}{\rightleftharpoons}} R\!\!\begin{array}{c}COO^-\\COO^-\\NH_3^+\end{array} \underset{}{\overset{K_3}{\rightleftharpoons}} R\!\!\begin{array}{c}COO^-\\COO^-\\NH_2\end{array}$$

Acid amino acid system

$$R\!\!\begin{array}{c}COOH\\NH_3^+\\NH_3^+\end{array} \underset{}{\overset{K_1}{\rightleftharpoons}} R\!\!\begin{array}{c}COO^-\\NH_3^+\\NH_3^+\end{array} \underset{}{\overset{K_2}{\rightleftharpoons}} R\!\!\begin{array}{c}COO^-\\NH_2\\NH_3^+\end{array} \underset{}{\overset{K_3}{\rightleftharpoons}} R\!\!\begin{array}{c}COO^-\\NH_2\\NH_2\end{array}$$

Basic amino acid system

An increase in the pH causes the first proton to be removed from the carboxyl group next to the $-NH_3^+$ group. Invariably, K_1 refers to the dissociation of the α-carboxyl group, and K_2 to an additional carboxyl group as in the acid amino acids. The α-amino group dissociates next after the carboxyl group except in the case of histidine. Here the imidazolium ion dissociates after the carboxyl group.

When the average net charge on the amino acid molecule is zero, this pH is referred to as the isoelectric point (pI). It may be calculated for neutral amino acids and dicarboxylic amino acids as shown in Table 3-2 as

$$pI = \frac{pK_{a_1} + pK_{a_2}}{2}$$

For basic amino acids the calculation is

$$pI = \frac{pK_{a_2} + pK_{a_3}}{2}$$

All of the buffers present intracellularly and extracellularly work together since the hydrogen ion concentration is common to all of the buffer systems. This related to the isohydric principle. The phenomenon occurs when a condition changes the hydrogen ion concentration; it causes a change in the balance of all of the buffer systems. The importance of this principle shows that the buffer systems actually buffer each other.

Suggested Reading—Chapter 3

Anfinsen, C. B. *The Molecular Basis of Evolution*. John Wiley & Sons, New York. 1959.
Bennett, T. P. and Frieden, E. *Modern Topics In Biochemistry*. The Macmillan Co. New York. 1966.
Bittar, E. E. *Cell pH*. Butterworth & Co. London. 1964.
Bourne, G. H. (editor). *Cytology and Cell Physiology*. 3rd edition. Academic Press. New York. 1964.
Buvat, R. Electron microscopy of plant protoplasm. Int. Rev. Cytol. *14:*41, 1963.
Cantarow, A. and Schepartz, B. *Biochemistry*. 4th edition. W. B. Saunders Co. Philadelphia. 1967.
Cohn, N. S. *Elements of Cytology*. Harcourt, Brace and World. New York. 1964.
Conn, E. C., and Stumpf, P. K. *Outlines of Biochemistry*. 2nd edition. John Wiley & Sons. New York. 1966.
Copenhaver, W. M. *Bailey's Textbook of Histology*. 15th edition. The Williams & Wilkins Co. Baltimore. 1964.
Davenport, H. W. *The ABC of Acid-Base Chemistry*. 4th edition. University of Chicago Press. Chicago. 1963.
DeRobertis, E. D. P., Nowinski, W. W. and Saez, F. A. *Cell Biology*. 4th edition. W. B. Saunders Co. Philadelphia. 1965.
Deuel, H. J., Jr. *The Lipids*. Vol. 1. Chemistry. Interscience Publishers. New York. 1951.
Dick, D. A. T. *Cell Water*. Butterworth & Co. London. 1965.
Giese, A. C. *Cell Physiology*. 3rd edition. W. B. Saunders Co. Philadelphia. 1968.
Green, D. E. and Goldberger, R. F. *Molecular Insights into the Living Process*. Academic Press. New York. 1967.
Guyton, A. C. *Testbook of Medical Physiology*. 3rd edition. W. B. Saunders Co. Philadelphia. 1966.
Harper, H. A. *Review of Physiological Chemistry*. Lange Medical Publications. Los Altos, Calif. 1967.
Karlson, P. *Introduction to Modern Biochemistry*. 3rd. edition. Academic Press. New York. 1968.
Kavanau, J. L. *Water and Solute-Water Interactions*. Holden-Day. San Francisco. 1964.
Kleiner, I. S. and Orten, J. M. *Biochemistry*. 7th edition. C. V. Mosby Co. St. Louis. 1966.
Leblond, C. P., Glegg, R. E., and Eidinger, D. Presence of carbohydrates with free 1,2-glycol groups in sites stained by the periodic acid-Schiff technique. J. Histochem. Cytochem. *5:*445, 1957.
Loewy, A. G. and Siekevitz, P. *Cell Structure and Function*, 2nd edition. Holt, Rinehart and Winston. New York. 1969.
Lucke, B. and McCutcheon, M. The living cell as an osmotic system and its permeability to water. Physiol. Rev. *12:*68, 1932.
Mahler, H. R. and Cordes, E. H. *Biological Chemistry*. Harper & Row. New York. 1966.
Neil, M. W. *Vertebrate Biochemistry in Preparation for Medicine*. 2nd edition. J. B. Lippincott Co. Philadelphia. 1965.

Oser, B. L. (editor). *Hawk's Physiological Chemistry*. 14th edition. McGraw-Hill Book Co. New York. 1965.
Perutz, M. F. *Proteins and Nucleic Acids*. Elsevier Publishing Co. New York. 1962.
Podolsky, R. J. The structure of water and electrolyte solutions. Circulation *21*:818, 1960.
Porter, K. R. The ground substance: observations from electron microscopy. In *The Cell*. Vol. 2. Brachet, J. and Mirsky, A. E. (editors). Academic Press, New York. 1961.
Schachman, H. K. Considerations on the tertiary structure of proteins. Cold Spring Harbor Symp. Quant. Biol. *28:*409, 1963.
Sebrell, W. H., Jr. and Harris, R. S. (editors). *The Vitamins*. 2nd edition. Academic Press. New York. 1967.
Toporek, M. *Basic Chemistry of Life*. Appleton-Century-Crofts. New York. 1968.
Truex, R. C. and Carpenter, M. B. *Human Neuroanatomy*. 6th edition. Williams & Wilkins Co. Baltimore. 1969.
Wagner, A. F. and Folkers, K. *Vitamins and Coenzymes*. Interscience Publishers. New York. 1964.
Weiss, P. The cell as a unit. J. Theor. Biol. 5:389, 1963.
West, E. S, Todd, W. R., Mason, H. S., and Van Bruggen, J. T. *Testbook of Biochemistry*. 4th edition. The Macmillan Co. New York. 1966.

Chapter Four

Transport Across Membranes

A. Permeability of Membranes

B. Osmosis and Diffusion

C. Active Transport

D. Donnan Equilibrium

E. Pinocytosis and Phagocytosis

F. The Nerve Impulse

G. Impulse-Transmission in Muscle Cells

The cell is the most important and unique entity in nature. As such, it has the capacity to maintain an internal environment by exchanging chemical substances with its external environment without disturbing its fundamental equilibrium. The cell is isolated, but not totally, from its environment by its surface coating—a membrane.

The term membrane can refer to practically any thin barrier which separates two phases or compartments. In biological terminology, many so-called membranes are no more than mechanical screens composed of collagen or polysaccharides secreted as extracellular substances by cells. Among these inert, acellular structures are basement membranes on which epithelial cells rest, Descemet's membrane of the cornea, Bruch's membrane of the uvea, and the cuticular membrane of an insect. Even frog skin and frog urinary bladder have an important place in membrane-model systems. The membranes of particular importance to the cell are the membrane covering the surface of the cell, and the membranes associated with cytoplasmic organelles.

The membrane at the cell surface is variously called the cell membrane, plasmalemma, and plasma membrane. The latter term should take precedence by right of priority since the cell surface covering was first called plasma membrane by Nägeli in 1855. He distinguished the plasma membrane from cytoplasm in that it had a greater density and viscosity. Nägeli also concluded that the plasma membrane endows the cell with its osmotic properties. It is almost unavoidable today to use, interchangeably, the term plasma membrane with cell membrane.

The plasma membrane is the limiting barrier between the cytoplasmic matrix and its extracellular environment. It is this cell membrane which ultimately separates an organism from its environment. Upon destruction of the plasma membrane, protoplasm flows to become mixed with its environment and death to the cell results. But some plasma membranes are plastic enough to stretch and even pass transplanted nuclei without being destroyed.

A cell is not an isolated entity resulting from the surface barrier of a plasma membrane. The nature of the membrane itself allows some substances to pass through it while restricting others. By 1895 Overton had noted two major trends in the permeability of membranes. One was that compounds which dissolved lipids passed through the cell membrane faster than those compounds which were water-soluble. The other trend was that small molecules diffused through membranes faster than large molecules, and often the larger molecules did not penetrate the cell at all. This led to the two fundamental features of membrane structure. One was that the membrane contained much lipid, and the other was that membrane contained pores of molecular dimensions through which solutes could traverse the membrane. Selective permeability is not only characteristic of the plasma membrane, but also of the membranous, intracellular organelles.

There is a distinct morphological similarity between the cell membrane and the organelle membranes. Although the similarity is not resolved by light microscopy, it shows very clearly in electron micrographs. The unit model of these membranes indicats a barrier of two dark, osmiophilic layers separated by a light, osmiophobic layer. The physical appearance of this trilaminar unit model has led to the term unit membrane. It has already been stated, and should be repeated here, that the unit membrane concept implies a trilaminar appearance, but does not explain the chemical nature of the membrane. This has been deduced from indirect evidence, and the actual conformation and chemistry is still an open question.

The traditional model of membrane structure has it that protein is layered in a β-configuration over the polar, hydrophilic surfaces of a lipid bilayer whose hydrophobic, apolar ends adjoin each other. Some recent studies, however, on the conformation of proteins in membranes indicate that the spatial relations of proteins and lipids do not lend support to the "butter sandwich" model concept. These studies are consistent with extensive, apolar, lipid-protein interactions. As a result, a membrane model is conjectured in which various subunits are assembled to form a protein lattice which is penetrated by cylinders of lipids. Thus, contact would occur between protein subunits and between protein and lipid in both polar and apolar regions of the membrane. Perhaps these subunits form the micelles of membranes indicated in recent electron micrographs. Membrane models are usually postulated to contain pores since the selective permeability of membranes allows such rapid transfer of some substances across membranes. A red cell membrane model has been proposed which incorporates both physical and chemical data as shown in Figure 4-1. This model includes an outer glycoprotein layer, plaques, a phospholipid bimolecular leaflet lined on each surface with Ca^{++}, and a protein layer in direct contact with hemoglobin. Polar pores might be lined with protein to give a net positive charge allowing anion exchange, but excluding Na^+ and K^+.

The morphological appearance of the unit membrane is similar in the cell membrane of animals, plants and bacteria; endoplasmic reticulum; Golgi complex; lysosomes; mitochondria; chloroplasts; and nuclei. The unit membrane appears as a single trilaminar structure in cell membranes, endoplasmic reticulum, Golgi complex, and lysosomes. In mitochondria, chloroplasts, and nuclei the organelles have an outer

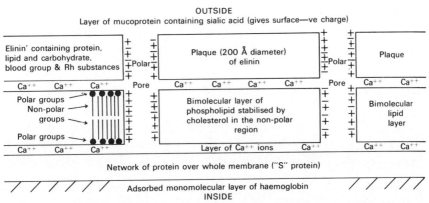

FIG. 4-1. Whittam model of red cell membrane. From Whittam, R. 1964. *Transport and Diffusion in Red Blood Cells*. Edward Arnold. London. p. 34.

and an inner membrane, each with the unit membrane trilaminar structural plan. There can be some variation in the thickness of the unit membrane as a result of varying thicknesses of the osmiophobic and osmiophilic layers. But in most cells examined, small variations only exist in membrane thickness from one species to another. Thus, the trilaminar appearance exhibited by cell membranes studied by electron microscopy leads one to think of the ubiquitous unit membrane as a universal concept.

A. Permeability of Membranes

The membranes of a cell pass small ions and molecules through them. The passage of ions or molecules may occur as passive diffusion, or active transport involving the expenditure of energy. In passive diffusion, membranes may be classed according to their degree of permeability.

1. Impermeable

A membrane of this kind allows nothing to pass through it. Certain unfertilized fish eggs, such as trout, are permeable only to gases; water labeled with deuterium does not penetrate the egg.

2. Semipermeable

No cell membranes are in this category. A model membrane may be constructed to allow passage of water molecules, but not solute particles.

3. Selectively permeable

Most membranes of the cell belong in this category. Such membranes allow water and certain selected ions and small molecules to pass through, but prohibit other ions as well as small and large molecules.

4. Dialyzing membranes

The endothelial cells and their basement membranes of the capillaries and nephron can act as a dialyzer. In this way hydrostatic pressure forces water molecules and crystalloids across the membrane down their concentration gradients while restricting the passage colloids.

B. Osmosis and Diffusion

A plasma membrane is freely permeable to water molecules. Although water passes in and out of a cell, it can do so because of a difference in concentrations of solutes on either side of the membrane. Precisely, the process whereby water passes from a region of higher water concentration through a membrane to a region of lower water concentration is called osmosis. The entry of water into a cell is termed endosmosis; water exit is called exosmosis. The concept of osmosis is often used also to include the passage of solutes in solution through a plasma membrane. In plants, exosmosis from cells cause plasmolysis in which the cell shrinks away from the cell wall. Deplasmolysis is brought about by endosmosis in plant cells.

There are a variety of solutes within a cell. For example, a mammalian erythrocyte contains K^+, Cl^-, Mg^{++}, $PO_4^=$, dissolved hemoglobin, and many other substances. If a red cell is placed in a salt solution containing 0.9% NaCl, it neither shrinks nor swells. A solution of this kind is called an isotonic solution. If the concentration of the NaCl solution is increased over 0.9% to form a hypertonic solution, the cell shrinks as exosmosis occurs. However, endosmosis occurs in a cell placed in a hypotonic solution in which the NaCl concentration is below 0.9%. In fact, in about 0.4% NaCl or less a cell usually hemolyzes.

As water diffuses into a cell in response to a concentration gradient, hydrostatic pressure increases. Since a plasma membrane is not a rigid, but a plastic membrane, a pressure develops within the cell as endosmosis occurs. This pressure is called osmotic pressure. It may be calculated by

$$\text{O. P.} = \text{RTM}_i \qquad (4\text{-}1)$$

where

O. P. = osmotic pressure (in atmospheres)
RTM = 0.08206 × 273 × molarity
i = van't Hoff factor (considered 1 for nonelectrolytes)

The selectivity of the plasma membrane is such that water traverses it in response to a concentration gradient by passive diffusion. The selectively permeable membrane of a cell will also pass certain solutes. For any membrane and any solute a permeability constant may be calculated by

$$K = \frac{\Delta V}{\Delta T} \times \frac{1}{A(P_1 - P_0)} \qquad (4\text{-}2)$$

where

K = permeability constant
$\Delta V/\Delta T$ = observed change in cell volume per unit time
A = surface area of cell

P_i = osmotic pressure inside the cell
P_0 = osmotic pressure outside the cell

Although a cell does not swell or shrink in an isotonic solution, from radioactive tracer studies it is known that water and solutes do exchange across the plasma membrane, particularly Na^+ and K^+. In fact, Na^+ are actively extruded metabolically by a cell placed in an isotonic solution. A 0.9% NaCl isotonic solution corresponds to an osmolar concentration of 0.154 M NaCl which has an osmotic pressure of about 5400 mm Hg. In hypotonic solutions even larger osmotic pressures are developed with correspondingly large osmotic gradients between the interior and exterior of a cell. It appears, therefore, that the greatest movement of water in and out of a cell is in response to osmotic gradients. Osmotic gradients are developed not only by passive diffusion of water and solutes through a selectively permeable membrane, but also by active transport mechanisms whereby the cell utilizes its own energy to pump solutes across its membrane.

C. Active Transport

Active transport is a generic term designating a process in which a cell continously expends metabolic energy to transport a substance across its membrane in a direction opposite that in which the substance tends to diffuse as a result of gradient differences. Or, to put it another way, the Rosenberg definition of an active transport process "is one that results in the net movement of a substance from a region of lower electrochemical potential to a region of higher electrochemical potential." An example is active Na^+-K^+ transport which has received much attention, but just how metabolic energy is used to extrude Na^+ from the cell is not known at present.

Generally, a cellular environment contains interstitial fluid with a high concentration of Na^+ and Cl^-. The intracellular fluid is high in K^+ and organic anions (A^-). Both Na^+ and A^- tend to diffuse, but the plasma membrane seems impermeable to A^-. Although Na^+ continues to leak into the cell, its concentration of Na^+ is low. And the extrusion of Na^+ is accompanied by an uptake by the cell of K^+ which diffuses more freely than Na^+. In fact, the membrane is about 100 times more permeable to K^+ than it is to Na^+. As a result, a transmembrane potential could develop as shown in Figure 4-2. Since A^- does not diffuse and the outward diffusion of K^+ is not replaced by an inward diffusion of Na^+, the inside acquires a net negative charge, and the outside a net

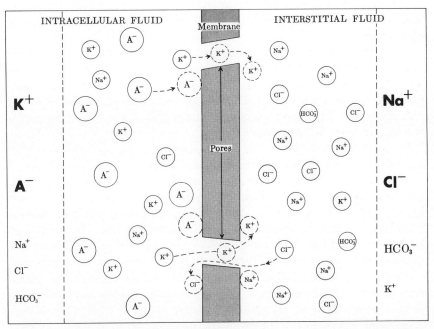

FIG. 4–2. Development of transmembrane voltage by an ion concentration gradient. Diagram of an intracellular fluid–membrane–interstitial fluid system. Membrane shown has some, but not all, properties of a real cell membrane. Hypothetical membrane is pierced by pores of such size that K^+ and Cl^- can move through them easily, Na^+ with difficulty, and A^- not at all. Sizes of symbols in left- and right-hand columns indicate relative concentrations of ions in fluids bathing the membrane. Dashed arrows and circles show paths taken by K^+, A^-, Na^+, and Cl^- as a K^+ or Cl^- travels through a pore. Penetration of the pore by a K^+ or Cl^- follows a collision between the K^+ or Cl^- and water molecules (not shown), giving the K^+ or Cl^- the necessary kinetic energy and proper direction. An A^- or Na^+ unable to cross the membrane is left behind when a K^+ or Cl^-, respectively, diffuses through a pore. Because K^+ is more concentrated on left than on right, more K^+ diffuses from left to right than from right to left, and conversely for Cl^-. Therefore, right-hand border of membrane becomes positively charged (K^+, Na^+) and left-hand negatively charged (Cl^-, A^-). Fluids away from the membrane are electrically neutral because of attraction between + and − charges. Charges separated by membrane stay near it because of their attraction. From Ruch, T. C. and H. D. Patton. 1965. 19th edition. *Physiology and Biophysics*. W. B. Saunders Co. Philadelphia. p. 9. Fig. 3.

positive charge as the K^+ is attracted by A^- back inside the cell. Although the concentration of Cl^- is high outside the cell, its tendency for diffusion is balanced by the electric forces inside the cell. Therefore, the outside and inside Cl^- concentrations are in electrochemical equilibrium.

The plasma membrane is postulated to contain pores of a diameter of about 7 A, or less. Although the Na^+ is smaller than the K^+, the former ion attracts and holds more dipolar water molecules than K^+ and, therefore, the Na^+ has a hydrated diameter of about 5 A as compared to the 4 A of the K^+. This condition of hydration and size may aid the explanation of greater K^+ permeability.

It also appears that the Na^+-K^+ transport system is linked to ATP since ATP hydrolysis is accelerated when both Na^+ and K^+ are present.

The enzyme adenosine triphosphatase which hydrolyzes ATP is intimately associated with cell membranes. The enzyme appears to be a lipoprotein since enzymatic activity is destroyed by lipolytic agents. Membrane ATPase, sometimes called pump ATPase, is stimulated by Na^+, K^+, and Mg^{++}; it is inhibited by Ca^{++}, F^-, digitalis, protamines, and histones. Membrane ATPase seems to be correlated with potassium concentration. In red cell membranes of dogs and cats with low potassium, there is low membrane ATPase; in marrow cells from the same species potassium and membrane ATPase are both high. During embryological development, organs with less potassium develop ATPase when potassium appears. The following scheme has been proposed to explain the lysis of ATP by membrane ATPase:

$$\begin{array}{l} \text{Enzyme (E)} + ATP^{32} \xrightarrow{xNa^+} [E \sim P^{32}]\, Nax^+ + ADP \\ [E \sim P^{32}]\, Nax \xrightarrow{yK^+} E + P_i^{32} + xNa^+ + yK^+ \\ \hline E + ATP^{32} \xrightarrow{xNa^+ + yK^+} E + ADP + P_i^{32} \end{array} \quad (4\text{-}3)$$

It is interesting to note that although ATPase has been identified by histochemical localization in a variety of epithelia noted for active transport including kidney tubules, frog skin, toad urinary bladder, and ciliary epithelium, ATPase activity has also been demonstrated histochemically in bulbar conjunctiva of the eye and palpebral epidermis of the eyelid where no active transport is known to occur. Histochemical identification of ATPase is lacking in the secretory epithelium of the avian salt gland in which active transport has been shown and identified chemically. It seems that a histochemical identification of ATPase does not necessarily mean a localized active transport occurrence. The

Na^+-K^+ activated ATPase has been isolated, however, from many tissues; these include red blood cell (RBC) membranes, brain, nerve, kidney, muscle, liver, intestine, thyroid, parotid gland, frog skin, toad bladder, lens, ciliary body, retina, and electric organ. Other enzymes have been found in association with cell membrane fractions such as acid and alkaline phosphodiesterases, acetylcholinesterase, diglyceride kinase, phosphatidic acid phosphatase, catalase, acid and alkaline phosphatases, glucose 6-phosphatase, NADH-cytochrome c reductase, and triose phosphate dehydrogenase.

There are other active transport pumping systems besides one for Na^+ and K^+ in cells which are specific for certain solutes. Some cells specifically pump amino acids; other glucose, but the same pump does not pump both macromolecules. Some cells have pumps which are specific for H^+, or Ca^+, or $PO_4^=$. *E. coli* contains a permease which allows the accumulation of galactosides.

Intraceullular active transport also occurs in such organelles as nuclei, mitochondria, and chloroplasts which are capable of accumulating or secreting solutes. In a mitochondrion, large amounts of Ca^{++}, Mg^{++}, and $PO_4^=$ can be accumulated. Ca^{++} accumulation can be prevented by 2,4-dinitrophenol which blocks the formation of ATP.

D. Donnan Equilibrium

A maintenance of cell volume for normal activity depends not only on osmotic gradients, but on other gradients also. One of these is the chemical concentration gradient; another is the electrical gradient. These electrochemical gradients greatly affect the distribution of ions permeating the plasma membrane. Although the plasma membrane maintains itself in a steady state by constant expenditure of energy by the cell, the membrane is not in just a state of equilibrium. However, the Donnan equilibrium has been used to demonstrate ion asymmetry on either side of the membrane. A Donnan model of a hypothetical cell may be postulated. The interior of this cell has nondiffusible anions which maintain a permanent internal negative. It also happens that cells do have an internal negative charge with respect to the outer plasma membrane which maintains a positive charge. If the cell model is immersed in a solution of KCl, both the chemical concentration gradient and the electrical gradient will drive the K^+ to the inside of the cell. The concentration gradient will drive the Cl^- to the inside of the cell, but the electrical gradient will drive the Cl^- outside the cell.

At equilibrium, there will be a higher concentration of K^+ inside than outside, but a lower concentration of Cl^- inside than outside the cell. By applying the law of mass action, Donnan showed that, at equilibrium,

$$[K^+]_{in} [Cl^-]_{in} = [K^+]_{out} [Cl^-]_{out} \qquad (4\text{-}4)$$

where the product of the concentration of the diffusible ions inside equals the product of the concentration of the diffusible ions outside the cell.

Actually, many cells do have more K^+ and less Cl^- inside than outside the cell. And most cells are negative inside, but positive on the outside. The voltage of many resting cells has been measured, and the resting potential generally falls in the range of 50–100 mV.

E. Pinocytosis and Phagocytosis

Substances pass into a cell other than through the plasma membrane. It is by the process of pinocytosis or phagocytosis that fluid globules, or food particles, are taken in by the cell. This action of endocytosis is brought about by the plasma membrane. Originally pinocytosis was described by Lewis as a process whereby a cell ingests particle-free globules. These globules were called pinosomes. Pinocytosis can be induced by certain solutes such as proteins, basic dyes, and charged ions. Temperature and pH will also induce pinocytosis.

The endocytic process for phagocytosis is essentially the same as it is for pinocytosis. In phagocytosis, food particles are engulfed by the cell.

In endocytosis, external substances are captured by an area of the plasma membrane which pinches itself off from the cell surface to form a vacuole or vesicle. The membrane-bound packet leaves the periphery of the cell, and migrates toward the cell interior. The packets so-formed may then fuse and assume various sizes, or they may fragment.

Endocytic vacuolization has been demonstrated in a number of cell types, but in some cells it has not been observed, or induced. Besides amebas and leukocytes which have been studied intensely for details of endocytosis, it has been demonstrated in fat cells, fibroblasts, the epithelial cells of gall bladder, epithelial cells of the intestine and kidney tubules which contain microvilli, macrophages, muscle cells, neurons, and osteoclasts as well as others. Time-lapse photography of the plasma membrane during the process of endocytosis indicates an active undulation of the plasma membrane. This undulation has been inter-

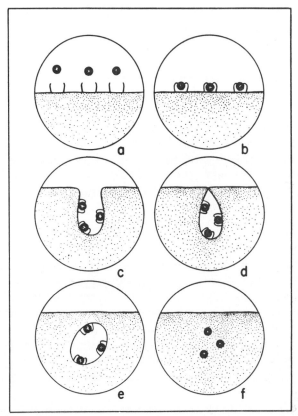

Fig. 4-3. The concept of membrane vesiculation during pinocytosis. In a and b, solute particles attach to the plasma membrane; the vesicle forms in c and d; the pinched-off vesicle in e liberates its contents to the cytoplasm in f. From H. S. Bennett. 1956. The concepts of membrane flow and membrane vesiculation as mechanisms for active transport and ion pumping. J. Biophys. Biochem. Cytol. 2: 99. Fig. 2.

preted as aiding the process in which the vacuole is formed. Another important act in the vacuolization process is the removal of water from the vacuole with the resulting concentration of the vacuolar contents.

The mechanism of endocytosis was first viewed conceptively as a transport by membrane vesiculation with selective uptake by induction as shown in Figure 4-3 by Bennett. A solute would bind to the plasma membrane with a concentration of solute-inducing molecules at the point of binding. The plasma membrane would then be stimulated to activity by forming a vacuole or vesicle which migrates to the interior of the cell. Indeed, this postulation was later confirmed by experimental evidence by Chapman-Andresen and Holter.

It now seems that the process of endocytosis is associated with the digestive apparatus of the cell involving lysosomes. Newly formed vacuoles enclosing extracellular substances fuse with some form of the pre-existing lysosome and acquire acid hydrolases. The residual body thus formed may be expelled from the cell by exocytosis.

In some cells, such as pancreatic gland cells, it has been shown that a reverse of endocytosis occurs. The process is one in which vacuoles in the cytoplasm move to the cell surface, attach to the plasma membrane, and discharge the secretory product to the cell environment. The process has been called emeiocytosis (cell-vomiting), but perhaps a better term for it would be exocytosis.

F. The Nerve Impulse

The conduction of impulses is an important characteristic of excitable cells such as nerve and muscle cells. Conduction is an active, self-propagating process, requiring an expenditure of energy by the cell. Since an exchange of Na^+ for K^+ occurs, a transmembrane potential develops. As K^+ diffuses out faster than Na^+ can be pumped into the cell, the outer membrane is positively charged with a steady-state membrane voltage.

A nerve cell has a low threshold for excitation. The threshold is the minimum strength of an electrical, chemical, or mechanical stimulus exciting a cell. When excited, there is usually a transient depolarization brought about in the cell as the membrane becomes more permeable to Na^+, or Na^+ and K^+. The Na^+ influx depolarizes the membrane which increases Na^+ permeability leading to still further depolarization. Depolarization is followed quickly by repolarization as K^+ outflow exceeds Na^+ inflow. The voltage change accompanying the depolarization-repolarization action is called an action potential. But regardless of the strength of the stimulus, the action potential of a nerve cell follows an all or none behavior in that the response is fixed in size, shape, duration, and conduction speed.

The conduction of the nerve impulse may be likened to a burning line of gun-powder in which each powder particle ignites and fires the next particle in line. At rest the nerve cell membrane is polarized so that the positive charges are lined up along the outside of the membrane and the negative charges on the inside. In Figure 4-4 is represented conductivity of the impulse in both nonmyelinated and myeli-

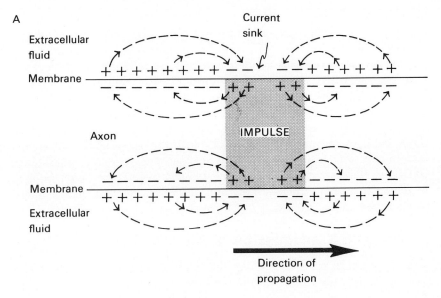

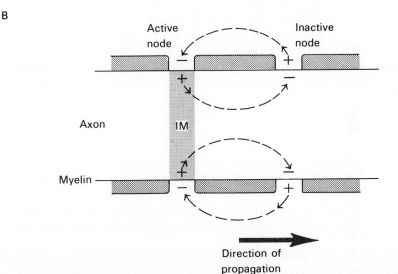

Fig. 4-4. Local current flow around an impulse in an axon. Note that current flow is represented as movement of positive charges. A represents the situation in nonmyelinated nerves, B that in myelinated nerves (saltatory conduction). IM = impulse. From Ganong, W. F. 1967. *Review of Medical Physiology*. Lange Medical Publishers. Los Altos, Calif. p. 26. Fig. 2-11.

nated nerve fibers. At the point of stimulus, the membrane becomes depolarized as it becomes more permeable to Na^+. As the membrane potential approaches the Na^+ equilibrium potential, a potential difference occurs between this local region and an adjacent inactive membrane region where the potential is near the K^+ equilibrium potential. As a result of the potential difference between these two regions, a local current flows from the active region to the inactive region through the intracellular fluid to discharge the membrane. The return current flows back to the active regions through the extracellular fluid and membrane. The local circular current flow acts to reduce membrane charge in the inactive region. Thus, propagation of the nerve impulse is achieved by local current flow and successive depolarizations ahead of the action potential.

G. Impulse-Transmission in Muscle Cells

As an impulse progresses along the fiber of the nerve cell, the transmission of the excitation to another nerve cell occurs at the synapse. The synapse is an area about 200 A wide between two neurons where an impulse is transmitted from one to another. The nerve fiber or axon may end in tiny swellings which is an arborization called telodendria. Each variety of neuron has its peculiar kind of synpatic terminal. There appears to be a transmitter substance involved in the transmission of an impulse at the synpase. One of these is acetylcholine; another is γ-aminobutyric acid (GABA). At synaptic junctions where apposing cell surfaces are in intimate contact, electrical energy serves as the chief agent of transmission instead of a chemical mediator.

Where the axon of the nerve cell approaches the muscle cell, the sheath of Schwann and the myelin sheath are lost. The endoneurium appears to become a part of the endomysium. The axon contacts the sarcoplasm of the muscle cell at the soleplasm. At end points of the terminal arborization are many small synaptic vesicles which contain a transmitter substance. Thus, a myoneural junction is formed at the motor end-plate.

The muscle cell has a modified type of endoplasmic reticulum called a sarcoplasmic reticulum. This system of membranes is specialized to transmit impulses reaching the muscle cell quickly to various areas of the cell. The sarcoplasmic reticulum surrounds the myofibrils of the muscle cell as shown in Figure 4-5. At various points along the surface of the muscle cell, specifically over the Z discs forming the boundaries of the sarcomere—the unit of contraction—there is a sarcolemmal in-

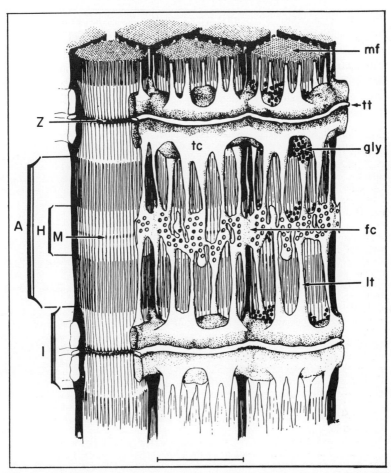

FIG. 4-5. Three-dimensional reconstruction of the sarcoplasmic reticulum (SR) associated with several myofibrils of frog sartorius muscle. The myofibrils (mf) show transverse banding pattern with the Z lines at the center of the light I bands (I). The A band (A) has a central light H zone (H) and denser outer regions where the thick and thin myofilaments overlap, and a dark M line (M) at its center. The fenestrated collar (fc), or H band cisterna, of the SR connects to the terminal cisternae (tc) by longitudinal tubules (lt). Transverse tubule (tt). Glycogen granules (gly). Dimensions: sarcomere, Z line to Z line, 265 μ; A band (thick myofibrils) $\sim$ 1.6 μ; thin myofibrils $\sim$ 1 μ; bar $\sim$ 1 μ. From (1) Peachey, L. D. 1965. The sarcoplasmic reticulum and transverse tubules of the frog's sartorius. J. Cell Biol. 25: 209. Fig. 8. (2) Street, S. F., Sheridan, M. N., and R. W. Ramsey. 1966. Some effects of extreme shortening on frog skeletal muscle. Med. Coll. Va. Quart., 2: 90. Fig. 1 (for the labels).

vagination. This leads into the transverse tubule which courses over the Z line. The T tubule is continuous laterally with the rest of the sarcoplasmic reticulum. It is possible, then, to conceive that an impulse reaching a myoneural junction could cause a wave of depolarization to spread deep into a muscle cell via the tubular sarcoplasmic reticulum so that a sudden muscle contraction could be effected in each sarcomere.

Suggested Reading—Chapter 4

Albers, R. W. Biochemical aspects of active transport. Ann. Rev. Biochem. *36:*727, 1967.

Baker, R. F. Ultrastructure of the red blood cell. Fed. Proc. *26:*1785, 1967.

Bennett, H. S. The concepts of membrane flow and membrane vesiculation as mechanisms for active transport and ion pumping. J. Biophys. Biochem. Cytol. (suppl.) *2:*99, 1956.

Branton, D. and Park, R. B. *Papers on Biological Membranes.* Little, Brown & Co. Boston. 1968.

Chapman-Andresen, C. and Holter, H. Differential uptake of protein and glucose by pinocytosis in *Amoeba proteus*. C. R. Lab. Carlsberg *34:*211, 1964.

Charnock, J. S. and Post, R. L. Evidence of the mechanism of oubain inhibition of cation activated adenosine triphosphatase. Nature (London) *199:*910, 1963.

Cole, Kenneth S. *Membranes, Ions and Impulses.* University of California Press. Berkeley. 1968.

Csáky, T. Z. Transport through biological membranes. Ann. Rev. Physiol. *27:*415, 1965.

Dalton, A J. and Haguenau, F. (editors). *The Membranes.* Academic Press. New York. 1968.

Danielli, J. F. and Davson, H. A contribution to the theory of permeability of thin films. J. Cell. Comp. Physiol. *5:*495, 1934–1935.

Donnan, F. G. The theory of membrane equilibrium in the presence of a non-dialyzable electrolyte. Z. Elektrochem. *17:*572, 1911.

Dowben, R. M. (editor). *Biological Membranes*. Little, Brown & Co. Boston. 1969.

Fawcett, D. W. The membranes of the cytoplasm. Lab. Invest. *10:*1162, 1961.

Fawcett, D. W. Physiologically significant specializations of the cell surface. Circulation *26:*1105, 1962.

Fawcett, D. W. and Revel, J. P. The sarcoplasmic reticulum of a fast-acting fish muscle. J. Biophys. Biochem. Cytol. (suppl.) *10:*89, 1961.

Fishman, A. P. (editor). Symposium on the plasma membrane. Circulation *26:*983, 1962.

Ganong, W. F. *Review of Medical Physiology.* Lange Medical Publishers. Los Altos, Calif. 1967.

Glynn, I. M. Membrane adenosine triphosphatase and cation transport. Brit. Med. Bull. *24:*165, 1968.

Gorter, E. and Grendel, F. On bimolecular layers of lipoids on the chromocytes of the blood. J. Exp. Med. *41:*439, 1925.

Grundfest, H. Some comparative aspects of membrane permeability control. Fed. Proc. *26:*1613, 1967.

Holter, H. Pinocytosis. Int. Rev. Cytol. *8:*481, 1959.

Holter, H. How things get into cells. Sci. Amer. *205(3):*167, 1961.

Hoskin, M. R. Studies on a $Na^+ + K^+$ dependent oubain sensitive adenosine triphosphatase in the avian salt gland. Biochem. Biophys. Acta *77:*108, 1963.

Kanno, Y. and Lowenstein, W. R. A study of the nucleus and cell membranes of ooctyes with an intracellular electrode. Exp. Cell Res. *31:*149, 1963.

Karrer, H. E. Electron microscopic study of the phagocytosis process in lung. J. Biophys. Biochem. Cytol. *7:*357, 1960.

Kernan, R. P. *Cell K*. Butterworth & Co. London. 1965.
Lewis, W. H. Pinocytosis. Bull. Johns Hopkins Hosp. *49:*17, 1931.
Locke, M. (editor). *Cellular Membranes in Development*. Academic Press. New York. 1964.
Lucy, J. A. and Glauert, A. M. Structure and assembly of macromolecular lipid complexes of globular micelles. J. Molec. Biol. *8:*727, 1964.
Maddy, A. H. The chemical organization of the plasma membrane of animal cells. Int. Rev. Cytol. *20:*1, 1966.
Menke, W. Structure and chemistry of plastids. Ann. Rev. Plant Physiol. *13:*27, 1962.
Muscatello, U., Andersson-Cedergren, E., Azzone, G. F., and von der Decken, A. The sarcotubular system of frog skeletal muscle. A morphological and biochemical study. J. Biophys. Biochem. Cytol. (suppl.) *10:*201, 1961.
Northcote, D. H. Structure and function of plant cell membranes. Brit. Med. Bull. *24:* 107, 1968.
Page, S. Structure of the sarcoplasmic reticulum in vertebrate muscle. Brit. Med. Bull. *24:*170, 1968.
Peachey, L. D. The sarcoplasmic reticulum and transverse tubules of the frog's sartorius. J. Cell Biol. *25:*209, 1965.
Policard, A. and Bessis, M. Micropinocytosis and rhopheocytosis. Nature (London) *194:* 110, 1962.
Ponder, E. The cell membrane and its properties. In *The Cell*. Vol. 2. Brachet, J. and Mirsky, A. E. (editors). Academic Press. New York. 1961.
Porter, E. R. The sarcoplasmic reticulum. Its recent history and present status. J. Biophys. Biochem. Ctyol. *10:*219, 1961.
Robertson, J. D. The unit membrane. In *Electron Microscopy in Anatomy*. Boyd, J. D., Johnson, F. R., and Lever, J. D. (editors). Williams & Wilkins Co. Baltimore. 1961.
Robertson, J. D. The membrane of the living cell. Sci. Amer. *206(4):*64, 1962.
Robertson, J. D. Unit membranes. In *Cellular Membranes in Development*. Locke, M. (editor). Academic Press. New York. 1964.
Robers, H. J. and Perkins, H. R. *Cell Walls and Membranes*. E. & F. N. Spon. London. 1968.
Rosenberg, T. A. The concept and definition of active transport. Sympos. Soc. Exp. Biol. *8:*27, 1954.
Rustad, R. C. Pinocytosis. Sci. Amer. *204(4):*120, 1961.
Schoffeniels, E. *Cellular Aspects of Membrane Permeability*. Pergamon Press. New York. 1968.
Simpson, F. O. and Oertelis, S. J. The fine structure of sheep myocardial cells; sarcolemmal invaginations and the transverse tubular system. J. Cell Biol. *12:*91, 1962.
Solomon, A. K. Pores in the cell membrane. Sci. Amer. *203(6):*146, 1960.
Stein, W. D. *The Movement of Molecules Across Cell Membranes*. Academic Press. New York. 1967.
Street, S. F., Sheridan, M. N., and Ramsey, R. W. Some effects of extreme shortening on frog skeletal muscle. Med. Coll. Va. Quart. *2:*90, 1966.
Tormey, J. M. Significance of the histochemical demonstration of ATPase in epithelia noted for active transport. Nature (London) *210:*820, 1966.
Troshin, A. S. *The Problem of Cell Permeability*. Pergamon Press. New York. 1966.
Wallach, D. F. H. and Gordon, A. Lipid protein interactions in cellular membranes. Fed. Proc. *27:*1263, 1968.
Whaley, W. G. Dynamics of cytoplasmic membranes during development. Soc. Study Develop. Growth *22:*135, 1964.
Whittaker, V. P. Structure and function of animal cell membranes. Brit. Med. Bull. *24:* 101, 1968.
Whittam, R. *Transport and Diffusion in Red Blood Cells*. Edward Arnold. London. 1964.

Section II.
Cell Metabolism

Chapter Five

Bioenergetics, Enzymes, and Cofactors

A. BIOENERGETICS
1. FREE ENERGY
2. REDOX REACTIONS
3. HIGH-ENERGY COMPOUNDS

B. ENZYMES
1. THEORY OF ACTION
2. KINETICS
3. INHIBITORS AND ANTIMETABOLITES
4. NOMENCLATURE
5. ISOZYMES

C. COFACTORS

A. Bioenergetics

Bioenergetics is the study of energy transformation in the living organism. There are certain basic processes occurring in cells which yield energy. One of these processes is photosynthesis; another is respiration.

Energy flows throughout the plant and animal world. In photosynthesis, chlorophyll absorbs radiant energy and converts it into chemical energy. It is this chemical energy which is utilized to reduce CO_2 from the atmosphere to ultimately form glucose. Autotrophic cells containing chlorophyll can synthesize organic molecules of carbohydrates, lipids, proteins, and nucleic acids from CO_2, H_2O, and NH_3. Other heterotrophic cells of both plants and animals use these food groups in the combustion processes of respiration. Of course, autotrophic cells also perform respiration.

There are other functions occurring in cells which require energy. Active transport is one and biosynthesis of molecules is another. From the processes yielding it, energy is shuttled by ATP to those processes which require energy. As physical and chemical reactions occur, they are accompanied by either an absorption or release of heat to the environment. Heat is a means of transferring energy, and occurs only when there is a difference in temperature between a system and its environment. Heat is absorbed in an endothermic process, and in an exothermic reaction heat is lost to the environment. Cells, however, are essentially isothermic in that there is very little temperature difference in various parts of cells or between cells and their environment.

1. Free Energy

How cells function under isothermal conditions is related to free energy (F). Although the first law of thermodynamics is essentially the conservation of energy, there is a direction in spontaneous physical and chemical reactions not explained by the first law, but is explained by the second law. All systems tend to approach states of equilibrium in which all measurable parameter states (such as temperature and pressure) become invariant. The second law states also, in effect, that the randomness of a system tends to increase so that entropy (S) becomes maximal. At this point equilibrium is reached, and both ΔS and ΔF are zero.

The internal energy (ΔE) of a system is affected by the gain or loss of heat (ΔH) by that system. And if the reaction occurs at a constant pressure and volume, then for condensed phases

$$\Delta H \simeq \Delta E \tag{5-1}$$

In a chemical reaction, ΔH (enthalpy, or the change of internal energy) can be measured by direct calorimetry. Organic compounds have a heat of combustion, *i.e.*, the calories of heat liberated to the environment when 1 gm of molecular weight is completely oxidized. A calorie is that amount of heat required to raise the temperature of 1 gm of water 1°C (14.5°–15.5°C). For example, in the combustion of 1 gm of molecular weight of glucose in a calorimeter, $\Delta H = -673{,}000$ cal/mole. The relation of heat and entropy to free energy was developed by Gibbs as

$$\Delta F = \Delta H - T\,\Delta S \tag{5-2}$$

where
  ΔF = change of free energy in the system
  ΔH = heat transferred between system and environment (enthalpy)
  T = absolute temperature
  ΔS = entropy change in the system.
In the chemical reaction,

$$aA \rightleftharpoons bB \tag{5-3}$$

the free energy content of substance A cannot be measured experimentally, but one can speak of the change in free energy, or the maximum amount of energy made available as "a" moles of A is converted to "b" moles of B. When the reaction occurs with a decrease in free energy, ΔF is negative. If the reverse of the reaction occurs, there will be an increase in free energy and ΔF will be positive. In spontaneous reactions, ΔF is negative. Reactions in which energy is taken up are termed endergonic; those which liberate energy to their surroundings are termed exergonic.

For any reaction such as

$$aA + bB = cC + dD + \cdots$$

$$K = \frac{aC^c aD^d \cdots}{aA^a aB^b \cdots}$$

the equilibrium constant (K) is a mathematic function of ΔF of the components of the reaction: As a function of the activities of the reactants and the activities of the products when equilibrium has been established, ΔF is related to K as

$$\Delta F^\circ = -RT \ln K \qquad (5\text{-}4)$$

where

$\Delta F^\circ =$ standard state, or standard change in free energy when the activities of both reactants and products are present in unit concentrations

R = gas constant (1.987 cal/mole- degree)
T = absolute temperature
ln K = natural logarithm of the equilibrium constant.

At 37°C,

$$\Delta F^\circ = -RT \ln K \qquad (5\text{-}5)$$

$$= (1.987)\,(273 + 37)\,(2.303)\,\log_{10} K$$

$$= -1419 \log_{10} K$$

With any arbitrary concentration other than the standard state, the value for free energy change may be calculated from

$$\Delta F = \Delta F^\circ + RT \ln K \qquad (5\text{-}6)$$

Most biological reactions occur in the cell at pH 7.0 or close to it. The symbol $\Delta F'$ is used to indicate the standard free energy change at pH

7.0. If a proton is neither formed nor utilized in the reaction, $\Delta F°$ will equal $\Delta F'$ since $\Delta F'$ will be independent of pH.

In photosynthesis, radioactive tracer studies have shown that the oxygen released to the atmosphere comes from water. In the formation of a glucose molecule from CO_2 and H_2O, there are. over a hundred chemically known sequences. Each reaction is catalyzed by a specific enzyme. The over-all aspects of photosynthesis may be represented by

$$6\ CO_2 + 12\ H_2O \xrightarrow[\text{energy}]{\text{radiant}} C_6H_{12}O_6 + 6\ O_2 + 6\ H_2O$$

$$\Delta F° = +\ 686{,}000\ \text{cal/mole}$$
$$\Delta H = +\ 673{,}000\ \text{cal/mole}$$
$$\Delta S = -\ 43.6\ \text{cal/mole.}$$

Clearly, the thermodynamics of the over-all process of photosynthesis represents an endergonic reaction with a large increase in free energy and a decrease in entropy.

Respiration represents the next large process in plant and animal cells where energy flows. The combustion of 1 gm of molecular weight of glucose, for example, may be identified thermodynamically as

$$C_6H_{12}O_6 + 6\ O_2 \rightarrow 6\ CO_2 + 6\ H_2O$$

$$\Delta F° = -\ 686{,}000\ \text{cal/mole}$$
$$\Delta H = -\ 673{,}000\ \text{cal/mole}$$
$$\Delta S = +\ 43.6\ \text{cal/mole.}$$

Respiration then represents an exergonic process with a decrease in free energy and an increase in entropy.

Free energy change and equilibrium may be calculated from a chemical reaction. During the breakdown of muscle glycogen to lactic acid, glucose 1-phosphate is catalyzed to glucose 6-phosphate enzymatically by phosphoglucomutase. The reaction may be initiated and completed as

$$\underset{(0.020\ M\ \rightarrow\ 0.001\ M)}{\text{Gl 1-PO}_4} \xrightarrow[\substack{25°C \\ pH\ =\ 7.0}]{\text{phosphoglucomutase}} \underset{(0.019\ M)}{\text{Gl 6-PO}_4}$$

The standard free energy may be calculated as
$$\Delta F' = -\ RT \ln K$$
$$= -\ (1.987)(298) \ln 0.019/0.001$$

BIOENERGETICS, ENZYMES, AND COFACTORS

$$\begin{aligned}
&= -(1.987)(298) \ln 19 \\
&= -(1.987)(298)(2.303) \log_{10} 19 \\
&= -1363 \times 1.28 \\
&= -1745 \text{ cal/mole.}
\end{aligned} \quad (5\text{-}7)$$

In this reaction $\Delta F' = \Delta F°$ since no acid is formed. It is most important to remember that an enzyme only accelerates a reaction; it does not influence the equilibrium point reached.

2. Redox Reactions

In chemical reactions in the cell involving oxidation-reduction processes, the change in free energy (ΔF) may be related to the difference in oxidation-reduction potentials (ΔE). The redox potential is that electric potential (volts or millivolts) measured against the hydrogen reference electrode. Some reactions involve the transfer of only one electron as in iron-porphyrin enzymes.

$$\underset{\text{reductant}}{Fe^{++}} \underset{\text{reduction}}{\overset{\text{oxidation}}{\rightleftarrows}} \underset{\text{oxidant}}{Fe^{+++} + e^-}$$

or electrons may be transferred in pairs as in most chemical reactions as

$$\underset{\text{hydroquinone}}{H_2Q} \underset{\text{reduction}}{\overset{\text{oxidation}}{\rightleftarrows}} \underset{\text{quinone}}{Q + 2H^+ + 2\ e^-}.$$

From these reactions it can be seen that a reducing agent (reductant) furnishes electrons and will be oxidized, and an oxidizing agent (oxidant) accepts electrons to become reduced.

In a reaction yielding an electromotive force (E), and supplying a quantity of electricity (Q), the electrical work performed is QE. Since Q is equal to the faraday (F) for each equivalent reacting, then the n equivalents reaction would be $Q = nF$. The electrical work for any reaction supplying nF coulombs of electricity at a potential ΔE is $nF\Delta E$. As work is accomplished with a decrease in free energy, then $-\Delta F$ must equal the net electrical work performed as

$$\Delta F = -nF\Delta E \quad (5\text{-}8)$$

Likewise,

$$\Delta F° = -nF\Delta E°. \quad (5\text{-}9)$$

Since, from equation 5-4

$$\Delta F° = -RT \ln K$$

and equating equations 5-4 and 5-9

$$-nF\Delta E° = -RT \ln K$$

then

$$\Delta E^0 = \frac{RT}{nF} \ln K \qquad (5\text{-}10)$$

where
ΔE^0 = the difference in the reduction potential between the oxidizing and reducing agents
R = gas constant
T = absolute temperature
ln K = natural logarithm of the equilibrium constant

In equation 5-9, $F\Delta E°$ units are coulomb-volts or joules. These are readily converted to free energy units since 4.18 joules equal 1 gm-cal. A common reaction in cells is the oxidation of NADH by molecular oxygen as

$$NADH + H^+ + \tfrac{1}{2} O_2 \rightarrow NAD^+ + H_2O.$$

The free energy of this reaction may be calculated from

$$\Delta F' = -nF\Delta E' \qquad 5\text{-}11$$

as

$$\Delta F' = \frac{-2 \times 96{,}500 \times [0.816 - (-0.320)]}{4.18}$$

$$= -52{,}400 \text{ cal.}$$

3. High-energy Compounds

All plant and animal cells use the same fundamental processes in transferring chemical energy. In oxidative reactions, the free energy available may be used to drive energy-requiring or endergonic reactions. A group of high energy phosphate compounds function largely to link exergonic to endergonic processes. These phosphate compounds are termed high-energy or energy-rich compounds since, upon hydrolysis, their decrease in free energy can be larger.

The most ubiquitous compound in which the energy from oxidative reactions is conserved as chemical energy is ATP. It is ATP which is the carrier and immediate source of chemical energy to those reactions in the cell which do not occur spontaneously, and can occur only if chemical energy is supplied. In Chapter 3 it was mentioned that purines and pyrimidines are universally distributed in cells in the form of nucleotides. Adenine is a purine derivative and a component of ATP. Attached to the base adenine are other components of ATP which are D-ribose and three phosphate groups. If ATP loses a phosphate group, it becomes adenine diphosphate (ADP). A loss of two phosphates converts ATP to adenine monophosphate (AMP). The structures of AMP, ADP, and ATP are given in Figure 5-1. All three of these adenine nucleotides occur within the cell.

It has already been mentioned that most reactions in the cell are carried out at about a pH of 7.0. At a pH of 7.0, each of the three phosphate groups of ATP are completely ionized. These four negative charges are concentrated around the linear polyphosphate chain, and lying close to each other, repel one another strongly. As a result, the ATP molecule has a high free energy of hydrolysis when ATP is hydrolyzed to ADP ($\Delta F' = -7$ kcal/mole). When the end-terminal phosphate of ATP is hydrolyzed, some of this electrostatic stress is reduced, and the ADP and phosphate will be electrostatically repulsed by each other. Another factor contributing to the high free energy of hydrolysis

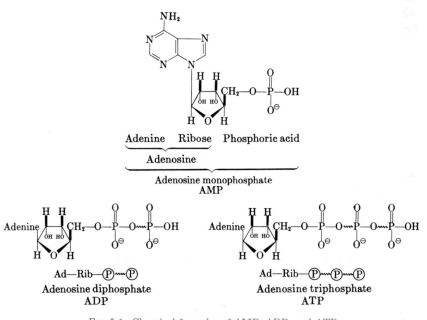

FIG. 5-1. Chemical formulas of AMP, ADP, and ATP

of ATP is that ADP and phosphate, as formed products, have a lower energy content than when they were part of the ATP molecule. On hydrolysis of ADP to AMP and inorganic phosphate, $\Delta F' = -6.5$ kcal/mole; hydrolysis of AMP to adenosine and phosphoric acid is less, $\Delta F' = -2.2$ kcal/mole.

The free energy of hydrolysis of some phosphate compounds can be very high in comparison with other compounds. Those with a high free energy of hydrolysis have a high phosphate transfer potential; those compounds with a low free energy of hydrolysis tend to keep phosphate groups, and thus have a lower potential. The $\Delta F'$ in kcal/mole of some of these compounds are phosphoenolpyruvate (-12.8), 1,3-diphosphoglycerate (-11.8), phosphocreatine (-10.5), acetyl phosphate (-10.1), ATP (-7), glucose 1-phosphate (-5), and glucose 6-phosphate (-3.3). The standard free energies of hydrolysis thus can vary in phosphate compounds with ATP having somewhat of an intermediate phosphate transfer group potential.

There seems to be a tendency to refer to those compounds with a high free energy of hydrolysis as having high energy bonds and designating them as $\sim$ P. Actually, this representation is a misnomer since it implies that the energy is in the bond, and that this energy is set free when the bond is split. Bond energy is actually the energy required to break a bond between two atoms. Therefore, the free energy of hydrolysis is not the chemical bond, but is a measure of the difference in energy content between the initial reactants and final products as indicated in equation 5-4.

ATP functions in the cell as the major intermediate in metabolic energy transformation. As a common intermediate, it links exergonic and endergonic reactions. ATP carries chemical energy from high energy donors to low energy phosphate acceptors. ATP can function in a cyclic process and in coupled reactions. For example, in the enzymatical controlled reaction,

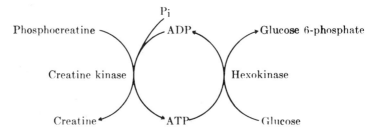

Here ATP is consumed, but regenerated through a coupled reaction. This type of reaction is common in the study of metabolism.

B. Enzymes

Cells do not function at or near chemical equilibrium, *i.e.*, where $\Delta F = 0$. In order for chemical systems in cells to function properly, they must be in a dynamic equilibrium or steady state. In other words, reactions in cells must tend toward equilibrium, but must not be at equilibrium. Chemical reactions occurring constantly which tend toward equilibrium provide the energy necessary for vital processes in cells. Steady state concentrations are maintained differently in dynamic equilibrium than in chemical equilibrium. Cells also have transport phenomena to aid steady state concentrations such as the uptake of foodstuff and oxygen and the elimination of CO_2 and waste products. The steady state concentrations of reacting substances are regulated enzymatically.

An enzyme is a specialized protein molecule having the capacity to act as a catalyst in a chemical reaction. A catalyst is any substance which influences the rate of a chemical reaction while remaining unchanged chemically at the end of the reaction. Thus, catalysis is the process of either positive or negative rate acceleration. The enzyme neither changes the ΔF of a reaction nor modifies the equilibrium constant. The equilibrium constant is high as the chemical reaction tends to completion. In exergonic reactions, the standard free energy change is negative, *i.e.*, there is a decline of free energy as the reaction proceeds. If the reaction does not proceed in the direction of completion, the equilibrium constant is low.

The greater the energy possessed by reactants, the less reactive they are. In other words, substances possess an energy of activation. According to this concept, reactants must first acquire enough energy to pass over an energy activation barrier before products are formed. An enzyme can function to lower the energy of activation of the reaction it catalyzes, and the depression of the energy of activation for equilibrium is exclusive for that particular reaction. Consequently, an enzyme has a reaction specificity. For example, if glucose is mixed with ATP very little glucose 6-phosphate and ADP are formed. But it hexokinase is added the enzyme greatly facilitates the reaction. As shown in Figure 5-2, it appears as if a molecule of glucose and a molecule of ATP are bound to the hexokinase at adjacent sites on the surface of the enzyme.

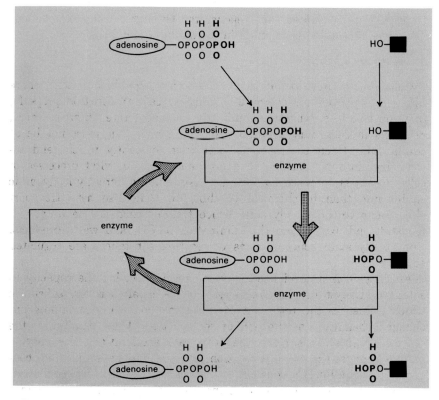

FIG. 5-2. Schematic representation of the action of hexokinase. The two reactants shown at the top of the figure are attached to the enzyme; the phosphoryl transfer then takes place; and the two products are released. The enzyme can then combine with a fresh pair of reactants. (The black square represents the glucose molecule.) From Bell, G. H., J. N. Davidson, and H. Scarborough. *Textbook of Physiology and Biochemistry.* 7th edition. E. & S. Livingstone. London. 1968. p. 138. Fig. 8.32.

After transfer of the phosphoryl group occurs the two products are released.

1. Theory of Action

Although it is not known just how enzymes operate chemically in the cell, enough data have been accumulated to form a working hypothesis. Apparently the enzyme enters into a chemical combination with its substrate to form a complex which then breaks down to give the products of the reaction. A model would be indicated as

$$\text{enzyme} + \text{substrate} \rightleftharpoons \text{complex}$$
$$E + S \rightleftharpoons ES$$

$$\text{complex} \rightarrow \text{enzyme} + \text{products}$$
$$ES \rightarrow E + P$$

Unless there is a physical combination of ES, it would be difficult to explain the high specificity of enzymes. But the most striking evidence is mathematical where the effect of substrate concentration on the rate of an enzyme-catalyzed reaction can be predicted.

2. Kinetics

Enzyme activity is greatly affected by temperature, hydrogen ion concentration, and other ionic concentrations. Generally, increasing the temperature 10° C doubles the rate (Q_{10}) of most chemical reactions. Since an enzyme is a protein, it has an upper limit at which loss of activity occurs as thermal denaturation commences at about 40°–60° C. Enzymes usually are most active at an optimum pH although the effective pH range of about 4–9 covers most enzymes. As with other proteins, enzymes are denatured at higher and lower pH limits away from their optimum pH. The pH effect on enzymes may be a cause-effect relation because of the presence of ionizable groups. Activity may be affected because certain groups need to be ionized while other groups are not. For activity, some enzymes require a loosely-bound cation (such as Mg^{++}).

The velocity or reaction rate of enzyme reactions is affected by enzyme concentration. This velocity-enzyme concentration is represented graphically in Figure 5-3. In the presence of an excess of the substrate, an increase in the concentration of enzyme results in an increase in reaction velocity.

At a constant enzyme concentration, the reaction rate is also dependent on the concentration of the substrate. This relation is illustrated in Figure 5-4. The Michaelis-Menten hypothesis explains this substrate dependency curve on the assumption that an intermediate enzyme-substrate complex is formed. Furthermore, it was hypothesized that the rate of conversion of the substrate to the reaction products is determined by the rate of conversion of the enzyme-substrate complex to reaction products and the enzyme. If, at a constant enzyme

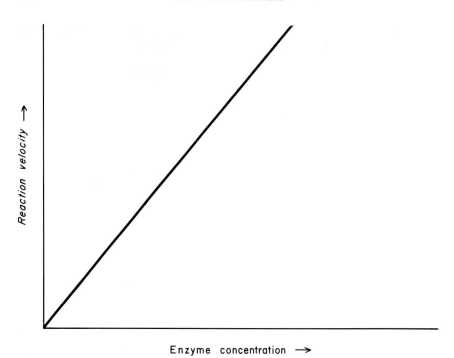

FIG. 5-3. The effect of increasing enzyme concentration on the velocity of an enzyme-catalyzed reaction.

concentration, the rate of formation of products varied directly with the substrate, a linear relation would be expected between velocity and substrate concentration. As shown in Figure 5-4 the relation, however, is not linear; to explain the experimental curve, Michaelis and Menten derived an expression

$$V = \frac{V_{max}[S]}{K_m + [S]} \qquad (5\text{-}12)$$

where

V = reaction velocity
V_{max} = point at which reaction velocity is maximal since virtually all enzyme exists as enzyme-substrate complex and saturated

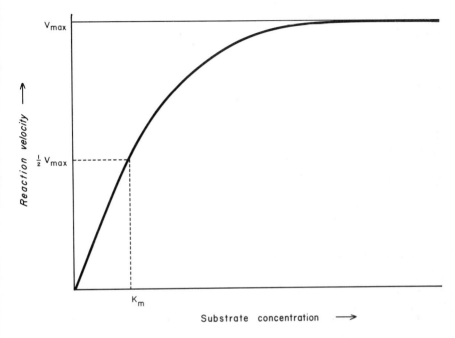

FIG. 5-4. The effect of increasing substrate concentration on the velocity of an enzyme-catalyzed reaction.

S = substrate concentration
K_m = the concentration (moles/l) of the substrate which gives half the numerical maximum velocity (Michaelis constant).

The Michaelis-Menten equation above contains two easily measurable variables:

(1) the reaction rate (V) at a given substrate concentration;
(2) the rate V_{max} is achieved at saturation of the enzyme.

These measurements can then be used to measure K_m. K_m values are important in that they provide a comparison of the affinity of an enzyme for different substrates. Along with V_{max}, K_m provides a useful parameter in describing the action of an enzyme.

Both V_{max} and K_m can be obtained from a single graphic presentation of the reaction velocity at various substrate concentrations. This evaluation is simplified by using the procedure of Lineweaver and Burk which is a reciprocal of equation 5-12.

$$\frac{1}{V} = \left(\frac{K_m}{V_{max}}\right)\left(\frac{1}{[S]}\right) + \frac{1}{V_{max}} \qquad (5\text{-}13)$$

Since 1/V and 1/[S] are variables, the equation represents the function of a straight line, $y = ax + b$. K_m can then be evaluated from the coordinate intercepts when 1/V (ordinate) is plotted *versus* 1/[S] (abscissa) as shown in Figure 5-5. In the plot shown

$$\frac{K_m}{V_{max}} = \text{the slope, and}$$

$$\frac{1}{V_{max}} = \text{the intercept on the } \frac{1}{V} \text{ axis}$$

When 1/V is plotted *versus* 1/[S] the intercept of the curve with 1/V is

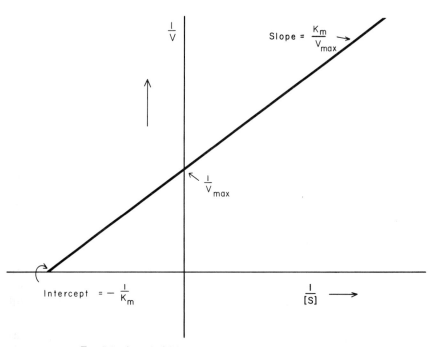

FIG. 5-5. A typical Lineweaver-Burk plot of equation 5-13

$1/V_{max}$; therefore, V_{max} can be calculated. The slope of the curve is obtained from the graph; it is equal to K_m/V_{max}. Since V_{max} is known, K_m is obtained as

$$K_m = V_{max} \times \text{the slope}$$

where the slope can be calculated from two points on the line as

$$\text{slope} = \frac{d\left(\frac{1}{V}\right)}{d\left(\frac{1}{[S]}\right)}.$$

3. Inhibitors and Antimetabolites

Since many drugs and poisons act to inhibit enzyme action, enzyme inhibition is important in cell activity. Inhibition of enzymes is quite different from denaturation since there is no large scale disruption of the protein molecule. Inhibition occurs when a compound combines with an enzyme, but does not serve as a substrate, and consequently blocks catalysis by the enzyme. Most inhibitors fall in one of two classes; either they are noncompetitive or competitive. Noncompetitive inhibition cannot be reversed by increasing the concentration of the substrate as it can in competitive inhibition.

In noncompetitive inhibition, many enzymes can be poisoned (inhibited or completely inactivated), e.g., by reagents which combine with free sulfhydryl (—SH) groups such as Cu^{++} ions, or mercury and arsenic compounds. Many copper- or iron-containing enzymes are inhibited by the cyanide ion (CN^-). Cyanide poisoning results because of the inhibition of cytochrome oxidase which is essential for nearly all mammalian cells.

When competitive inhibition occurs, another organic molecule competes with the substrate for binding at the active site of the enzyme. Usually such inhibitors show a close structural resemblance to the substrates of the enzymes they inhibit. In competing for the active site of the enzyme, an inhibitor can displace the substrate if it is present in a sufficiently high concentration. The inhibitor may, on the other hand, be displaced by an increase of substrate concentration.

A Lineweaver-Burk plot can be used to demonstrate whether an inhibition is competitive or noncompetitive. For example, in Figure 5-6 where the plot of inhibited and noninhibited systems are shown, it may be seen that the noncompetitive system has a different inter-

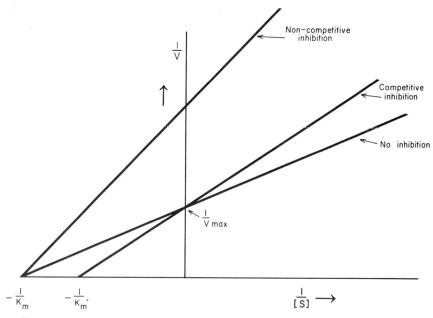

Fig. 5-6. A Lineweaver-Burk plot of V and [S] in the presence of a competitive and noncompetitive inhibitor.

cept on the 1/V axis with a different slope, but the competitive systems have the same intercept with different slopes.

A classic example of competitive inhibition is succinic acid dehydrogenase which oxidizes succinic acid to fumaric acid. If increasing concentrations of malonic acid, (or malic acid, and oxaloacetic acid) are added, the activity of succinic acid dehydrogenase decreases rapidly. Presumably, all of these acids act as competitive inhibitors since they have a structural resemblance to succinic acid. Interestingly, by increasing the concentration of the substrate, succinic acid, the inhibition can be reversed.

In the living cell, the essential substances aiding in maintaining the steady state by enzymatic action are termed metabolites. As in the phenomenon of competitive inhibition of an isolated enzyme, certain substances may be introduced into a cell and compete for the active site of the enzyme in a particular metabolic pathway. These substances are called antimetabolites. Normally they have a similar structure to the metabolite with which they compete. Antimetabolites are particularly studied as inhibitors of pathogenic microorganisms and neoplasms. Those antimetabolites which have been isolated from living cells are called antibiotics. For example, it has been mentioned that p-aminobenzoic acid (PABA) is a vitamin. Those organisms requiring PABA for growth utilize it for folic acid synthesis. The sulfonamides can inhibit the growth of these organisms requiring PABA, and this inhibition can be reversed by supplying PABA.

Few enzymes produce a radical change in their substrate. Consequently, products tend to resemble their substrates and therefore might be expected to inhibit reactions as they are formed. Indeed, it is quite often true that an end-product inhibition occurs which is the result of the inhibitor binding to the surface of the enzyme. End-product inhibition is a form of competitive inhibition and not just a general inhibition resulting from an accumulation of products. End-product inhibition acts as a control in an individual reaction. As long as there is an excess of substrate over product, inhibition is minimal. As a result, reactions in metabolic pathways can occur with formation of intermediates which continue until the end-product level rises and exerts a negative feedback on the system.

There seems to be two sites on the enzyme involved in end-product inhibition. One is the site for catalytic substrate binding, and the other is for end-product binding. In the latter case, inhibitory binding apparently alters the tertiary structure of the enzyme molecule which subsequently affects its catalytic reactivity. This binding at the non-catalytic site is called the allosteric site, and enzymes which possess a capability for end-product inhibition are thus called allosteric enzymes.

4. Nomenclature

Enzymes have long been known by the trivial names they possess, i.e., pepsin, trypsin, etc. Mostly, enzymes have been named according to the reaction they catalyzed simply by adding the suffix -ase to the substrate, e.g., arginase which acts on arginine. There are also groups of enzymes which have generic names since they catalyze similar reactions. Examples of these are lipases, proteases, etc. With a rapid growth in enzymology during the last several years, difficulties in terminology arose. As more enzymes were discovered and as individual workers continued to name their own enzymes, duplicate terminology appeared inevitable. And with no general agreement on nomenclature

of coenzymes, a nomenclature began to appear in which the same name was sometimes given to different enzymes, different names were given to the same enzyme, and similar names were given to enzymes of different types. Subsequently, an International Commission on Enzymes was formed. By 1961 the commission had devised a complete and complex system for naming, classifying, and coding enzymes systematically. The six main divisions adopted by the commission were:
1. oxidoreductases
2. transferases
3. hydrolases
4. lyases
5. isomerases
6. ligases (synthetases)

Oxidoreductases catalyze redox reactions. They transfer electrons and hydrogen ions. Dehydrogenases, oxidases, oxygenases, and peroxidases are included in this division.

Transferases catalyze the transfer of a group of atoms from one molecule to another. The enzymes of this division transfer the following groups:
1. one carbon
2. aldehydic or ketonic residues
3. acyl
4. glycosyl
5. alkyl
6. nitrogenous
7. phosphorus-containing
8. sulfur-containing

Hydrolases catalyze the hydrolysis of a complex molecule into two components by adding the elements of water across the bond which is cleaved. These enzymes act on the following bonds:
1. ester
2. glycosyl
3. ether
4. peptide
5. other C—N bonds
6. acid anhydride
7. C—C
8. halide
9. P—N

Lyases catalyze either the removal of a group of atoms from their

substrate leaving double bonds or add groups to double bonds without hydrolysis, oxidation, or reduction. These enzymes act on the following bonds:
1. C—C
2. C—O
3. C—N
4. C—S
5. C—halide

Isomerases catalyze the intramolecular rearrangement of atoms in their substrates. These include a large variety of enzymes:
1. racemases and epimerases
2. cis-trans isomerases
3. intramolecular oxidoreductases
4. intramolecular transferases
5. intramolecular lyases

Ligases, or synthetases, catalyze the linkage of two molecules, coupled with the breakdown of a pyrophosphate bond in ATP or a similar triphosphate. These enzymes form the following bonds:
1. C—O
2. C—S
3. C—N
4. C—C

Lactate dehydrogenase may be used as an example of the systematic classification of enzymes by the Enzyme Commission: EC 1.1.1.27 (L-lactate:NAD oxidoreductase). The trivial name would be retained as lactate dehydrogenase. At cell pH, organic acids can be ionized and exist in the salt form. In this instance, lactic acid exists as lactate, and its enzyme is lactate dehydrogenase.

In many texts and papers today most authors refer to the enzyme classification based on a derived enzyme name from its substrate with the suffix -ase added. For convenience of discussion, some of these generic names are given.
1. *Carbohydrases*—hydrolyze glycosidic linkages of glycosides, oligosaccharides, and polysaccharides.
2. *Proteases*
   a. endopeptidases—hydrolyze peptide bonds in interior of proteins.
   b. exopeptidases—hydrolyze peptide bonds adjacent to free amino or carboxyl group.
3. *Amidases*—hydrolyze nonpeptide C—N groups.

4. *Esterases*—hydrolyze ester linkages, *e.g.*, phosphatases, lipases.
5. *Dehydrogenases*—act on substrate to make possible removal of hydrogen.
6. *Oxidases*—activate molecular oxygen so that it can serve as a hydrogen acceptor.
7. *Decarboxylases*—remove carbon dioxide from carboxylic acids.
8. *Hydrases*—add or remove water from substrate.
9. *Transferases*—catalyze a group transfer from one substance to another.
10. *Isomerases*—catalyze intramolecular rearrangement.

5. Isozymes

Enzymes may exist in more than one form called isozymes. Two isozymes which have been identified are malate dehydrogenase (MDH) and lactate dehydrogenase (LDH). The isozyme most thoroughly investigated is LDH which has been demonstrated in nearly all vertebrate organs with starch-gel electrophoresis. Different molecular forms of LDH exist. For example, in mammals, the class mostly studied, LDH appears as five tetromers resulting from two different monomers. One is predominately associated with heart and the other with skeletal muscle, *e.g.*, (1) H_4, (2) H_3M, (3) H_2M_2, (4) HM_3, and (5) M_4. Heart LDH is strongly inhibited by pyruvate and muscle LDH, weakly inhibited. Heart LDH has a different amino acid composition from muscle LDH. And different species may have a different number of major and minor LDH isozymes. In fishes the major patterns can vary from 1 to 5; the fluke has only one, but both shad and herring have five major LDH isozymes. A variety of minor isozymes may also be present in various species of fishes.

There seems to be an interesting relation between LDH and the physiological role of flight in birds. In those birds grouped as short flyers, over 99% muscle LDH is present in breast muscle. In birds characterized as sustained and long flyers, breast muscle contains over 95% heart LDH which suggests a possible relation to sustained heart contraction. Those birds classed as intermediate flyers contain mixtures of both heart LDH and muscle LDH. In Figure 5-7 is shown an isozymogram of LDH in the house sparrow for heart, breast muscle, and cerebrum.

C. Cofactors

In order for some chemical reactions to occur, some other substance may be necessary other than a substrate and an enzyme. These sub-

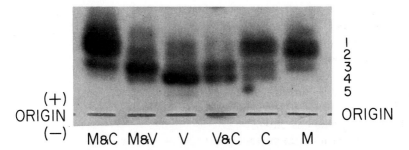

FIG. 5-7. Isozymogram of lactate dehydrogenase (LDH) in house sparrow pectoralis muscle (M), heart ventricle (V), and cerebrum (C) extracts prepared separately and which are mixed equally with one another for dissociation-reassociation. From Bush, F. M. and W. W. Farrar. 1969. Dissociation-reassociation of lactate dehydrogenase: reversed isozyme migration and kinetic properties. Proc. Soc. Exp. Biol. Med. *131*:13. Fig. 1.

---

stances seem appropriately called cofactors since they help make the reaction go, and are found unchanged at the end of a catalyzed reaction. They are not enzymes since all enzymes which have been described are proteins. In fact, most enzymes are conjugated proteins called holoenzymes. A holoenzyme is composed of a protein part, the apoenzyme, and a nonprotein portion, the prosthetic group. Often the prosthetic group is not removed when the enzyme is purified so that it appears to be very tightly bound. This group can, however, often be removed and added back to the apoprotein to reconstitute the enzyme. There are other cases in which the nonprotein moiety is only loosely bound, dissociates easily, and sometimes is not isolated with the enzyme. Therefore, some authors do not distinguish between prosthetic groups and coenzymes. In this sense, both prosthetic groups and coenzymes are considered as cofactors. In fact, a prosthetic group can be a generic term indicating any group attached to a protein whether or not the protein has a catalytic effect. The distinction made here is that a prosthetic group becomes oxidized and reduced while it remains attached to the enzyme molecule. Conversely, a coenzyme is a carrier substrate which undergoes a reduction while attached to one enzyme molecule, but migrates to another enzyme molecule by which it can be oxidized. Accordingly, nicotinamide adenine dinucleotide (NAD), nicotinamide adenine dinucleotide phosphate (NADP), and coenzyme A are considered coenzyme carrier substrates, but hemes, flavins, biotin, and pyridoxal phosphate are prosthetic groups. In some cases enzymes may be brought to an active catalytic state by simple substances called activators. They may be no more than inorganic ions. A metal ion may form a complex with the substrate which then reacts with the enzyme. Thus, activators may also be considered as cofactors. Metal cations which have been found to participate in one or more enzymatic reactions are Na^+, Cs^+, Rb^+, K^+, Ca^{++}, Co^{++}, Cu^{++}, Mg^{++}, Mn^{++}, Cd^{++}, Fe^{++}, Cr^{+++}, and Al^{+++}. At ordinary concentrations, the effect of most anions on enzymes is negligible.

Two of the most important reactions occurring in cells are concerned with hydrogen transport and phosphate transport. The former is involved with the production of energy whereas the latter is concerned with transfer of energy from one process to another. ATP is the main carrier of phosphate. The enzymes which catalyze transphosphorylation processes are kinases. Hydrogen and electron transport carriers are important in biological oxidations which are catalyzed by enzymes. Each enzyme functions in conjunction with a cofactor. Many cofactors are closely related to vitamins. Generally, a vitamin is the main component of a coenzyme, or a vitamin may have a prosthetic group as a part of its molecule. The essentialness of a vitamin seems related to its association with cofactors.

There are various coenzymes and prosthetic groups which act as hydrogen and electron carriers, and other groups (Table 5-1). Of particular importance here are the two pyridine nucleotides, NAD and NADP. These two are coenzymes for the enzymes known as dehydrogenases. NAD and NADP are the coenzyme forms of the pellagra-preventing vitamin, nicotinamide. It is the vitamin portion of the coenzyme which can be reduced. The reduction may be described as the transfer of a hydride ion from a dehydrogenase-substrate complex (metabolite) to the nicotinamide portion of the pyridine nucleotide with a concomitant release of a proton to the medium:

$$MH_2 + NAD^+ \rightarrow M + NADH + H^+$$

The pyridine nucleotides can, once reduced, function in the reduction of flavin adenine mononucleotide (FMN) and flavin adenine dinucleotide (FAD). These are the prosthetic groups of enzymes which carry out the oxidation-reduction of organic substrates and are derivatives of the vitamin riboflavin. Therefore, the pyridine nucleotides, reduced by the dehydrogenase-substrate complex, carries hydrogen to the flavoproteins which become reduced:

$$H^+ + NADH + FAD \rightleftharpoons NAD^+ + FADH_2$$

TABLE 5-1
Coenzymes and Prosthetic Groups

Coenzyme or Prosthetic Group	Symbol	Enzymatic Function or Group Transferred	Related Vitamin
Nicotinamide adenine dinucleotide	NAD	Hydrogen carrier	Nicotinamide
Nicotinamide adenine dinucleotide phosphate	NADP	Hydrogen carrier	Nicotinamide
Flavin mononucleotide	FMN	Hydrogen carrier	Riboflavin
Flavin adenine dinucleotide	FAD	Hydrogen carrier	Riboflavin
Ubiquinone (coenzyme Q)	Q	Hydrogen carrier	
Lipoic acid	LIP(S_2)	Hydrogen and acyl carrier; oxidative decarboxylation	
Iron-protoporphyrin		Electrons	
Adenine triphosphate	ATP	Transphosphorylation	
Pyridoxal phosphate	PALP	Amino group; decarboxylation	Pyridoxine
Tetrahydrofolic acid	CoF	Formyl group	Folic acid
Adenosylmethionine		Methyl group	Methionine
Biotin		Carboxyl group	Biotin
Coenzyme A	CoA	Acyl group	Pantothenic acid
Thiamine pyrophosphate	TPP	Oxidative decarboxylation; active aldehyde carrier	Thiamine
Uridine diphosphate	UDP	Active aldehyde carrier	
B_{12} coenzyme		Carboxyl intramolecular shift	Cobalamine

Whereas the NAD and NADP dissociate readily, the FMN and FAD are tightly bound as flavoproteins. Most flavoproteins do not contain FMN, but rather FAD. Over 60 flavoproteins are known at present.

The flavoproteins are a group of enzymes which catalyze oxidation-reduction reactions. The prosthetic groups of these enzymes are FAD and FMN, both of which contain riboflavin. Hence, the term flavoproteins. There are two well known chemical reactions in which flavoproteins participate. One flavoprotein, called NADH dehydrogenase (with FMN as its prosthetic group), accepts hydrogen from reduced pyridine nucleotides. The other flavoprotein is succinic acid dehydrogenase which catalyzes the oxidation of succinic acid to fumaric acid in the TCA cycle. Both flavoproteins, in either case, transfer their electrons to ubiquinone in the electron transport chain. Succinic acid dehydrogenase has FAD as its prosthetic group. It acts on its substrate without the mediation of pyridine nucleotides. In

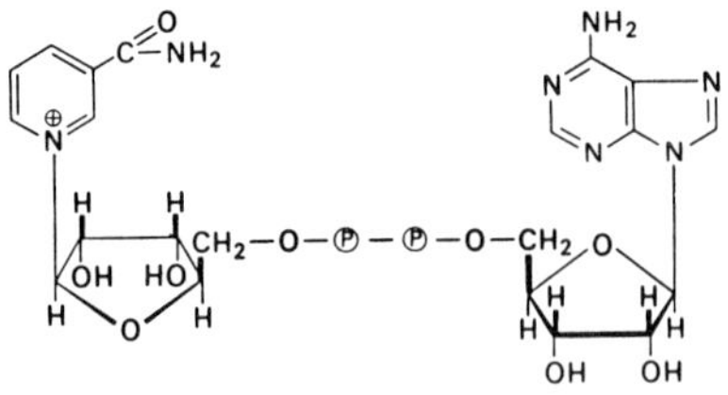

FIG. 5-8. Chemical formulas of NAD, NADP, riboflavin, FMN, and FAD

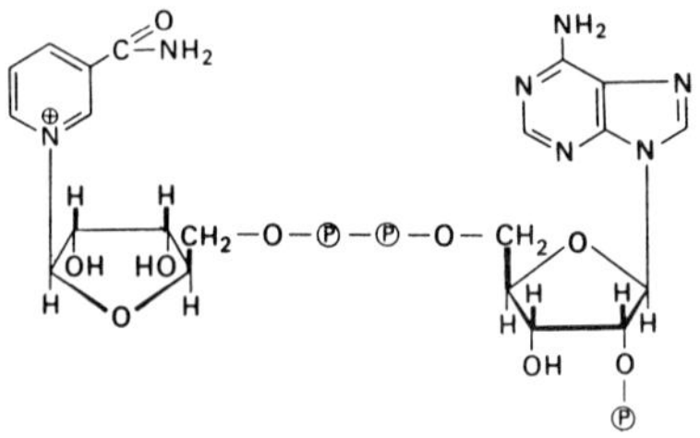

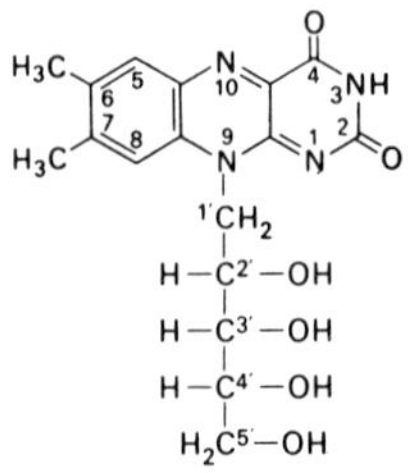

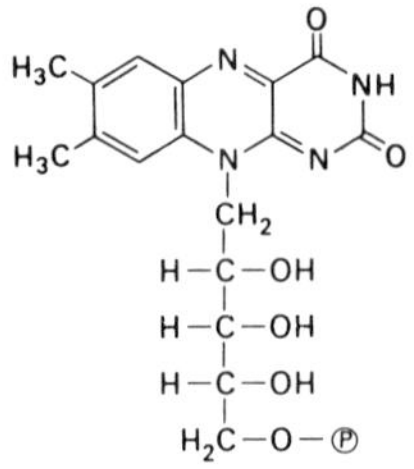

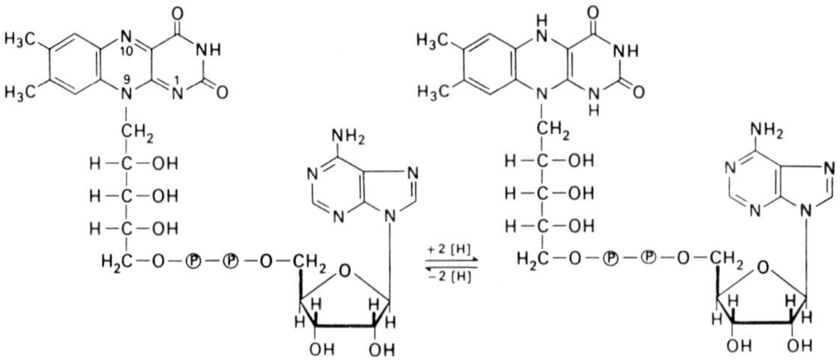

Figure 5-8 are shown formulas for NAD, NADP, riboflavin, FMN, and the oxidized and reduced forms of FAD.

Suggested Reading—Chapter 5

Bell, G. H., Davidson, J. N., and Scarborough, H *Textbook of Physiology and Biochemistry.* 7th edition. E. & S. Livingstone. London. 1968.
Bernhard, S. A. *The Structure and Function of Enzymes.* W. A. Benjamin. New York. 1968.
Boyer, P. D., Lardy, A. H., and Myerbäck, K. *The Enzymes.* 2nd edition. Academic Press. New York. 1960.
Bush, F. M. and Farrar, W. W. Dissociation-reassociation of lactate dehydrogenase: reversed isozyme migration and kinetic properties. Proc. Soc. Exp. Biol. Med. *131:*13, 1969.
Conn, E. C. and Stumpf, P. F. *Outlines of Biochemistry.* 2nd edition. John Wiley & Sons. New York. 1966.
Diem, K. (editor). *Documenta Geigy.* Geigy Pharmaceuticals. Ardsley. New York. 1962.
Dixon, M. and Webb, E. C. *Enzymes.* 2nd edition. Academic Press. New York. 1964.
Giese, A. C. *Cell Physiology.* 3rd edition. W. B. Saunders Co. Philadelphia. 1968.
International Union of Biochemistry. *Enzyme Nomenclature.* Elsevier. New York. 1965.
Karlson, P. *Introduction to Modern Biochemistry.* 3rd edition. Academic Press. New York. 1968.
Lehninger, A. L. *Bioenergetics.* W. A. Benjamin. New York. 1965.
Markert, C. L. The molecular basis for isozymes. Ann. N. Y. Acad. Sci. *151:*14, 1968.
Markert, C. L. and Faulhaber, I. Lactate dehydrogenase isozyme patterns of fish. J. Exp. Zool. *159:*319, 1965.
Maron, S. H. and Prutton, C. F. *Principles of Physical Chemistry.* 4th edition. The Macmillan Co. New York. 1965.
Michaelis, L. and Menten, M. L. Kinetics of invertase action. Biochem. Z. *49:*333, 1913.
Moyed, H. S. and Umbarger, H. E. Regulation of biosynthetic pathways. Physiol. Rev. *42:*444, 1962.
Roodyn, D. B. The classification and partial tabulation of enzyme studies on subcellular fractions isolated by differential centrifuging. Int. Rev. Cytol. *18:*99, 1965.
White, A. Handler, P. and Smith, E. L. *Principles of Biochemistry.* 4th edition. McGraw-Hill Book Co. New York. 1968.
Racker, E. *Mechanisms in Bioenergetics.* Academic Press. New York. 1965.
Segel, I. H. Phosphate bond energy. In *Encyclopedia of Biochemistry.* Lansford, E. M., Jr. (editor). Reinhold. New York. 1966.
Shaw, C. R. Electrophoretic variation in enzymes. Science *149:*936, 1965.
Shows, T. B. and Ruddle, F. H. Malate dehydrogenase: Evidence for tetrameric structure in Mus musculus. Science *160:*1356, 1968.
Smith, D. J. The structure of flight muscle sarcomeres in the blowfly *Calliphora erythrocephala* (Diptera). J. Cell Biol. *19:*135, 1963.
Stadtmen, E. R. Allosteric regulation of enzyme chemistry. Advances Enzymol. *28:*41, 1966.

Chapter Six

Biological Oxidations and Mitochondria

A. MITOCHONDRIAL PATHWAYS

B. THE TRICARBOXYLIC ACID CYCLE

C. ELECTRON TRANSPORT AND OXIDATIVE PHOSPHORYLATION

A. Mitochondrial Pathways

The carrier pyridine nucleotides and flavoproteins are associated with biological oxidations occurring in mitochondria. When the major foodstuffs of carbohydrates, proteins, and lipids from intermediary metabolism undergo oxidation, hydrogen atoms and electrons pass along certain pathways in mitochondria. In Figure 6-1 these pathways are summarized in a generalized scheme. If one looks at the scheme, the relation of the TCA cycle to electron transport and oxidative phosphorylation becomes more apparent. Later, more specific reactions involved in oxidation, energy flow and energy conservation, can be related to particular points in the scheme. As hydrogen and electrons are removed in the TCA cycle by oxidation processes, they pass via the carriers to the electron transport chain. As electrons flow from one reaction to another along the respiratory chain, energy tends to be lost. But at certain points in the chain some energy is conserved in ATP in phosphorylating reactions coupled to oxidations. Subsequently, electrons are transferred to react with hydrogen and oxygen to form water. These chemical reactions functioning in mitochondria are related to mitochondrial structure.

The mitochondrion has been studied for over 100 years. In no other

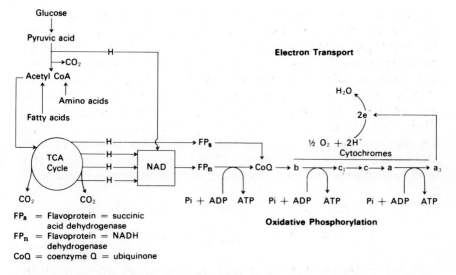

Fig. 6-1. A generalized summary of the TCA cycle, electron transport and oxidative phosphorylation in the oxidation of the major foodstuffs to CO_2 and H_2O in mitochondria.

cell organelle has there been demonstrated such remarkable confluences between structure and function. Several major events occurred over the past century to aid in our present-day knowledge of mitochondria. It was Kölliker in 1888 who demonstrated that granules could be separated from intact insect muscle cells, that they would swell in water, and that they possessed a membrane. Ten years later Benda coined the term mitochondrion and devised a rather specific crystal violet stain for them. In another 10 years, Regaud demonstrated a mitochondrial staining response to phospholipid and protein which differentiated them from other cellular granules. Although some workers believed mitochondria were related to genetics, Kingsbury in 1912 postulated they were associated with cellular oxidations. Indeed, Warburg in the following year confirmed this view experimentally. A new approach was added to the study of mitochondrial behavior in 1914 by Lewis and Lewis with tissue culture techniques. Warburg was continuing his respiration studies, and observed that in oxidation reactions oxygen utilized by cells occurred with the aid of an iron-containing catalyst

(Atmungsferment = cytochrome oxidase) which was inhibited by cyanide and carbon monoxide. Concomitant with Warburg's studies, Keilin was investigating McMunn's myohematin of 1886. Keilin observed microspectroscopically these hemoproteins underwent changes in oxidation states in intact bee-wing muscles. Later, he named them cytochromes. In the 1930's Warburg and Theorell provided evidence of the action of the pyridine nucleotides and their action with dehydrogenases. Their discovery of the flavoproteins and flavin nucleotides led to the present-day concept of flavoproteins as mediators between pyridine nucleotides and cytochromes. By 1930 Keilin isolated cytochrome c from heart muscle and later formulated a view for electron transport.

$$MH_2 + C \xrightarrow{\text{dehydrogenase}} CH_2 + M$$
$$CH_2 + \tfrac{1}{2} O_2 \xrightarrow{\text{cytochrome oxidase}} C + H_2O$$

In 1937 Krebs postulated a cycle of reactions which accounted for the oxidation of pyruvic acid to carbon dioxide and water. It was known at this time that the combustion of carbohydrates produced a large amount of heat. Although Lohmann discovered ATP in 1931, it was not until 1938 that Warburg demonstrated formation, from enzymatic oxidation, of glyceraldehyde phosphate, and Meyerhof demonstrated ATP formation from phosphopyruvate. By the beginning of the 1940's Kalckar had described that oxidative phosphorylation was coupled to respiration. And shortly thereafter Belitser hypothesized that oxidative phosphorylation was associated with electron transport through flavoproteins and cytochromes. By 1949 the tricarboxylic acid cycle was shown to be the oxidation pathway for 2-carbon units from fatty acids by Lehninger.

Isolating mitochondria seemed feasible from Bensley's work as early as 1934. But it wasn't until 1946 that Claude's work on differentially centrifuged cells showed promise. As a result, in 1948 Hogeboom, Schneider, and Palade were able to isolate intact mitochondria by differential centrifugation in sucrose solution. In the following few years, with ultrathin sectioning and improved electron microscopes, both Palade and Sjöstrand in 1952 and 1953 were able, independently, to produce electron micrographs of a mitochondrion showing its outer membrane and inner membrane.

The outer and inner membranes of a mitochondrion appear to differ both structurally and chemically. High resolution electron micrographs by negative staining indicate the presence of elementary particles on the inner cristal membranes. On the inner membrane the elementary particles appear spherical or ellipsoidal, and connected to the cristae by a stalk as first shown in beef heart by Fernández-Morán in 1962. These particles have been described in such diverse cells as *Neurospora*, rat liver, and insect flight muscles. Parsons evidence (Figure 6-2) confirmed the elementary particle figuration of Fernádez-Morán. At least these submitochondrial particles now seem to consist of an intramembranous basepiece which may or may not be pronounced, and connected by a stalk to a somewhat spherical headpiece. Treating the elementary particles with trypsin and urea produces trypsin-urea particles without spheres. These TU particles have been found to contain components of the entire electron transport chain. The spheres are the morphological expression of an ATPase oxidative phosphorylation coupling factor (F_1) necessary for coupling phosphorylation to oxidation. Passing the elementary particles on cristal fragments through a sephadex column (less drastic than trypsin) and subsequently treating them with urea prepares sephadex-urea particles without spheres, and loss of ATPase activity. On addition of the soluble ATPase to the sephadex-urea particles, the characteristic shape of the elementary particle with stalk and sphere is reconstituted and attached to the cristae. Even so, the presence of the particles have been questioned as artifacts of negative staining, or perhaps represent globular miscelles of lipids.

The chemical composition of the membranes differ from the mitochondrial matrix as well as from each other. Membranes contain about one third lipids, mostly phospholipids, and about two thirds protein. Actomyosin has been extracted from mitochondrial membranes; it is presumed to be a membrane component, and similar to myofibrillar actomyosin. The dense granules in mitochondria are insoluble inorganic salts. Their composition depends on the salts to which the mitochondria have been subjected. The enzyme concentration of membranes from mitochondria of different cells is similar.

Structurally, mitochondria are similar in different cell types. Most variation occurs only in the pattern of the cristae. Mitochondria from invertebrates show the same fundamental structure as those from vertebrates.

Solutes and ions also pass through mitochondrial membranes. Cloudy swelling, sometimes indicative of diseased cells, in liver cells and proximal convoluted tubule cells of the kidney, results from swelling of mitochondria. The outer membrane resembles the smooth membranes of

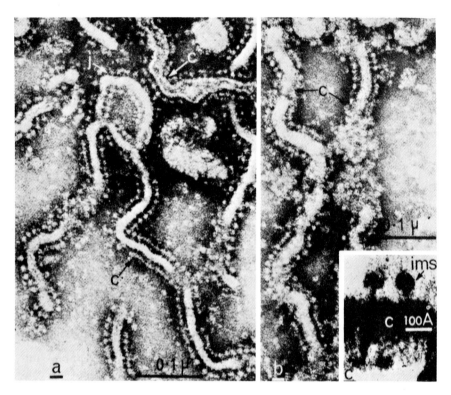

FIG. 6-2. The subunit associated with the inner membranes or cristae. (a) Part of a mitochondrion from a negatively stained preparation of mouse liver. A few of the cristae (c) are shown. The cristae consist of long filaments which branch at some points (j). The surfaces of the cristae are covered with projecting subunits ($\times$ 186,000). (b) Negatively stained cristae (c) prepared by spreading isolated lysed rat liver mitochondria. The subunits on the cristae appear similar to those of Fig. a ($\times$ 186,000). (c) Higher magnification—a few subunits from the same preparation as Fig. b. The spherical heads are 75 to 80 A diameter and the stems 30 to 35 A wide and 45 to 50 A long. The center-to-center spacing is 100 A. Reversed print ($\times$ 770,000). From Parsons, D. F. 1963. Mitochondrial structure: Two types subunits on negatively stained mitochondrial membranes. Science 140:985. Fig. 1.

the endoplasmic reticulum. The phospholipids appear to be abundant here along with cardiolipin, the structural protein. This protein forms about half of the total mitochondrial protein. It has a molecular weight of about 22,000, and polymerizes rapidly in solution at neutral pH. There are present NADH-cytochrome c reductase, NADH-cytochrome b_5 reductase, and glucose 6-phosphatase. These enzymes are also present in endoplasmic reticulum, but not the inner membrane of mitochondria.

When mitochondria are disrupted by any means, cytochromes and flavoproteins are found in the insoluble portions and are presumed to be derived from the membranes. The membrane fragments also contain the enzymes for electron transport and the capacity to couple phosphorylation of ADP to electron transport. Succinic acid, β-hydroxybutyrate, and α-glycerophosphate dehydrogenases are also found as part of membrane fragments. The enzymes which catalyze respiration and the conservation of respiratory energy as ATP do not occur singly in the free form in the soluble portion of mitochondria. They are fixed in specific areas and arrays. It is the inner membrane which contains the enzymes for the electron transfer system and the mechanism for oxidative phosphorylation. The soluble portion, however, contains the enzymes for the TCA cycle, and seemingly, they exist in the mitochondrial matrix or between the inner and outer membranes. Or it may be that these enzymes exist in loose complexes in, or near, the cristae. Some specific TCA cycle enzymes have been found to exist in molar concentrations to each other as well as cytochromes. And spectral analyses of flavoproteins and cytochromes show that they exist in simple concentration ratios to each other. Therefore, it seems justifiable to presume that a respiratory assembly chain exists in a rather specified sequence. It is the hydrogen and electrons obtained from the TCA cycle which feeds into the respiratory assembly chain.

B. The Tricarboxylic Acid Cycle

In intermediary metabolism, products are formed from carbohydrates, fats, and proteins which can enter the TCA cycle. One of the most important, of these products is acetyl coenzyme A. It is derived from two major sources; glucose metabolism and fatty acid metabolism. In glycolysis, a glucose molecule is transformed by a series of reactions resulting in the formation of two molecules of pyruvic acid which enters mitochondria. The relation of pyruvic acid and acetyl CoA to the TCA cycle is indicated in Figure 6-3. It will be helpful to refer at times to Figure 6-3 in reading the following.

The over-all reaction in which acetyl CoA is formed from pyruvic acid is called an oxidative decarboxylation since it involves both oxidation and loss of carbon dioxide. Involved in this complex process are five cofactors—coenzyme A, NAD, lipoic acid, Mg^{++}, and thiamin pyrophosphate; and three enzymes—pyruvic acid decarboxylase, dihydroxylipoyl transacetylase, and dihydrolipoyl dehydrogenase. The whole complex is called pyruvic acid dehydrogenase. Acetyl CoA is formed as the acetyl moiety of pyruvic acid is transferred to coenzyme A; concomitantly, the carbon of the carboxyl group is liberated as carbon dioxide. Two hydrogen atoms remain; one from the pyruvate carboxyl

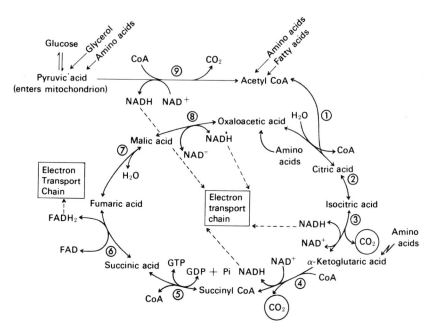

Fig. 6-3. The tricarboxylic acid (Krebs citric acid) cycle. Reaction steps (involving enzymes, coenzymes, and prosthetic groups): 1. citrate synthetase; 2. aconitase; Fe^{++}; 3. isocitric acid dehydrogenase; Mg^{++}; 4. α-ketoglutaric acid dehydrogenase; 5. succinic acid thiokinase; 6. succinic acid dehydrogenase; 7. fumarase; 8. malic acid dehydrogenase; 9. pyruvic acid dehydrogenase.

group and one from the sulfhydryl group of coenzyme A. These hydrogens are transferred to NAD.

The TCA cycle first begins to turn properly when the acetyl CoA gives its acetyl group to oxaloacetic acid to form citric acid. Citrate synthetase catalyzes this reaction which is also called condensing enzyme. A molecule of water is required to hydrolyze the linkage between the acetyl group and coenzyme A which is liberated to react with more pyruvic acid.

In the second reaction of the cycle, citric acid is catalyzed by aconitase to from isocitric acid by isomerization. Fe^{++} is also required for the internal rearrangement. This occurs as the hydroxyl group of the citric acid migrates to an adjacent carbon to form isocitric acid.

Isocitric acid, in reaction three, is catalyzed by one of two isocitric dehydrogenases to α-ketoglutaric acid and carbon dioxide in the presence of a pyridine nucleotide. In mitochondria, the enzyme is NAD specific and Mg^{++} dependent. The isocitric dehydrogenase in cytoplasm is specific for NADP and dependent on the cofactor Mn^{++}. The latter reaction involves the formation of an enzyme-bound intermediate oxalosuccinic acid. Evidence, however, is lacking for oxalosuccinate as an intermediate in the reaction catalyzed by isocitric acid dehydrogenase in the mitochondrion. The concentration of NADP in mitochondria varies according to the tissue, e.g., low in muscle, if present, and high in liver. NADP occurs mostly in the reduced form whereas NAD is found mostly oxidized. Apparently there is no direct pathway for electron transport in mitochondria from NADP to oxygen. The oxidation of NADP can occur through such reducing biosyntheses as fatty acid formation, or transfer to NAD in the presence of transhydrogenase as

$$NADPH + NAD \leftrightarrows NADP + NADH.$$

Amino acids can enter the cycle at this point. By oxidation or transamination, glutamic acid can form α-ketoglutaric acid. Thus, amino acids such as ornithine, proline, glutamine, and histidine which metabolically yield glutamic acid are potential sources of α-ketoglutaric acid.

In the fourth step of the TCA cycle, succinyl-CoA is formed from α-ketoglutaric acid by oxidative decarboxylation. The reaction is analogous to that of pyruvic acid described above. The reaction is catalyzed by the α-ketoglutaric acid dehydrogenase complex requiring cofactors NAD, thiamin pyrophosphate, lipoic acid, coenzyme A, and Mg^{++}. Succinyl CoA may condense with glycine to initiate porphyrin synthesis.

But most of the succinyl CoA forms succinic acid in a fifth reaction of the cycle. This step is catalyzed by succinic acid thiokinase. The thioester energy of succinyl CoA is conserved by formation of guanosine triphosphate (GTP). As coenzyme A is released in the reaction, GTP can in turn react with ADP to form guanosine diphosphate (GDP) and ATP in the presence of Mg^{++} and nucleoside diphosphate kinase, or be used directly for the activation of fatty acids.

Step six of the TCA cycle represents the only dehydrogenation which does not involve a pyridine nucleotide. Succinate dehydrogenase catalyzes the removal of two hydrogen atoms from succinic acid to form fumaric acid. The immediate acceptor of the hydrogens is the flavin prosthetic group (FAD) of the flavoprotein enzyme, succinic acid dehydrogenase. As indicated in Figure 6-1, this enzyme is a flavoprotein which picks up hydrogen from the substrate and feeds it directly into the electron transport chain, probably at the site of ubiquinone.

Fumarase catalyzes the reversible hydration of fumaric acid to form malic acid in the seventh step of the cycle.

In the eighth step oxaloacetic acid is regenerated from malic acid by malic acid dehydrogenase in the presence of NAD. At this point, the cycle is completed. The oxaloacetic acid formed is now available to condense with another molecule of acetyl CoA for another turn of the cycle. Oxaloacetic acid can also be formed by transamination from aspartic acid.

Several points can be mentioned about the TCA cycle. As the cycle completes one turn, one mole of acetate (in the form of acetyl CoA), is oxidized to carbon dioxide and water. With the exception of the oxidative decarboxylation of α-ketoglutaric acid, all reactions in the TCA cycle are reversible. Some cycle intermediates, such as oxaloacetic acid and α-ketoglutaric acid, can undergo other reactions in the cell. Of course, other biological oxidations occur in the cell outside of the mitochondria.

Not all hydrogen feeding into the electron transport chain derives from the TCA cycle. Some hydrogen can be transported into mitochondria via shuttle, or shunt, systems. Since the pyridine nucleotides outside of the mitochondria in the cytoplasm do not readily penetrate the mitochondrial membranes, it is the NADH in the mitochondria that is oxidized. An active system for the transfer of reducing power between the cytoplasm and the mitochondrial pyridine nucleotides is illustrated in Figure 6-4.

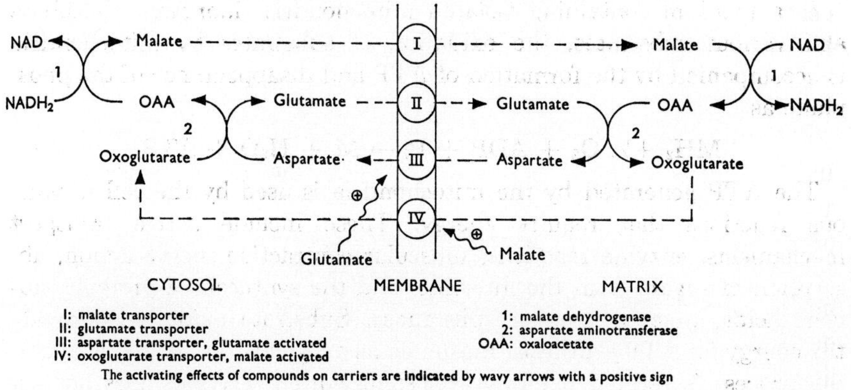

FIG. 6-4. The role of some transporting systems in the oxidation of cytoplasmic $NADH_2$ by mitochondria. From Chappell, J. B. 1968. Systems used for the transport of substrates into mitochondria. Brit. Med, Bull. 24:150. Fig. 4.

C. Electron Transport and Oxidative Phosphorylation

As the TCA cycle turns, two decarboxylations and four dehydrogenations occur. Subsequent to the two decarboxylations, the 4-carbon oxaloacetic acid condenses with the 2-carbon acetyl CoA to form 6-carbon citric acid, a tricarboxylic acid. Thus, regeneration of the cycle occurs. As the dehydrogenation reactions occur, hydrogen atoms and their electrons are carried to the electron transport chain.

The components of the chain have been isolated from fragments of mitochondrial membranes, *i.e.*, the cristae. Two these membrane-bound components are iron-flavoproteins: NADH dehydrogenase and succinic acid dehydrogenase. In addition, there are the iron-containing cytochromes. Spectroscopic analysis has been used to identify cytochromes b, c_1, c, a, and a_3. Cytochromes a and a_3 function as cytochrome oxidase. Cytochrome c is easily extracted, but the others are firmly bound to the membranes. Ubiquinone is also present in larger amounts than cytochromes and is easily extracted. If intermediates of the TCA cycle are incubated anaerobically with mitochondria or cristae, all of the above mentioned, membrane-bound components FP_n, FP_s, CoQ, nontheme, and cytochromes b, c_1, c, a, and a_3, are reduced.

In a medium containing isolated mitochondria, inorganic phosphate and various substrates, the oxidation of substrates by mitochondria is accompanied by the formation of ATP and disappearance of the phosphate as

$$MH_2 + \tfrac{1}{2} O_2 + ADP + P_i \rightarrow M + H_2O + ATP$$

The ATP generated by the mitochondria is used by the cell in various reactions that require energy. These include active transport mechanisms, enzyme reactions, muscular contraction, nerve action, absorption of sugars from the intestine, and the synthesis of proteins, nucleic acids, urea, and other substances. Substrate oxidations provide the energy for ATP synthesis. Inasmuch as oxygen is the oxidizing agent, the process is termed oxidative phosphorylation. Oxygen does not act directly on substrates, but through a series of reactions apparently in a sequence referred to as the electron transport, or respiratory, chain. The electrons are transported by dehydrogenases, flavoproteins, and cytochromes arranged in an order of increasing oxidation-reduction potential to oxygen which is reduced to water. A higher redox potential indicates that the substance has a greater affinity for electrons.

Oxidative phosphorylating mechanisms are linked to the respiratory chain at three sites. Although the generalities of oxidative phosphosphorylation are known, it is the specifics which are lacking: one, the identification of the intermediates, and two, the elucidation of the energy-conversion mechanism. Consequently, a definitive formulation of the respiratory chain cannot be given at the present time that is completely free from controversy.

There are two main views regarding the mechanism of oxidative phosphorylation. One of these is the chemical hypothesis; the other is the chemosmotic hypothesis. There are two essential differences between these two hypotheses. The first is concerned with the nature of the high-energy state. In the chemical hypothesis the high-energy state is a defined chemical intermediate which includes $A \sim X$, a member of the electron transport chain. In the chemosmotic hypothesis, the high-energy state is an electrical membrane potential which is created during electron transport. The second essential difference between the chemical and chemosmotic hypotheses for oxidative phosphorylation concerns the membrane. The chemical hypothesis has the role of the membrane as an organizer of the multienzyme complex. According the chemosmotic hypothesis, the membrane must have an inside and an outside. Yet the two views are somewhat similar. In both hypotheses there is a high-energy intermediate, $X \sim Y$, for generating ATP, and both are closely associated with ion transport as indicated in the following.

$$\begin{array}{c} \text{Ion transport} \\ \uparrow \\ \sim \\ AH_2 \nearrow \; x \sim y \; \xrightleftharpoons[P_i]{ADP} \; ATP \\ \searrow \; \updownarrow \\ A \sim x \\ \downarrow \\ \text{Ion transport} \end{array}$$

Be that as it may, the scheme of electron transport through the respiratory chain and the points in the chain where coupled phosphorylation is believed to occur is indicated in Figure 6-5.

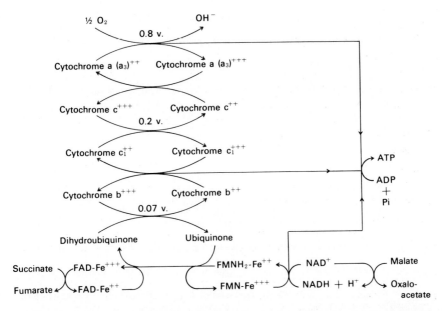

FIG. 6-5. Apparent organization of electron flow and coupled phosphorylation in the mitochondrial electron transport system. Approximate values for E' are shown. Modified from White, A., Handler, P., and Smith, E. L. 1968. *Principles of Biochemistry*. 4th edition. McGraw-Hill Book Co. New York. p. 340. Fig. 16.8.

The nicotinamide adenine dinucleotide, NADH, delivers hydrogen from the TCA cycle substrate to a flavoprotein, NADH dehydrogenase (FP_n = (FMN). Succinic acid dehydrogenase (FP_s = FAD), another flavoprotein, also functions as a hydrogen carrier. Both flavoproteins, it is believed, transport hydrogen to ubiquinone (coenzyme Q); it may be reoxidized by the Fe^{+++} of cytochrome b. Hydrogen ions are formed here, and then only electrons are transported through the cytochrome chain by a change in valence of iron. Ultimately, cytochrome oxidase transfers two electrons onto oxygen forming an unstable ion, O^-. It immediately combines with two H^+ in the media to form water.

Establishing the order of electron transport among the components identified with cristae has been aided by the use of specific inhibitors. Likewise, the tight coupling of oxidative phosphorylation to electron transport has been aided by the use of uncoupling agents acting at specific sites in the chain.

Suggested Reading—Chapter 6

Benda, C. Weitere Mitteilungen über die Mitochondrien. Verh. Physiol. Ges. *1899*:376, 1899.
Bittar, E. E. and Bittar, N. (editors). *The Biological Basis of Medicine*. Academic Press. New York. 1968.
Chance, B., Lee, C., and Mela, L. Control and conservation of energy in the cytochrome chain. Fed. Proc. *26*:1341, 1967.
Chance, B., Estabrook, R. W., and Lee, C. Electron transport in oxysome. Science *140*:379, 1963.
Chance, B., Parsons, D. F., and Willams, G. R. Cytochrome content of mitochondria stripped of inner membrane structure. Science *143*:136, 1964.
Chappell, J. B. Systems used for the transport of substrates into mitochondria. Brit. Med. Bull. *24*:150, 1968.
Cowdry, E. V. The mitochondrial constituents of protoplasm. Carnegie Contr. Embryol. *8*:39, 1918.
Cowdry, E. V. Historical background of research on mitochondria. J. Histochem. Cytochem. *1*:183, 1953.
Ernster, L. and Lindberg, O. Animal mitochondria. Ann. Rev. Physiol. *20*:13, 1958.
Fernández-Morán, H. Correlation of ultrastructure and function in mitochondrial membranes. Circulation *26*:1047, 1962.
Green, D. E. Mechanism of energy transformations in biological membranes. Med. Coll. Va. Quart. *4*:96, 1968.
Green, D. E. and Silman, I. Structure of the mitochondrial electron transfer chain. Ann. Rev. Plant Physiol. *18*:147, 1967.
Hoffman, H. and Gregg, G. An electron microscope study of mitochondrial formation. Exp. Cell Res. *15*:118, 1958.
Hogeboom, G. H., Schneider, W. C., and Palade, G. H. Isolation of intact mitochondria from rat liver: some biochemical properties of mitochondria and submicroscopic particulate material. J. Biol. Chem. *172*:619, 1948.

Kennedy, E. P. and Lehninger, A. L. Intracellular structures and the fatty acid oxidase system of rat liver. J. Biol. Chem. *172*:847, 1948.

Klingenberg, M. and Kröger, A. On the role of ubiquinone in the respiratory chain. In *Biochemistry of Mitochondria*. Slater, E. C., Kaniuga, Z., and Wojtczak, L. (editors). Academic Press. New York. 1967.

Krebs, H. A. and Johnson, W. A. The role of citric acid in intermediate metabolism in animal tissues. Enzymologia *4*:148, 1937.

Lehninger, A. L. *The Mitochondrion*. W. A. Benjamin. New York. 1965.

Lehninger, A. L. Molecular basis of mitochondrial structure and function. In *Molecular Organization and Biological Function*. Allen, J. M. (editor). Harper & Row. New York. 1967.

Luck, D. J. L. and Reich, E. DNA in mitochondria of *Neurospora crassa*. Proc. Nat. Acad. Sci. USA *52*:931, 1964.

Mattisson, A. G. M. and Birch-Anderson, A. On the fine structure of the mitochondria and its relation to oxidative capacity in muscles in various invertebrates. J. Ultrastruct. Res. *6*:205, 1962.

Michaelis, L. Die vitale Färbung, eine Darstellungsmethode der Zellgranule. Arch. Mikroskop. Anat. *55*:558, 1900.

Millerd, H. and Bonner, J. The biology of plant mitochondria. J. Histochem. Cytochem. *1*:254, 1953.

Mitchell, P. Coupling of phosphorylation to electron and hydrogen transfer by a chemiosmotic type of mechanism. Nature (London) *191*:144, 1961.

Mitchell, P. Proton-translocation phosphorylation in mitochondria, chloroplasts and bacteria: natural fuel cells and solar cells. Fed. Proc. *26*:1370, 1967.

Nass, M. M. K. The circularity of mitochondrial DNA. Proc. Nat. Acad. Sci. USA *56*:1215, 1966.

Nass, M. M. K., Nass, S., and Afzelius, B. A. The general occurrence of mitochondrial DNA. Exp. Cell Res. *37*:516, 1965.

Passmore, R. and Robson, J. S. (editors). *A Companion to Medical Studies*. F. A. Davis. Philadelphia. 1968.

Peachey, L. D. Electron microscope observations on the accumulation of divalent cations in intramitochondrial granules. J. Cell Biol. *20*:95, 1964.

Parsons, D. F. Mitochondrial structure: two types of subunits on negatively stained mitochondrial membranes. Science *140*:985, 1963.

Parsons, D. F. Recent advances correlating structure and function in mitochondria. Int. Rev. Exp. Path. *4*:1, 1965.

Pressman, B. C. Induced active transport of ions in mitochondria. Proc. Nat. Acad. Sci. USA *53*:1076, 1965.

Pullman, M. E. Mitochondrial oxidations and energy coupling. Ann. Rev. Biochem. *36*:539, 1967.

Rabinowitz, M., Sinclair, J., DeSalle, L., Haselkorn, R., and Swift, H. H. Isolation of deoxyribonucleic acid from mitochondria of chick embryo heart and liver. Proc. Nat. Acad. Sci. USA *53*:1126, 1964.

Racker, E. Resolution and reconstitution of the inner mitochondrial membrane. Fed. Proc. *26*:1335, 1967.

Racker, E. The membrane of the mitochondrion. Sci. Amer. *218*(2):32, 1968.

Roodyn, D. B. *Enzyme Cytology*. Academic Press. New York. 1967.

Roodyn, D. B. The mitochondrion. In *The Biological Basis of Medicine*. Bittar, E. E. and Bittar, N. (editors). Academic Press. New York. 1968.

Roodyn, D. B. and Wilkie, D. *The Biogenesis of Mitochondria*. Methuen & Co. London. 1968.

Rouiller, C. Physiological and pathological changes in mitochondrial morphology. Int. Rev. Cytol. 9:227, 1960.

Sanadi, D. R. Energy-linked reactions in mitochondria. Ann. Rev. Biochem. 34:21, 1965.

Sjöstrand, F. S. A new ultrastructural element of the membranes in mitochondria and of some cytoplasmic membranes. J. Ultrastruct. Res. 9:340, 1963.

Slater, E. C., Kaniuga, Z. and Wojtczak, L. *Biochemistry of Mitochondria.* Academic Press. New York. 1967.

Stoeckenius, W. Some observations on negatively stained mitochondria. J. Cell Biol. 17:443, 1963.

Swift, H., Kislev, N., and Bogorad, L. Evidence for DNA and RNA in mitochondria and chloroplasts. J. Cell Biol. 23(2):91A, 1964.

Chapter Seven

Cellular Respiration

A. CALORIMETRY AND RESPIRATION QUOTIENT
B. HEMOGLOBINS AND GAS TRANSPORT
C. CARBOHYDRATE FORMATION AND CONVERSIONS
   1. LIGHT REACTIONS
   2. DARK REACTIONS

Living cells maintain a dynamic steady state with their environment. Although cells are not in equilibrium with their environment, they operate in a constant exchange through the membrane. As a result, cells continually undergo chemical changes within their matrix. These changes reflect a state of metabolism in which enzymes participate in the chemical reactions of the metabolite molecules. Regulatory mechanisms operate in converting one metabolite to another which may involve a series of reactions. The sequences form metabolic pathways. Different pathways involving different metabolites can liberate, over-all, various amounts of energy. Some of this energy is conserved in ATP, and some of it is lost to the environment as heat. As an average for all foodstuffs, about 60% of the energy present becomes heat when ATP is formed. More heat is liberated as ATP transfers its energy so that all of the functional systems utilize not more than about 25% of all the energy formed from foodstuffs.

A. Calorimetry and Respiratory Quotient

Heat is usually expressed in terms of the kilocalorie (=1000 calories) which is represented by the large Calorie. The major foodstuffs have been measured for their caloric values. On the average, when 1 gm of carbohydrate is burned, it liberates about 4.1 Cal. Lipids yield about 9.3 Cal/gm, and proteins about 4.1 Cal/gm. These values represent averages since individual members of each class of different foodstuffs yield somewhat different figures. For example, glucose yields 3.75 Cal/gm, but glycogen liberates 4.3 Cal/gm. Most foodstuffs of animals seem to yield caloric value higher than those of plants.

Cellular respiration may be considered as aerobic metabolism. During the process of respiration in a cell, not only is energy liberated but carbon dioxide is released. The amount of oxygen used by the cell and the amount of carbon dioxide released depends on the type of foodstuff being oxidized. In fact, the amount of oxygen consumed and the amount of carbon dioxide produced can be calculated from the stoichiometry of the equations for the oxidations of different foodstuffs. The theoretical relation between the gaseous exchange is a ratio, $i.e.$, the volume of carbon dioxide produced divided by the volume of oxygen consumed is a ratio defined as the respiratory quotient (R.Q.).

The R.Q. varies as shown for the following compounds:

$$C_6H_{12}O_6 + 6\ O_2 \rightarrow 6\ CO_2 + 6\ H_2O$$
Glucose   R.Q. = 6/6 = 1.00

$$C_{57}H_{104}O_6 + 80\ O_2 \rightarrow 57\ CO_2 + 52\ H_2O$$
Triolein   R.Q. = 57/80 = 0.71

$$2\ C_3H_7O_2\ N + 6\ O_2 \rightarrow CO(NH_2)_2 + 5\ CO_2 + 5\ H_2O$$
Alanine   R.Q. = 5/6 = 0.83

The R.Q. for all carbohydrates is 1.00. In each carbohydrate molecule there is exactly enough oxygen to oxidize only the intramolecular hydrogen, and each carbon present requires only one molecule of atmospheric oxygen. The R.Q. for dietary fat averages about 0.70 and for dietary protein it is about 0.80. It is more difficult to calculate a theoretical R.Q. for proteins than fats and carbohydrates because oxidation of proteins does not proceed completely to water and carbon dioxide formation; also nitrogenous end products are usually excreted in the urine.

Many methods have been developed for studying oxygen consumption in cells and organisms. These methods vary and include techniques for studying either (single) cells or whole organisms. Regardless of the size or whether or not unicellular or multicellular, the consumption of oxygen is expressed in terms of rate of respiration indicated by Q_{O_2}. The Q_{O_2} is the volume of oxygen consumed (corrected to standard temperature and pressure, STP) per unit weight per unit time. For cells such as erythrocytes, bacteria, algae, yeast, ova, spermatozoa, and molds the unit of measurement is ordinarily expressed in microliters (μl) per milligram per hour. For multicellular organisms the units of measurement are usually given as ml/gm/hr. Q_{O_2} values may be reported for either wet or dry weight. Dry weight is approximately 25% of wet weight.

Whereas cell respiration involves the utilization of atmospheric oxygen in aerobic processes, glycolysis and fermentation undergo anaerobic respiration. In these processes foodstuffs are oxidized in the absence of atmospheric oxygen.

In the various methods for determining oxygen consumption of cells, tissue slices or homogenates, or even whole organisms, procedures are given utilizing either aquatic or atmospheric techniques. In Figure 7-1 is shown the sketch of a set-up in which oxygen consumption can be measured chemically. As water flows through the chamber housing the organism, a sample of water is drawn from the upper needle into a syringe. Oxygen is determined by a modified Winkler method. By using the clamp to control the water flow, another sample of water is taken from the lower needle for oxygen analysis. From these data the Q_{O_2} may be calculated.

Most studies on cell respiration are made with the Barcroft-Warburg respirometer which is pictured in Figure 7-2. The respirometer assembly provides for manometer and flask support, shaker, and constant temperature bath. In Figure 7-3 are shown the flask and manometer as attached to the respirometer assembly unit; cells are in a suitable buffer in the flask which contains a center well with a little strong alkali for absorption of carbon dioxide produced. Pressure changes are therefore the result of oxygen consumption. The side-arm allows addition of substrates, inhibitors, activators, etc. during an experiment. The gas phase is kept at a constant volume in the flask and manometer. A fall in the pressure is a measure of oxygen consumption and is read on the manometer. The pressure readings are corrected for the flask constant. A thermobarometer is required to correct for temperature and pressure changes in the reaction flask.

Although oxygen consumption in cells can be measured directly, there is no convenient direct method for measuring carbon dioxide. Carbon dioxide is produced by oxidative decarboxylation of organic acids, and not by the direct oxidation of carbon compounds as is known now. Carbon dioxide values are given in terms of R.Q. as for

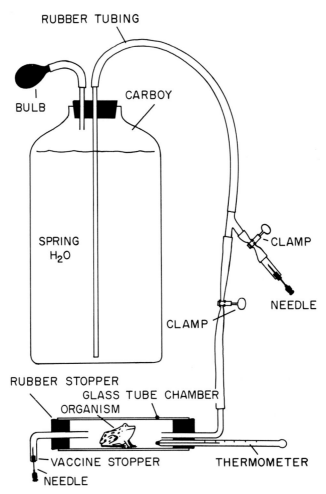

Fig. 7-1. An apparatus showing an animal in position for an oxygen consumption experiment. Oxygen determinations are made on water samples by a modified Winkler method. Syringe samples are analyzed for oxygen before, and after, passing through the chamber.

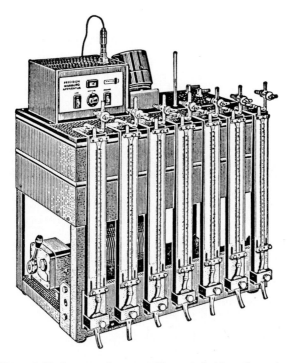

Fig. 7-2. A Barcroft-Warburg respirometer. The unit holds a thermobarometer and six Warburg, or three Barcroft differential, manometers. From Fisher Scientific Co., 7722 Fenton Street, Silver Spring, Maryland 20910.

oxygen. Differential manometers have been developed to permit measurements of carbon dioxide production as well as oxygen consumption in the presence of physiological tensions of carbon dioxide. The constant-volume differential manometer consists of two flasks attached to two manometers as introduced by Barcroft in 1908. In this type of apparatus the rate of photosynthesis, either the evolution of oxygen or oxygen evolution and carbon dioxide uptake simultaneously, and the rates of respiration and fermentation can be determined. The manometers are arranged so that pressure changes in the two flasks are opposed to each other on the manometer. Therefore, in an experiment, the measurement is of the difference in pressure between the two flasks rather than the total pressure in each flask. The pressure dif-

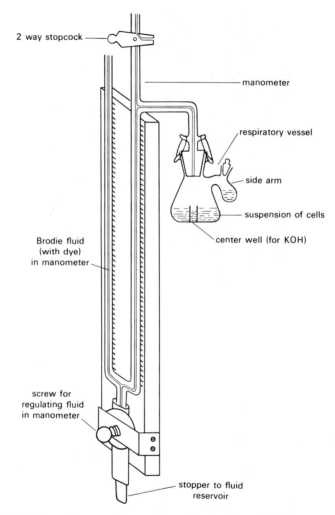

Fig. 7-3. A flask and manometer are shown for the Barcroft-Warburg respirometer. From Giese, A. C. 1968. *Cell Physiology*. 3rd edition. W. B. Saunders Co. Philadelphia. p. 422. Fig. 18-15.

ference results from both oxygen consumption and carbon dioxide production. The carbon dioxide in the gas phase can be measured by alkali absorption.

The differential manometer is used to measure aerobic glycolysis ($Q_G^{O_2}$), but the simpler Barcroft-Warburg manometer can be used to measure anaerobic glycolysis ($Q_G^{N_2}$). In Table 7-1 are given some values of various tissues, for Qo_2, R.Q., $Q_G^{O_2}$, and $Q_G^{N_2}$. Anaerobic glycolysis can be inhibited by oxygen and is called the Pasteur effect. The Crabtree effect is the inhibition of aerobic respiration by glucose.

Other respirometers which have been developed are modifications in one way or another of the Barcroft-Warburg type. One of these is depicted in Figure 7-4 which has been useful for measuring oxygen consumption in small organisms. As oxygen is consumed, pressure in the closed system falls. By manipulating the stopcock, air can be drawn into the syringe, and then directed into the container to equalize the

TABLE 7-1
Respiration and Glycolysis of Selected Animal Tissue; Average Values

Tissue	Qo_2	R.Q.	$Q_G^{O_2}$	$Q_G^{N_2}$
Muscle (dog)	2	0.95	0	4
Pancreas (dog)	3	...	0	4
Bone marrow (rabbit)	5	0.90	2	13
Rous sarcoma (chicken)	5	0.93	20	30
Liver (fetal, rat)	7	1.00	0.5	8
Testis (rat)	8	0.90	4	8
Jensen sarcoma (rat)	9	0.78	17	34
Liver (adult, rat)	10	0.5–1.0	0.5	3
Embryo (chicken)	11	1.00	2	18
Intestinal mucosa (rat)	12	0.85	2	4
Spleen (rat)	12	0.89	2	8
Brain (rat)	13	1.00	2	19
Thyroid gland (rat)	13	...	0	2
Retina (rat)	19	1.00	22	88
Kidney (rat)	21	0.83	0	3
Chorion (rat)	26	1.00	7	32

$Qo_2 = \mu l/hr/gm$ (dry wt.)
R.Q. $= CO_2/O_2$
$Q_G^{O_2}$ = aerobic glycolysis
$Q_G^{N_2}$ = anaerobic glycolysis

From Oser, B. L. (editor). 1965. *Hawk's Physiological Chemistry*. 14th edition. McGraw-Hill Book Co. New York. p. 449. Table 14-3.

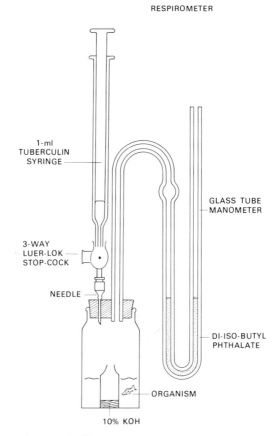

Fig. 7-4. A respirometer is illustrated used for oxygen consumption determinations in either aquatic or terrestrial organisms. An identical apparatus without organism serves as a thermobarometer.

---

pressure in the system to the zero mark. Oxygen consumption can be read off the syringe scale. Of course, a thermobarometer is used with this type of respirometer.

In Figure 7-5 is shown a microrespirometer which has been variously modified over a number of years. A graduated capillary tube connects an experimental flask to a control flask which serves as a compensator. A volume change is indicated by a kerosene drop colored with a dye. This respirometer is essentially a Barcroft differential respirometer

CELLULAR RESPIRATION

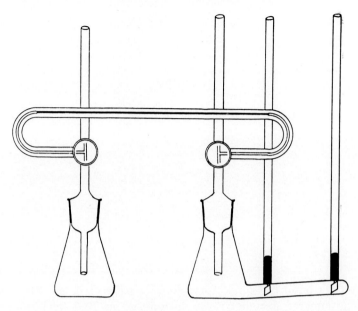

FIG. 7-5. A diagram of the original Fenn microrespirometer for determining oxygen volumetrically and carbon dioxide by conductivity. Essentially this apparatus is the same as a Barcroft differential manometer with the manometer raised to a horizontal position. There have been many variations of the microrespirometer shown here described in the literature. From Fenn, W. O. 1928. A new method for the simultaneous determination of minute amounts of carbon dioxide and oxygen. Amer. J. Physiol. *84:*110. Fig. 1.

---

with the manometer raised to a horizontal position. Not only can this microrespirometer be used to determine oxygen volumetrically, but can measure carbon dioxide by conductivity with platinum electrodes.

Ultramicrorespirometers have also been devised. The one diagrammed in Figure 7-6 is suitable for measurements of gas exchanges from 0.001 to 0.5 mm^3. The calculations required for converting pressure reading into volumes are exactly as those described for the Barcroft-Warburg respirometer. Individual readings can be made from 15 to 30 seconds. The apparatus is sensitive enough to measure oxygen uptake by developing shrimp embryos within egg membranes. Its sensitivity appears to be as great as that of a Cartesian diver ultramicrorespirometer.

The Cartesian diver apparently had nothing to do with Descartes.

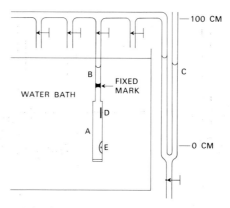

Fig. 7-6. A diagram of the Gregg ultramicrorespirometer. A = Respiratory chamber; B = Capillary tube; C = Manometer containing Brodie's solution; D = Alkali on filter paper; E = Respiratory material. From Gregg, J. R. 1947. A microrespirometer. Rev. Sci. Instrum. 18:514. Fig. 1.

Margiotti, a pupil of Galileo, first described the diver in 1648. Its first use in ultamicrorespirometry was introduced in 1937 by Linderstrøm-Lang. The Cartesian diver and micromanometer is represented in Figure 7-7. Single cells and unicellular organisms are studied in such a set-up for QO_2 analyses. The principle of the diver action is that a gas bubble is enclosed in the diver and the diver is immersed in fluid of a larger vessel. The buoyancy of the diver will vary according to rise and fall of the pressure on the fluid in the larger vessel. If the gas in the diver diminishes or increases, the pressure must be varied to maintain the diver at a constant level. Thus, the Cartesian diver manometer serves as a constant volume manometer. The glass tail on the diver is to maintain it in a vertical position.

Another type of ultramicrorespirometer is shown in Figure 7-8. It is a variation of the Cartesian diver type. In the Cartesian diver, the constancy of the gas volume is determined by floating the whole diving chamber. In this system, a reference diver is floated in the respiratory chamber containing the cell. The pressure is adjusted so that the bubble diver is suspended at a constant height in the respiratory chamber. Pressure changes in this constant volume system are read on the micromanometer.

There are other methods also for measuirng respiratory rates. One of these is by infrared absorption spectroscopy. Carbon dioxide absorbs specifically at 4.2 μ in the infrared. Another method which is par-

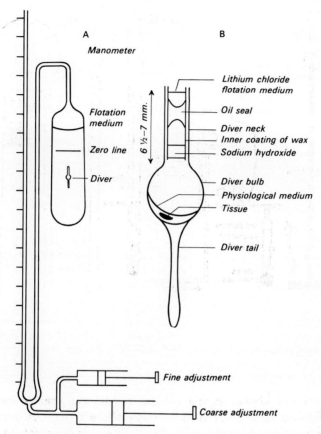

FIG. 7-7. The Cartesian diver ultramicrorespirometer for single cell measurement. From Boell, E. J. and Needham, J. 1939. Morphogenesis and metabolism: studies with the Cartesian diver ultramicromanometer. Proc. Roy. Soc. [Biol.] *127*:322. Fig. 1.

---

ticularly useful is the oxygen (platinum) electrode. The apparatus is standardized against a known oxygen content in a solution by the Winkler method, or against the oxygen content in the atmosphere. In any event, respiratory rate can be calculated from the known oxygen content, and the rate of change in current registered in the apparatus. Not only can oxygen consumption be measured, but oxygen tension can also be recorded. Results can be recorded continuously and instantaneously for long periods of time.

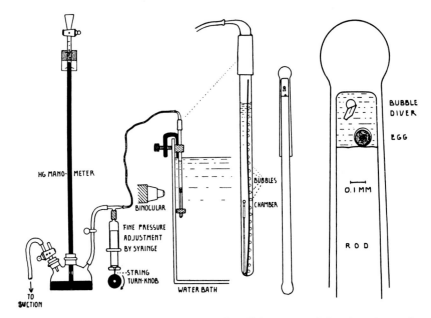

FIG. 7-8. Respiration measurement of single cell by means of the ultramicrorespirometer reference diver. From Scholander, P. F., Claff, C. L., and Sveinsson, S. L. 1952. Respiratory studies of single cells. I. Methods. Biol. Bull. *102:*159. Fig. 1.

As mentioned above, during cellular respiration heat is continually being produced as a by-product of metabolic activity. As heat is produced, some heat is continually lost to the environment as well as conserved. If the rate of heat production exceeds the rate of heat loss, there will be an increase in total body heat. Likewise, a rise in environmental temperature increases the respiratory rate of cells in suspension within the biokinetic zone limitations. For every 10° C rise in temperature, the rate of reaction increases 2–4 times. This temperature coefficient is referred to as Q_{10}.

Indirect calorimetry can be used to measure basal metabolism with a spirometer and recorder. It is possible to calculate heat production, from the volume of oxygen consumed (corrected to STP) by using the value 4.825. For an assumed R.Q. of 0.82, this figure represents the calorie equivalents of 1 liter of oxygen. A R.Q. of 0.82 is an average value for an individual (man) on a normal diet of carbohydrates, fats, and proteins.

B. Hemoglobins and Gas Transport

Gases diffuse independently of each other. The cause of the movement by diffusion is a pressure gradient. Gas diffusion in unicellular organisms is less of a problem than it is in multicellular organisms. As oxygen is utilized in cellular respiration, a pressure gradient (P_{O_2}) can restore the oxygen used. When the foodstuffs are oxidized the carbon dioxide pressure (P_{CO_2}) increases in the cell and the pressure gradient causes an extracellular diffusion. Although both oxygen and carbon dioxide are transported in plants and animals via fluid media, gas transport problems are greater in animals than plants.

Fortunately, the hemoglobin molecule evolved to aid gas transport in animals. In doing so, the hemoglobin molecule evolved genetic stability so that a species has one or more hemoglobins present. Genetic variability may also produce abnormal hemoglobins within a species.

It seems a well known fact that oxygen is transported by hemoglobin in the erythrocyte (about 97%) and the other 3% dissolved in the plasma. But it is not so well known that hemoglobin and the red blood cell carry carbon dioxide. As illustrated in Figure 7-9 carbon dioxide is transported in four different chemical forms: (1) dissolved; (2) carbonic acid;

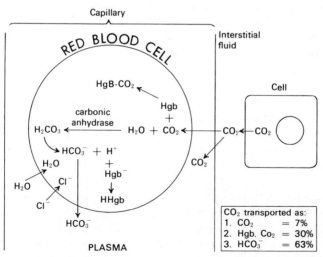

FIG. 7-9. Transport of carbon dioxide in the blood. From Guyton, A. C. 1966. *Textbook of Medical Physiology*. 3rd edition. W. B. Saunders Co. Philadelphia. p. 586. Fig. 468.

(3) bicarbonate ions; and (4) carbamino compounds. Gaseous carbon dioxide diffuses out of the cell into the blood of the capillary. Some carbon dioxide dissolves in the plasma and some slowly combines with water to form carbonic acid. Most carbon dioxide, however, diffuses into the red blood cell which contains carbonic anhydrase. This enzyme facilitates by 200–300 times the formation of carbonic acid in the red cell. The carbonic acid immediately dissociates into hydrogen and bicarbonate ions. Since hemoglobin is an acid-base buffer also, hydrogen ions are removed by combining with hemoglobin. The carbon dioxide is then eliminated from the organism by diffusion from the bicarbonate ion. A hemoglobin molecule can carry oxygen and carbon dioxide at the same time because the two gases have different attachment points in the molecule.

Not only can the pattern of hemoglobins be different in different species, but their capacity to carry oxygen can also vary. This in turn can have an effect on the distribution of a species as well as its activity. A hemoglobin solution can be prepared and subjected to electrophoresis for hemoglobin separation and indentification. In Figure 7-10 are shown electropherograms of hemoglobins in two species of basses, the smallmouth bass (*Micropterus dolomieui*) and the largemouth bass (*Micropterus salmoides*). The hemoglobin patterns of these two fishes are different. Typically, they reproduce in two different habitats. The smallmouth bass is associated with cold lakes and streams, but the largemouth bass is most often found in warm rivers, ponds, and lakes. Sometimes the largemouth bass inhabits and can exist in water holding breeding populations of smallmouth bass. Knowing that each bass species has different hemoglobins, that each ordinarily occupies a different habitat, and that each bass has a characteristic fighting habit, it might be suspected that each fish also has a different oxyhemoglobin affinity. Indeed this is so as seen in Figure 7-11 for the largemouth and Figure 7-12 for the smallmouth bass.

Oxyhemoglobin affinity curves can be constructed for whole blood or hemoglobin solution analysis. In either case the percentage of saturation of hemoglobin by oxygen is plotted against oxygen tension in millimeters of mercury. For convenience in comparing one curve with another, the value P^{50} is used. This figure is the pressure of oxygen in mm Hg which represents 50% saturation of hemoglobin with oxygen. Usually oxyhemoglobin affinity curves are determined at blood pH of 7.4 and at a lower pH, often a pH of 6.8.

A comparison of the data in Figures 7-11 and 7-12 seem to relate to

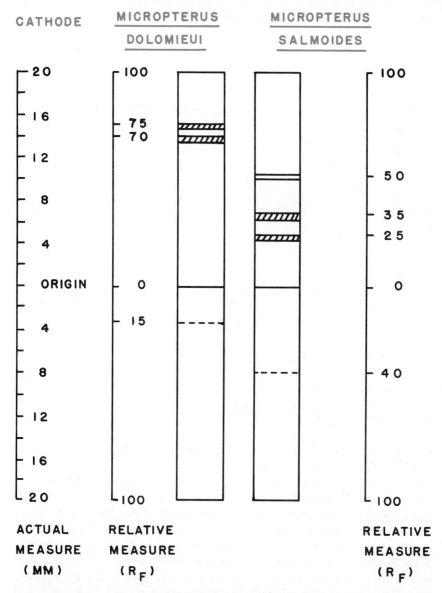

FIG. 7-10. Electrophoretic patterns of hemoglobins formed on cellulose acetate membranes are shown for the smallmouth bass (*Micropterus dolmieui*) and largemouth bass (*Micropterus salmoides*). Distance (in millimeters) of migration is shown as well as percentage of relative measure (R_F). Except for the origin line, the bands are drawn in relative density to each other. From Burke, J. D. 1965. Oxygen affinities and electrophoretic patterns of hemoglobins in trout and basses from Virginia. Med. Coll. Va. Quart. *1*:16. Fig. 2.

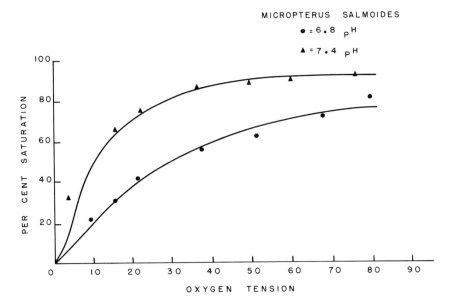

Fig. 7-11. Oxyhemoglobin affinity curves for largemouth bass (*Micropterus salmoides*) determined at 25°C. From Burke, J. D. 1965. Oxygen affinities and electrophoretic patterns of hemoglobins in trout and basses from Virginia. Med. Coll. Va. Quart. *1*:16. Fig. 7.

two ecophysiological factors differentiating the largemouth from the smallmouth bass. First, the hemoglobin of the largemouth has a greater affinity for oxygen than that of the smallmouth bass (P^{50} of 9 mm Hg *versus* P^{50} of 14 mm Hg). Assuming this to be a genetic difference, it would explain the ability of the largemouth bass to inhabit water with a lower oxygen content, lower turnover rate, and higher temperature. Conversely, since the smallmouth bass has hemoglobin with a lower affinity for oxygen, this species would tend to be restricted to waters where the temperature is lower and oxygen content higher. These conditions are associated with mountain streams and lakes, and fast flowing streams with rapids. The second factor is the Bohr effect, *i.e.*, the affinity of hemoglobin for oxygen decreases as acidity increases. The decreasing oxyhemoglobin affinity is noticeably associated with an increasing Bohr effect in the smallmouth bass. This relation is particularly advantageous to the organism when metabolic demands require

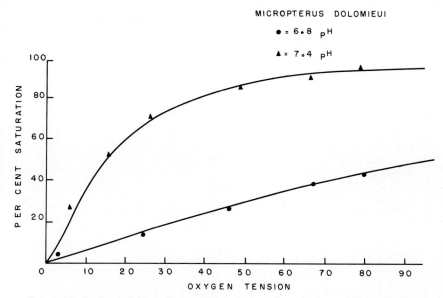

Fig. 7-12. Oxyhemoglobin affinity curves for smallmouth bass (*Micropterus dolomieui*) determined at 25°C. From Burke, J. D. 1965. Oxygen affinities and electrophoretic patterns of hemoglobins in trout and basses from Virginia. Med. Coll. Va. Qaurt. *1*:16. Fig. 6.

a high unloading tension to cells. It may also explain why it is that, as acid metabolites form and carbon dioxide increases, the hemoglobin is buffered at a lower pH and more oxygen is, therefore, unloaded to cells. The smallmouth bass is noted for its characteristic fighting habit when hooked, and has been called "the gamest fish that swims." Perhaps the large Bohr effect ($R = \Delta \log P^{50}/\Delta$ pH) with a ratio of -1.37 as compared with R for the largemouth bass (-0.73) helps explain the vigor and strength of smallmouth bass in which an increasing supply of oxygen to the cells would be necessary to sustain great activity.

Several factors affect the affinity of hemoglobin for oxygen. One of these is pH as mentioned above. Others include temperature, ionic strength, and Po_2. As pH decreases the acidity increases; acidity increases with an increase in the concentration of carbon dioxide. This pH decrease shifts the oxyhemoglobin curve to the right (Fig. 7-13) which indicates the decrease in oxyhemoglobin affinity. An increase in tem-

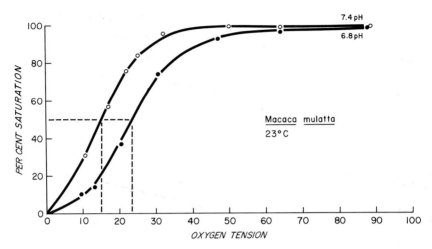

FIG. 7-13. Oxyhemoglobin affinity at pH 7.4 and 6.8 for the Rhesus monkey (*Macaca mulatta*). Data for both curves were obtained at 23 C. The P_{50} is 15 mm Hg at pH 7.4, and 23 mm Hg at pH 6.8. From Burke, J. D. 1966. A simple rapid method for determining oxyhemoglobin affinity: Illustration using blood from Rhesus monkey. Med. Coll. Va. Quart. 2:219. Fig. 2.

---

perature also decreases the affinity of hemoglobin for oxygen as does an increase in acidity. In very high and very low concentrations of salt, oxyhemoglobin affinity decreases. Therefore, in determining oxyhemoglobin affinity for construction curves, pH, temperature, and ionic strength must be controlled experimentally. Then Po_2 can be varied directly and the percentage of saturation of hemoglobin with oxygen can be measured at various points on the curve. In some studies it is necessary to determine oxyhemoglobin affinity on whole blood. A reinvestigation of the normal oxyhemoglobin affinity curve for man is shown in Figure 7-14.

As cellular respiration occurs, a continual supply of atmospheric oxygen must be available. In animals, the aerobic metabolic process is facilitated by gas transport by hemoglobin whether in the smallest (2.0 gm) mammal (*Pachyura etrusca*) or hummingbird (*Calypte helenae*) less than 1 gm, the smallest (2 mg) fish (*Schindleria praematurus*), or even the largest (100 tons) animal known, past or present, the blue whale (*Sibbaldus musculus*). But oxygen reaches the smallest cell (*Mycoplasma laidlaivii*), the pleuropneumonia-like organism (PPLO), by a diffusion gradient.

It has been long established and well documented that oxygen con-

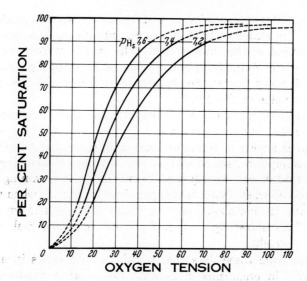

FIG. 7-14. Average O_2 dissociation curve from 14 human subjects at serum pH 7.6, 7.4, and 7.2. Ordinate: percentage of O_2 saturation; abscissa: pressure of O_2 in mm Hg 37 C. From Bartels, H., Betke, K., Hilpert, P., Niemeyer, G., and Riegel, K. 1961. Die sogenannte Standard-O_2-Dissoziationskurve des gesunden erwachsenen menschen. Arch. Ges. Physiol. 272:372. Fig. 1.

---

sumption varies with the body weight of an organism. The usual expression is in terms of metabolic rate, *i.e.*, the Q_{O_2}. Then the total metabolism may be calculated as some exponential of body weight. When metabolism is expressed as a power function of body weight, unicellular and multicellular organisms can be compared either intraspecifically or interspecifically. When large and small animals within a species, or when different animals of different sizes are compared, it is found that the total metabolism of larger animals is greater than that of smaller animals, but the Q_{O_2} of smaller animals exceeds that of larger animals. If the total metabolism, or amount of oxygen consumed per unit time equals M, and body weight is expressed by x, then

$$M = kx^n$$

and metabolic rate is

$$Q_{O_2} = \log \frac{M}{x} = \log k + n \log x$$

When a logarithmic regression of metabolic rate (Q_{O_2}) is made on body weight (x), the straight line has a slope (n) and log k is equivalent to the y—intercept of that line. If metabolic rate is proportional to body weight, n = 1.0. Actually, n is less than 1.0 for most organisms; exceptions include some insects. For homeotherms and poikilotherms in general, n has a value of about 0.73.

In Figure 7-15 is shown a semilogarithmic plot of metabolism per kilogram against body weight for different mammals. The average total metabolism is shown in terms of average heat production. The rate of oxygen consumption rises so steeply that a mammal weighing less than 2.0 gm could not eat enough food to sustain it. A hummingbird has a Q_{O_2} of about 85 when flying as compared with about 15 when not flying. To maintain its metabolic rate at night, its temperature drops. In Figure 7-16 are shown Q_{O_2} values for some small mammals as a function of body weight.

It is clear from these data that metabolic rate varies inversely with body weight. In organisms in general, the negative exponential value

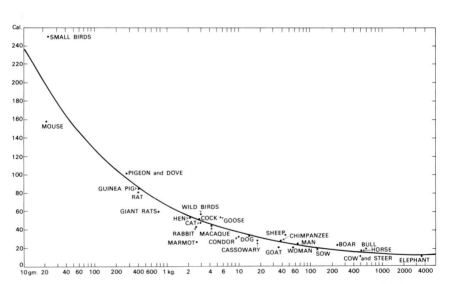

Fig. 7-15. Semilogarithmic chart showing the trend of the average heat production per kilogram of each animal species referred to the average body weight—weight range from 20 gm to 4000 kg. From Benedict, F. G. 1938. *Vital Energetics*. Carnegie Institute, Washington Publ. No. 503. p. 145. Fig. 37.

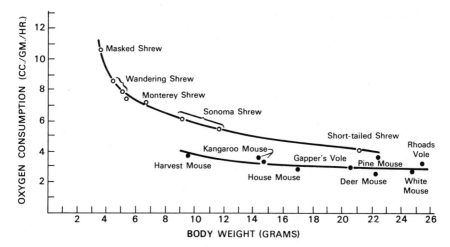

FIG. 7-16. Oxygen consumption of small mammals as a function of body weight. From Pearson, O. P. 1948. Metabolism of small mammals, with remarks on the lower limit of mammalian size. Science 108:44. Fig. 1.

---

may vary from 0.55 to 1.0. But the Qo_2 is not the only parameter varying inversely with body weight. Other parameters which have been related to Qo_2 are total blood oxygen capacity (BOC), blood volume, and oxyhemoglobin affinity.

The total BOC is correlated with metabolic rate since BOC sets a maximum limit on the amount of oxygen that the blood can take up in a unit time.

It was established in 1957 that BOC/gm varied inversely with body weight in albino rats (Fig. 7-17). Since that time, it has been also reported that a decrease in BOC/gm body weight is correlated with an increase in body weight in fishes, Ranid frogs, turtles, pigeons, and 25 different species of birds (Figs. 7-18 through 7-25, respectively). The coefficient of correlation is lower for turtles than for any other group of vertebrates represented. This may be explained, at least in part, by the principle of similitude. This principle is related to the fact that an increase in body weight can be accompanied by a disproportionate increase of inert metabolic tissue (such as shell). Indeed it has been found that metabolic rate is not correlated with body weight in turtles. Statistical data pertaining to Figures 7-17 through 7-25 are presented in Table 7-2. Variables in regression analyses include age, sex, phylogenetic varia-

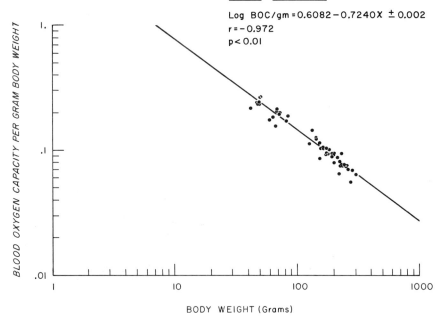

FIG. 7-17. Volumes per cent blood oxygen capacity per gram of body weight plotted against body weight in 64 albino rats, *Rattus norvegicus* (co-ordinates on logarithmic scale). From Burke, J. D. 1966. Vertebrate blood oxygen capacity and body weight. Nature (London) *212*:46. Fig. 1.

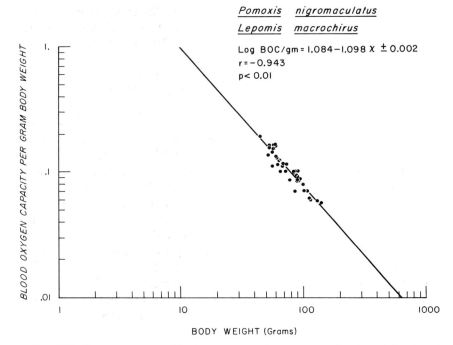

FIG. 7-18. Volumes per cent blood oxygen capacity per gram of body weight plotted against body weight in 40 fishes, *Lepomis macrochirus* and *Pomoxis nigromaculatus* (co-ordinates on logarithmic scale). From Burke, J. D. 1966. Vertebrate blood oxygen capacity and body weight. Nature (London) *212*:46. Fig. 2.

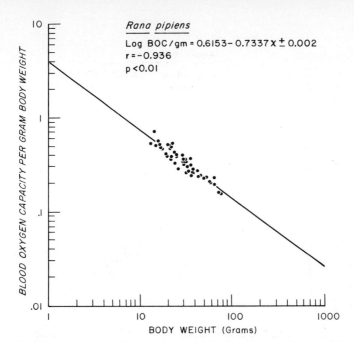

Fig. 7-19. Volumes per cent blood oxygen capacity per gram of body weight plotted against body weight in 50 meadow frogs, *Rana pipiens* (co-ordinates on logarithmic scale). From Burke, J. D. 1966. Vertebrate blood oxygen capacity and body weight. Nature (London) *212*:46. Fig. 3.

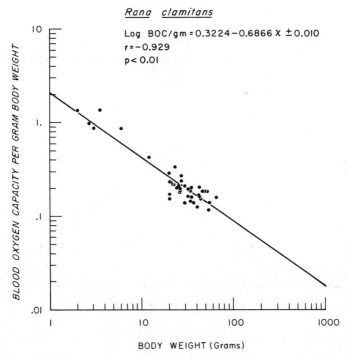

Fig. 7-20. Volumes per cent blood oxygen capacity per gram of body weight plotted against body weight in 43 green frogs, *Rana clamitans* (co-ordinates on logarithmic scale). From Burke, J. D. 1966. Vertebrate blood oxygen capacity and body weight. Nature (London) *212*:46. Fig. 4.

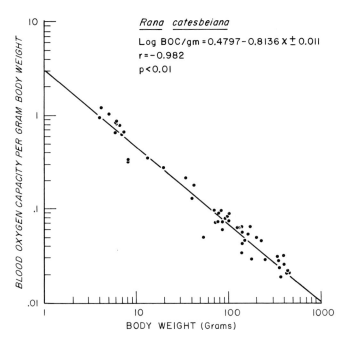

Fig. 7-21. Volumes per cent blood oxygen capacity per gram of body weight plotted against body weight in 52 bullfrogs, *Rana catesbeiana* (co-ordinates on logarithmic scale). From Burke, J. D. 1966. Vertebrate blood oxygen capacity and body weight. Nature (London) *212:*46. Fig. 5.

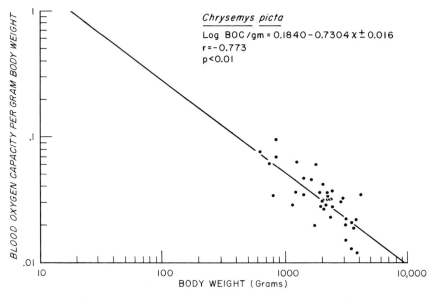

Fig. 7-22. Volumes per cent blood oxygen capacity per gram of body weight plotted against body weight in 40 painted turtles, *Chrysemys picta* (co-ordinates on logarithmic scale). From Burke, J. D. 1966. Vertebrate blood oxygen capacity and body weight. Nature (London) *212:*46. Fig. 6.

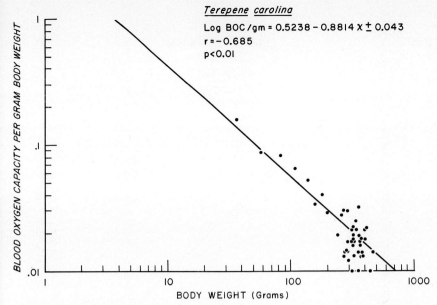

FIG. 7-23. Volumes per cent blood oxygen capacity per gram of body weight plotted against body weight in 48 box turtles, *Terrepene carolina* (co-ordinates on logarithmic scale). From Burke, J. D. 1966. Vertebrate blood oxygen capacity and body weight. Nature (London) 212:46. Fig. 7.

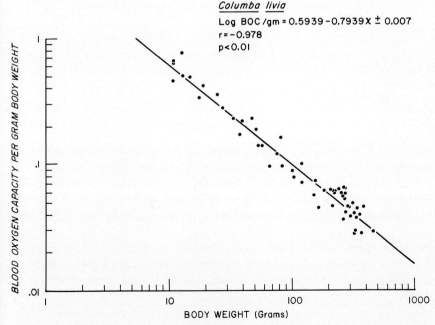

FIG. 7-24. Volumes per cent blood oxygen capacity per gram of body weight plotted against body weight in 56 domestic pigeons, *Columba livia* (co-ordinates on logarithmic scale). From Burke, J. D. 1966. Vertebrate blood oxygen capacity and body weight. Nature (London) 212:46. Fig. 8.

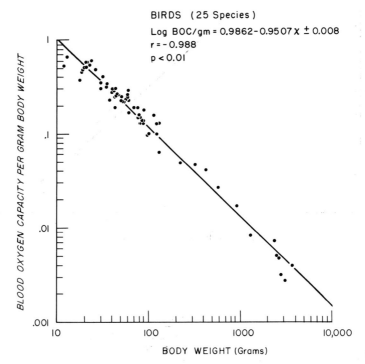

FIG. 7-25. Volumes per cent blood oxygen capacity per gram of body weight plotted against body weight in 67 individuals representing 25 species of birds (co-ordinates on logarithmic scale). From Burke, J. D. 1966. Vertebrate blood oxygen capacity and body weight. Nature (London) *212*:46. Fig. 9.

---

tion, and body weight. The excellent coefficient of correlations clearly suggests that body weight is the predominant factor influencing BOC.

The inverse relation of BOC (and Q_{O_2}) to body weight seems to be correlated with the amount of blood in an animal. It is known that blood volume per unit weight decreases with increasing body weight. Inverse relations between blood volume and body weight have been reported in fishes, frogs, turtles, birds, and mammals. In Figure 7-26 an example is shown for swine where blood volume per kilogram of body weight is plotted against body weight. Blood volume values were obtained in each animal from experiments made simultaneously with T-1824 (Evans blue) dye and ^{32}P methods. The correlation coefficient was statistically significant being 87% (T-1824) and 86% (^{32}P).

TABLE 7-2
Linear Regressions of Log Blood Oxygen Capacity on Log Body Weight in Various Vertebrates

Animals	Regression Equation*	Correlation Coefficient	Probability that Slope is Zero
Albino rats 64 *Rattus norvegicus*	$Y = 0.6082 - 0.7240\chi \quad \pm 0.002$	-0.972	$P < 0.01$
Birds (25 species) 67 individuals	$Y = 0.9862 - 0.9507\chi \quad \pm 0.008$	-0.988	$P < 0.01$
Pigeons 56 *Columba livia*	$Y = 0.5939 - 0.7939\chi \quad \pm 0.007$	-0.978	$P < 0.01$
Painted turtles 40 *Chrysemys picta*	$Y = 0.1840 - 0.7304\chi \quad \pm 0.016$	-0.773	$P < 0.01$
Box turtles 48 *Terrepene carolina*	$Y = 0.5238 - 0.8814\chi \quad \pm 0.043$	-0.685	$P < 0.01$
Meadow frogs 50 *Rana pipiens*	$Y = 0.6153 - 0.7337\chi \quad \pm 0.002$	-0.936	$P < 0.01$
Green frogs 43 *Rana clamitans*	$Y = 0.3224 - 0.6866\chi \quad \pm 0.010$	-0.929	$P < 0.01$
Bullfrogs 52 *Rana catesbeiana*	$Y = 0.4797 - 0.8136\chi \quad \pm 0.011$	-0.982	$P < 0.01$
Fishes (2 species) 20 *Lepomis macrochirus* 20 *Pomoxis nigromaculatus*	$Y = 1.084 - 1.098\chi \quad \pm 0.002$	-0.943	$P < 0.01$

* $\log \text{BOC/gm} = \log k + n \log \text{body wt. (gm)}$.
$Y = \log \text{BOC/g} = \log$ of blood oxygen capacity/gm body wt.
$\log k = y -$ intercept.
$n =$ slope of line.
$\chi = \log$ body wt. (gm).
Variance = standard error of BOC/gm from body wt. (gm).
From Burke, J. D. 1966. Vertebrate blood oxygen capacity and body weight. Nature (London) 212:46. Table 1.

An inverse relation between oxyhemoglobin affinity and body weight has been reported. The P^{50} (the oxygen pressure in mm Hg at which one half saturation of the blood occurs) decreases as body weight increases. In Figure 7-27 is shown that as various animals increase in body weight there is an increase in the affinity of blood hemoglobin for oxygen as indicated by a decreasing P^{50}.

Several assumptions can be made about the inverse relation of metabolic rate, total blood oxygen capacity, blood volume, and oxyhemoglobin affinity to body weight in vertebrates. It seems that vertebrates (either intraspecies and interspecies) of lesser weight can maintain a

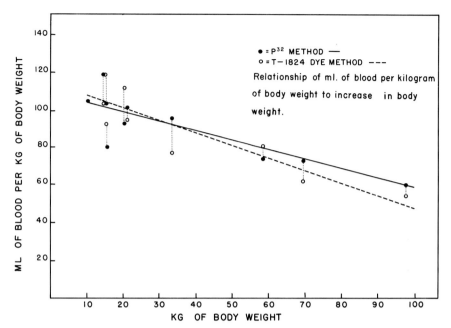

FIG. 7-26. From Burke, J. D. 1954. Blood volume in mammals. Physiol. Zool. *27*:1. Fig. 8.

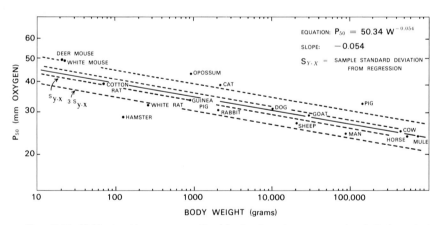

FIG. 7-27. Half-saturation pressure for blood of various mammals plotted against body weight. (Co-ordinates on logarithmic scale). From Schmidt-Neilsen, R. and Larimer, J. L. 1958. Oxygen dissociation curves of mammalian blood in relation to body size. Amer. J. Physiol. *195*:424. Fig. 2.

higher metabolic rate because: (1) smaller animals have a greater blood volume per unit body weight; (2) smaller animals have a greater total blood oxygen capacity per unit body weight; and (3) smaller animals have a higher unloading tension for oxygen which is more favorable for a rapid diffusion of oxygen to the cells for cellular respiration. It also appears that the related parameters of Qo_2, BOC/gm, and total blood volume per gram of body weight do not limit animals increase in body weight. On the contrary, it is apparent that an evolutionary mechanism developed to aid increasing body weight. This was, of course, the evolvement of a hemoglobin molecule which had an increasing affinity for oxygen thereby permitting evolving species with increasing body weights.

C. Carbohydrate Formation and Conversions

Chloroplasts alone carry out photosynthesis in higher plants, the chief site of the photosynthetic process is the leaf. In Figure 7-28 a cross-section of the leaf shows the tissues which are epidermis, meso-

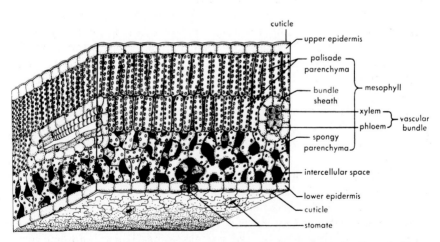

FIG. 7-28. Diagram of section view of a leaf, showing nature of cells and tissues. From Wilson, C. L and Loomis, W. E. 1962. *Botany*. 3rd edition. Holt, Rinehart, and Winston. New York. p. 77. Fig. 6.2.

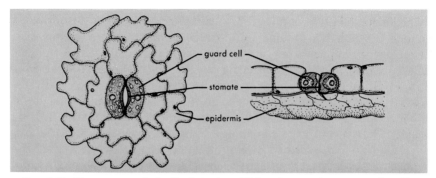

FIG. 7-29. The stomate. Left: Surface view. Right: As seen in cross section. From Wilson, C. L. and Loomis, W. E. 1962. *Botany*. 3rd edition. Holt, Rinehart, and Winston. New York. p. 77. Fig. 6.3.

---

phyll, and vascular bundles. Two specialized epidermal cells, called guard cells (Fig. 7-29) contain chloroplasts, unlike other epidermal cells. Because of the kidney-shape and thick medial borders, a change in turgor pressure causes the opening (stomate) between the two guard cells to open and close. It is through the stomate that the carbon dioxide, oxygen, and water vapor pass. Primarily, gas exchange occurs freely in response to a diffusion gradient.

Photosynthesis is localized mostly in the mesophyll of the leaf. Chloroplasts are located in both the upper palaside and lower spongy parenchyma. In Figure 7-28 it may be seen that the stomate forms an air passage space connecting with the intercellular air spaces extending throughout the mesophyll. The chloroplasts are usually more abundant in the palaside layer than in the spongy cells. Although the composition of chloroplasts varies in different cells, generally their percentages range for proteins, 50–70%; lipids, 20–35%; chlorophyll, 5–8%; carotenoids, 0.7–1.1%; RNA, 1.0–7.5%; DNA, 0.02–0.5%; chloroplasts, although generally in a fixed area of a cell, are phototaxic.

In higher plants, chlorophyll is located in grana of the chloroplast. The ultrastructure of grana and chloroplasts was discussed in Chapter 2. Grana are composed of a multilayer system of lamellae. The lamellae consist of stacked membranous sheets. On the sheet surface are regular, packed units. Each unit is termed a quantosome. The quantosome may be the photosynthetic unit containing an aggregation of all the components necessary for photosynthesis.

CELLULAR RESPIRATION 213

The absorption of light by chlorophyll is the fundamental event in photosynthesis. Green plants are characterized by having chlorophyll A and chloroplyll B. As indicated in Figure 7-30, in the upper right ring, the methyl group of chlorophyll is replaced by a formyl group (—CHO) in chlorophyll B. Otherwise, both chlorophylls A and B are identical. There are different chlorophylls in other cells. In brown algae and diatoms, chlorophylls A and C are present. Chlorophylls A and D are found in red algae cells. Chlorophylls differ from one another only in the side chains attached to the outer ends of the tetrapyrrole nucleus. Chlorophylls exist within chloroplasts as a lipoprotein complex. A chlorophyll called bacteriochlorophyll is found in purple bacteria. It differs from chlorophyll A in that the double bond at C_3—C_4 is replaced by a single bond, and an acetyl residue replaces the vinyl substituent at C_2. Absorption spectra for chlorophylls A and B and bacteriochlorophyll are shown in Figure 7-31. It appears on best evidence at present that chlorophyll is evenly distributed in grana and stromal lamellae.

The process of photosynthesis involves two reactions: one, in which

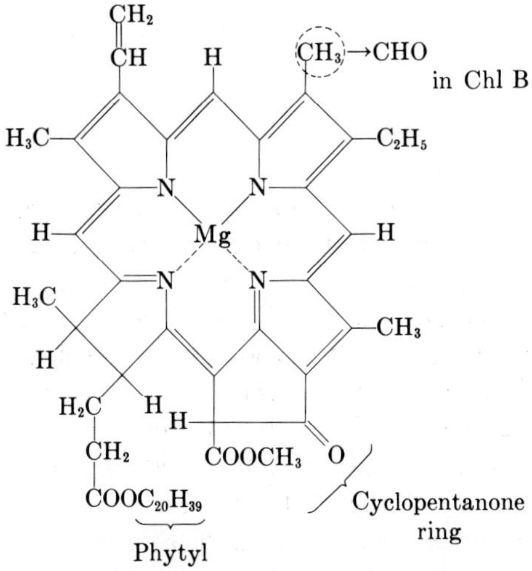

FIG. 7-30. The tetrapyrrole structure of chlorophyll. The difference in structure between chlorophyll A and chlorophyll B is at the top right of the figure. From Clayton, R. K. 1965. *Molecular Physics in Photosynthesis.* Blaisdell Publishing Co. Waltham, Mass. p. 195, Fig. A.1.

214 CELL METABOLISM

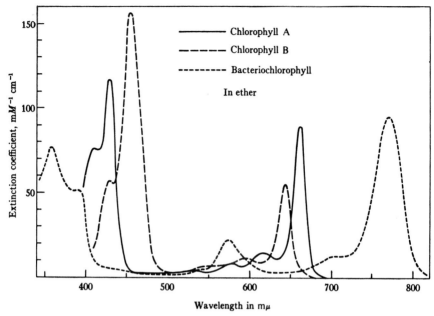

FIG. 7-31. Absorption spectra of various chlorophylls. Note that chlorophyll A absorbs at longer red wave lengths than chlorophyll B. From Clayton, R. K. 1965. *Molecular Physics in Photosynthesis*. Blaisdell Publishing Co. Waltham, Mass. p. 196, Fig. A. 2.

---

light is necessary; two, another reaction in which light is not required. The first reaction is called the light or Hill reaction and involves photophosphorylation and photolysis of water. The second reaction is known as the dark or Blackman reaction. In this reaction, carbon dioxide fixation leads to the formation of carbohydrates. Subsequently, chemical reactions in the cell convert some carbohydrates into other foodstuffs.

Since photosynthesis involves energy conversion and flow, several points ought to be kept in mind about it. When atoms bond together to form molecules, energy changes always accompany the bonding. In the formation of polar bonds, ions are formed. Electrons are transferred from one atom to another. This electron transfer leads to a net positive charge on one while the other atom acquires a net negative charge. In the formation of nonpolar, or covalent, bonds electrons are shared by two atoms. In either case, when either polar or nonpolar bonds are formed, energy is liberated. Conversely, when a bond is broken, energy must be absorbed. The greater the stability of the bond, the greater the amount of energy required for bond breakage. When an individual atom absorbs energy, the outer orbital may be excited *i.e.*, moved to an outer orbit of a higher energy level. When the excited electron returns to its orbit, this energy may be released in the form of heat or by emission of a light quantum as fluoresence. The unit of energy absorbed or released is called a quantum. The photon is a quantum of radiant energy (light). The quantum value of light is directly related to its wave length, *i.e.*, a longer wave length has a lower quantum value. Light quanta are absorbed by molecules only at specific wave lengths. In higher plants most of the light on a leaf is absorbed by chloroplylls A and B. In Figure 7-31 it may be seen that both chlorophylls A and B abosorb maximally in the blue-violet and red spectrum of visible light. Shorter wave lengths may also be utilized because accessory pigments, carotenoids, for example, absorb at wave lengths from about 400 to 550 mμ.

1. LIGHT REACTIONS

Life on this planet depends on the fact that light energy is converted to chemical energy. It is in the chloroplast where the transformation of light energy to chemical energy occurs during the process of photophosphorylation. There seems to be two types of photosynthetic phosphorylation. One of these types is called cyclic photophosphorylation. The other type is known as noncylic photophosphorylation. Both cyclic and noncyclic photophosphorylation seem essential in that ATP formed in one is not sufficient for carbon dioxide fixation in carbohydrate formation.

Cyclic photophosphorylation requires light and chlorophyll, and the only substrates needed are ADP and inorganic phosphate to form ATP and water. This most fundamental photosynthetic reaction has been demonstrated as occurring in bacteria, algae, and higher plants. Unlike oxidative phosphorylation, no oxygen is required in cyclic photophosphorylation. The over-all scheme for cyclic photophosphorylation may be indicated as

$$\text{ADP} + \text{P}_i \xrightarrow[\text{chlorophyll}]{\text{light}} \text{ATP} + \text{H}_2\text{O}$$

Noncyclic photophosphorylation produces ATP, oxygen, and NADPH or reduced ferrodoxin. A summary of noncyclic photophosphorylation shows that

$$4\ Fd_{ox} + 2\ ADP + 2\ P_i + 4\ H_2O \xrightarrow[\text{chlorophyll}]{\text{light}} 4\ Fd_{red} + 2\ ATP + 2\ H_2O + 4\ H^+ + O_2$$

and NADP is ultimately reduced by a transfer of electrons from reduced ferrodoxin, apparently via a flavoprotein (ferrodoxin-NADP reductase). Unlike oxidative phosphorylation, ATP formation here in this reaction is coupled to oxygen evolution.

There seem to be two complex pigment systems involved in photophosphorylation. Functionally, these two systems are closely related. They are called System I and System II. System I includes chlorophyll A having along wave length maximum absorption peak at 683 mμ, and another chlorophyll having a peak absorption at 700 mμ. Associated with System II is chlorophyll B (and chlorophyll A). In some plants carotenoids (*Chlorella*), fucoxanthin (diatoms and brown algae), and phycobilin pigments (phycocyanin and phycoerythrin) in the red and blue-green algae function efficiently as light-absorbers for photosynthesis.

When light energy is connected to chemical energy, a difference in redox potential occurs. A reference to Figure 7-32 shows that the main difference between System I and System II is the redox potential. It may also be seen in the figure that Systems I and II are linked to each other by a chain of redox catalysts; these are plastoquinone, cytochrome b_6, plastocyanine, and cytochrome f. In the following discussion it will be helpful to refer to Figure 7-32 at intervals.

In cyclic photophosphorylation, quantum absorption by chlorophyll A produces an excitation of orbital electrons. These excited electrons are raised to a higher energy level in the complex chlorophyll molecule. The excited electrons flow from chlorophyll A (via a primary electron acceptor, Z ?) to ferrodoxin, an iron-containing compound. From ferrodoxin electrons pass on to plastoquinone via a flavoprotein (?) with the probable formation of 1 ATP per 2 e^-. Electron transport proceeds to cytochrome f via cytochrome b_6 and plastocyanine with ATP formation. When cytochrome f returns electrons to chlorophyll A, the cycle is complete.

The most striking feature of noncyclic photophosphorylation is that the excited electrons of chlorophyll B do not return, but flow via NADPH into an electron sink, mostly for carbon dioxide reduction to carbohydrate. It is believed at present that the exogenous donor of elec-

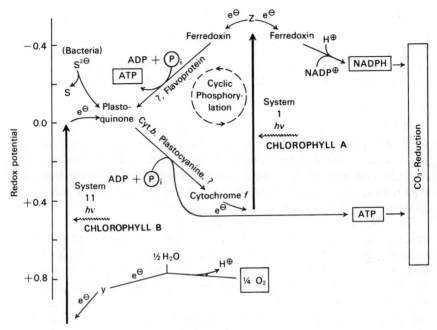

Fig. 7-32. Schematic diagram of photosynthesis. The boldface arrows represent the energy rise of the electrons due to absorbed light quanta. From Karlson, P. 1968. *Introduction to Modern Biochemistry*. 3rd edition. Academic Press. New York. p. 299. Fig. 43.

---

trons is the OH^- ions from the photolysis of water. Molecular oxygen is released from hydroxyl ions when electrons are removed as indicated in a Hill reaction shown in Chapter 2. Excited chlorophyll may effect the transfer of electrons from y (an unknown redox system) to plastoquinone. Oxidized y could take electrons from water and oxidize it to O_2 as

$$2 H_2O - 4 e^- \rightarrow O_2 + 4 e^-$$

These electrons are used for the ultimate reduction of NADP. As such the electrons would be replaced by the photolysis of water. On the basis of NADPH formation, the stoichiometry of noncyclic photophosphorylation is indicated as

$$2\text{ NADP} + 2\text{ H}_2\text{O} + 2\text{ ADP} + 2\text{ P}_i \xrightarrow{\text{light}} 2\text{ NADPH} + 2\text{ H}^+ + 2\text{ ATP} + O_2$$

Because ferrodoxin (an intermediate) is the most electronegative of naturally occurring electron carriers, it can reduce NADP to form NADPH as a stable transport metabolite for hydrogen. The NADPH can be utilized along with the ATP generated in both cyclic and noncyclic photophosphorylation for carbohydrate formation in photosynthesis.

2. DARK REACTIONS

The dark reactions of photosynthesis can occur without light if both ATP and NADPH are present. In these reactions carbon dioxide fixation occurs forming carbohydrates. As indicated in the discussion above on light reactions light energy is transformed into phosphate bond energy during photophosphorylation; this is the derived energy for carbon fixation in photosynthesis by ATP and NADPH.

The path of carbon in the dark reactions has been worked out with radioactive carbon-14. ^{14}C is introduced into suspensions of photosynthesizing algae cells. After brief periods of exposure the cells are killed, usually in hot alcohol, at timed intervals. The products formed are identified by chromatography and radioautography. After cells have been exposed only 1 second to light, ^{14}C is found incorporated in phosphoglyceric acid. After 5 seconds exposure, not only is labeled phosphoglyceric acid present, but ^{14}C-labeled alanine, aspartic acid, and malic acid are also present. After a full minute of illumination radioactive glucose phosphate and radioactive fructose phosphate are found. In addition to these labeled compounds at the end of 5 minutes exposure to light, other compounds such as lipids and proteins are labeled.

Finally, it was determined that in the carbon reduction cycle (see Fig. 7-33) the key reaction was carboxylation of a 5-carbon compound (ribulose 1,5-diphosphate) and not a 2-carbon compound. This reaction is catalyzed by diphosphoribulose carboxylase. Carbon dioxide adds to C-2 of the ribulose diphosphate to form an unstable β-keto acid which undergoes hydrolysis to form two molecules of phosphoglyceric acid. The phosphoglyceric acid is reduced to glyceraldehyde. As shown in Figure 7-33, hexose may be accumulated and pentose regenerated so that ribulose 1,5-diphosphate may be recovered for another carboxylation reaction. If ribulose diphosphate were not regenerated, photosynthesis would halt!

Ribulose diphosphate regeneration by dark reactions is often referred to as the Calvin cycle. The regeneration of this pentose proceeds via a pathway which includes many intermediates of the Embden-

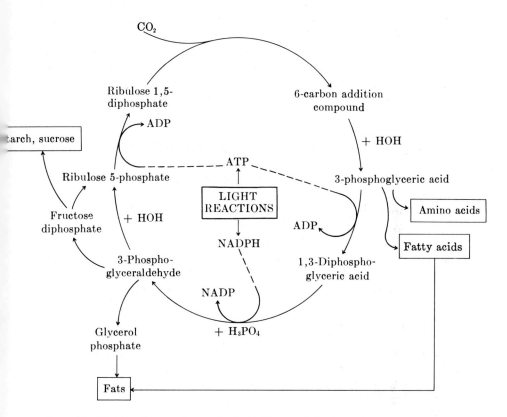

Fig. 7-33. Photosynthetic carbon cycle (or Calvin cycle). In these dark reactions illustrated, carbon dioxide is fixed by condensing with ribulose diphosphate. The relation of the light reactions to the dark reactions is indicated by both ATP and NADPH from the light reactions contributing to reduction of phosphoglyceric acid to phosphoglyceraldehyde. The carbon cycle shown here is greatly simplified. The end-products of photosynthesis are not limited to carbohydrates, but also include fats, fatty acids, and amino acids as well as other compounds.

Meyerhof and phosphogluconate oxidative (pentose phosphate) pathways; the enzymes involved in the dark reactions are identical to those of the glycolytic pathways. Glycolysis will be discussed in the following chapter on carbohydrate metabolism.

In Table 7-3 are shown the various steps of reactions in which hexose is accumulated and ribulose diphosphate is regenerated. A summarized diagram is given below which outlines by carbon relation the regeneration of ribulose diphosphate.

$$CO_2 \ (1 \ C)$$
$$+$$
phosphoglyceric acid (3 C) $\leftarrow$ ribulose diphosphate (5 C)
$+$
phosphoglyceric acid (3 C) $\rightarrow$ phosphoglyceraldehyde (3 C)
$$+$$
hexose monophosphate $\leftarrow$ phosphoglyceraldehyde (3 C)
(6 C)
$+$
phosphoglyceraldehyde $\rightarrow$ ribulose monophosphate (5 C)
(3 C)   and
phosphoglyceraldehyde $+$ tetrose monophosphate (4 C)
$\downarrow$
sedoheptulose (7 C)
$+$
phosphoglyceraldehyde $\rightarrow$ ribulose monophosphate (5 C)
(3 C)   and
ribulose diphosphate (5 C) $\leftarrow$ ATP $+$ ribulose monophosphate
                                           (5 C)

The light reactions of photophosphorylation are related to the dark reactions as shown in Figure 7-33. As ATP and NADPH are formed during photophosphorylation the phosphate bond energy contained in these compounds is used to drive the Calvin cycle. As the cycle turns ribulose diphosphate is regenerated, carbohydrates are formed as well as other foodstuffs from intermediates of the cycle.

Energy flows throughout the living systems of organisms. In this biosphere, light energy is captured during photosynthesis and released in anaerobic and aerobic metabolism. In Figure 7-34 is shown a diagram representing a biosphere. Matter and energy required from the environment flows in at the bottom of the boundary and the products released at the top boundary to the environment. The biosphere can represent

TABLE 7-3
Hexose Formation and Pentose Regeneration in Photosynthesis

Step	Enzyme	Reaction	Carbon Balance
a	Carboxylation enzyme	6 Ribulose 1,5-diphosphate + $6CO_2$ → 12 3-phosphoglyceric acid	6(5) + 6(1) → 12(3)
b	Phosphoglyceric acid kinase	12 3-Phosphoglyceric acid + 12ATP → 12 1,3-diphosphoglyceric acid + 12ADP	12(3) → 12(3)
	Phosphoglyceraldehyde dehydrogenase	12 1,3-Diphosphoglyceric acid + 12DPNH + $12H^+$ → 12 3-phosphoglyceraldehyde + $12DPN^+$ + $12P_i$	12(3) → 12(3)
c	Triose isomerase	5 3-Phosphoglyceraldehyde → 5 dihydroxyacetone phosphate	5(3) → 5(3)
	Aldolase	5 3-Phosphoglyceraldehyde + 5 dihydroxyacetone phosphate → 5 fructose 1,6-diphosphate	5(3) + 5(3) → 5(6)
d	Phosphatase	5 Fructose 1,6-diphosphate → 5 fructose 6-phosphate + $5P_i$	5(6) → 5(6)
e	Transketolase	2 Fructose 6-phosphate + 2 3-phosphoglyceraldehyde → 2 xylulose 5-phosphate + 2 erythrose 4-phosphate	2(6) + 2(3) → 2(5) + 2(4)
f	Transaldolase	2 Fructose 6-phosphate + 2 erythrose 4-phosphate → 2 sedoheptulose 7-phosphate + 2 3-phosphoglyceraldehyde	2(6) + 2(4) → 2(7) + 2(3)
g	Transketolase	2 Sedoheptulose 7-phosphate + 2 3-phosphoglyceraldehyde → 4 xylulose 5-phosphate	2(7) + 2(3) → 4(5)
h	Epimerase	6 Xylulose 5-phosphate → 6 ribulose 5-phosphate	6(5) → 6(5)
i	Phosphoribulokinase	6 Ribulose 5-phosphate + 6ATP → 6 ribulose 1,5-diphosphate + 6ADP	6(5) → 6(5)
Net: 6 Ribulose 1,5-diphosphate + $6CO_2$ + 18ATP + 12DPNH + $12H^+$ → 6 ribulose 1,5-diphosphate + 1 fructose 6-phosphate + $17P_i$ + 18 ADP + $12DPN^+$			6(5) + 6(1) → 6(5) + 1(6)

From White, A., Handler, P., and Smith, E. L. 1968. *Principles of Biochemistry*. 4th edition. McGraw-Hill Book Co. New York. p. 455. Table 20.1.

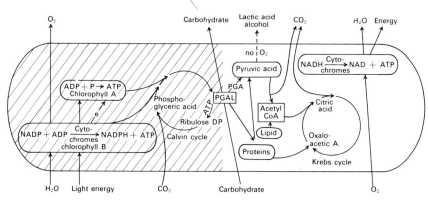

FIG. 7-34. Diagram illustrating the relation of photosynthetic reactions (left) to the reactions of oxidative and anaerobic metabolism (right). From DuPraw, E. J. 1968. *Cell and Molecular Biology.* Academic Press. New York. p. 146. Fig. 7-4.

a cell in which energy and matter flows through it to maintain the cell in a steady state.

Suggested Reading—Chapter 7

Arnon, D. I. Conversion of light into chemical energy in photosynthesis. Nature (London) *184*:10, 1959.
Arnon, D. I. The role of light in photosynthesis. Sci. Amer. *203(5)*:105, 1960.
Arnon, D. I. The chloroplast as a functional unit in photosynthesis. *Handbuch der Pflanzenphysiologie 5/1*:773, Springer Verlag. Berlin. 1960.
Barcroft, J. Differential method of blood-gas analysis. J. Physiol. (London) *37*:12, 1908.
Barrington, E. J. W. *The Chemical Basis of Physiological Regulation.* Scott, Foresman and Co. Glenview, Ill. 1968.
Bartels, H., Betke, K., Hilpert, P., Niemeyer, G., and Riegel, K. Die sogenannte Standard-O_2-Dissoziationskurve des gesunden erwachsenen menschen. Arch. Ges. Physiol. *272*:372, 1961.
Bassham, J. A. The path of carbon in photosynthesis. Sci. Amer. *206(6)*:89, 1962.
Bassham, J. A. and Calvin, M. *The Photosynthesis of Carbon Compounds.* W. A. Benjamin. New York. 1962.
Benedict, F. G. *Vital Energetics.* Carnegie Institute, Washington. Publ. No. 503. 1938.
Boell, E. J., Needham, J., and Rogers, V. Morphogenesis and metabolism: studies with the Cartesian diver ultramicromanometer. Proc. Roy. Soc. [Biol.] *127*:322, 1939.
Bogorad, L. The organization and development of chloroplasts. In *Molecular Organization and Biological Function.* Allen, J. M. (editor). Harper & Row. New York. 1967.
Bond, C. G. and Gilbert, P. W. Comparative study of blood volume in representative aquatic and nonaquatic birds. Amer. J. Physiol. *194*:519, 1958.
Bonner, J. T. *Morphogenesis.* Princeton University Press. Princeton, N. J. 1952.
Bonner, J. and Varner, J. E. *Plant Biochemistry.* Academic Press. New York. 1965.
Bruun, A. F. A study of a collection of the fish *Schindleria* from South Pacific waters. Carlsberg Found, Oceanogr. Exped. 1928-1930. Rept. *21*:1, 1940.

Burke, J. D. Oxygen capacity in mammals. Physiol. Zool. *26:*259, 1953.
Burke, J. D. Blood volume in mammals. Physiol. Zool. *27:*1, 1954.
Burke, J. D. Blood oxygen capacity of albino rats. Amer. J. Physiol. *188:*118, 1957.
Burke, J. D. The effects of different salt concentrations on the affinity of hemoglobin for oxygen. J. Cell. Comp. Physiol. *54:*126, 1959.
Burke, J. D. Determination of oxygen in water using a 10 ml. syringe. J. Elisha Mitchell Sci. Soc. *78:*145, 1962.
Burke, J. D. Oxygen affinities and electrophoretic patterns of hemoglobins in trout and basses from Virginia. Med. Coll. Va. Quart. *1:*16, 1965.
Burke, J. D. A simple rapid method for determining oxyhemoglobin affinity: Illustration using blood from Rhesus monkey. Med. Coll. Va. Quart. *2:*219, 1966.
Burke, J. D. Vertebrate blood oxygen capacity and body weight. Nature (London) *212:*46, 1966.
Burke, J. D. and Woolcott, W. S. A comparison of the blood oxygen capacity in the black crappie (*Pomoxis nigromaculatus*) and the bluegill (*Lepomis macrochirus*). Va. J. Sci. *8:*113, 1957.
Calvin, M. The path of carbon in photosynthesis. Science *135:*879, 1962.
Clayton, R. K. *Molecular Physics in Photosynthesis.* Blaisdell Publishing Co. Waltham, Mass. 1965.
DuPraw, E. J. *Cell and Molecular Biology.* Academic Press. New York. 1968.
Ellis, R. J. Chloroplast ribosomes: specificity of inhibition by chloramphenicol. Science *163:*477, 1969.
Fenn, W. O. A new method for the simultaneous determination of minute amounts of carbon dioxide and oxygen. Amer. J. Physiol. *84:*110, 1928.
Foreman, C. W. A comparative study of the oxygen dissociation of mammalian hemoglobin. A. Cell Comp. Physiol. *44:*421, 1954.
Gibbs, M. Photosynthesis. Ann. Rev. Biochem. *36:*757, 1967.
Giese, A. C. *Cell Physiology.* 3rd edition. W. S. Saunders Co. Philadelphia. 1968.
Glick, D. *Techniques of Histo- and Cytochemistry.* Interscience. New York. 1949.
Gluecksohn-Waelsch, S. The inheritance of hemoglobin types and other biochemical traits in mammals. J. Cell Comp. Physiol. (suppl. 1) *56:*89, 1960.
Goodwin, T. W. (editor). *Biochemistry of Chloroplasts.* Vols. 1 and 2. Academic Press. New York. 1966.
Granick, S. The chloroplasts: inheritance, structure, and function. In *The Cell.* Vol. 2. Brachet, J. and Mirsky, A. E. (editors). Academic Press. New York. 1961.
Gregg, J. R. A microrespirometer. Rev. Sci. Instrum. *18:*514, 1947.
Guyton, A. C. *Textbook of Medical Physiology.* 3rd edition. W. B. Saunders Co. Philadelphia. 1966.
Hall, F. G. Minimal utilizable oxygen and the oxygen dissociation curve of blood of rodents. J. Appl. Physiol. *21:*375, 1966.
Harlan, J. R. and Speaker, E. B. Iowa Fish and Fishing. Des Moines State Conservation Commission. 1951.
Harvey, E. N. The oxygen consumption of luminous bacteria. J. Gen. Physiol. *11:*469, 1928.
Hemmingsen, A. M. The relation of standard basal metabolism to total fresh weight of living organisms. Rep. Steno. Mem. Hosp. *4:*7, 1950.
Hutton, K. E. Blood volume, corpuscular constants, and shell weight in turtles. Amer. J. Physiol. *200:*1004, 1961.
Jagendorf, A. T. Acid-base transitions and phosphorylation by chloroplasts. Fed. Proc. *26:*1361, 1967.
Kabat, D. Organization of hemoglobin synthesis in chicken erythrocytes. J. Biol. Chem. *243:*2597, 1968.
Karlson, P. *Introduction to Modern Biochemistry.* 3rd edition. Academic Press. New York. 1968.

Kirk, J. T. O. and Tilney-Bassett, R. A. E. *The Plastids*. W. H. Freeman Co. San Francisco. 1967.
Kleiber, M. Body size and metabolic rate. Physiol. Rev. *27*:511, 1947.
Larimer, J. L. Hemoglobin concentration and oxygen capacity of mammalian blood. J. Elisha Mitchell Sci. Soc. *75*:174, 1959.
Leftwich, F. B. and Burke, J. D. Blood oxygen capacity of Ranid frogs. Amer. Mid. Nat. *72*:241, 1964.
Linderstrøm-Lang, K. Principle of the Cartesian diver applied to gasometric technique. Nature (London) *140*:108, 1937.
Martin, A. W. The blood volume of some elasmobranchs. Fed. Proc. *6*:164, 1942.
Morrison, P. R. Oxygen consumption in several mammals under basal conditions. J. Cell. Comp. Physiol. *31*:281, 1948.
Moudriankis, E. N. Structural and functional aspects of photosynthetic lamellae. Fed. Proc. *27*:1180, 1968.
Oser, B. L. (editor). *Hawk's Physiological Chemistry*. 14th edition. McGraw-Hill Book Co. New York. 1965.
Packer, L. and Siegenthaler, P. A. Control of chloroplast structure by light. Int. Rev. Cytol. *20*:97, 1966.
Park, R. B. Subunits of chloroplast structure and quantum conversion in photosynthesis. Int. Rev. Cytol. *20*:67, 1966.
Park, R. B. and Biggins, J. Quantasome: size and composition. Science *144*:1009, 1964.
Payne, H. J. and Burke, J. D. Blood oxygen capacity in turtles. Amer. Mid. Nat. *71*:460, 1964.
Pearson, O. P. Metabolism of small mammals, with remarks on the lower limit of mammalian size. Science *108*:44, 1948.
Powell, J. R. and Burke, J. D. Avian blood oxygen capacity. Amer. Mid. Nat. *75*:425, 1966.
Prosser, C. L. and Brown, F. A. *Comparative Animal Physiology*. W. B. Saunders Co. Philadelphia. 1961.
Prosser, C. L. and Weinstein, S. J. F. Comparison of blood volume in animals with open and closed circulatory systems. Physiol. Zool. *23*:113, 1950.
Robinson, J. R. *Fundamentals of Acid-Base Regulation*. 3rd edition. F. A. Davis Co. Philadelphia. 1967.
Ronald, K., MacNab, H. C., Stewart, J. E., and Beaton, B. Blood properties of aquatic vertebrates. I. Total blood volume of the Atlantic cod, *Gadus morhua L*. Canad. J. Zool. *42*:1127, 1964.
Schmidt-Neilsen, K. and Larimer, J. L. Oxygen dissociation curves of mammalian blood in relation to body size. Amer. J. Physiol. *195*:424, 1958.
Scholander, P. F., Claff, C. L., and Sveinsson, S. L. Respiratory studies of single cells. I. Methods. Biol. Bull. *102*:159, 1952.
Sealander, J. A. The influence of body size, season, sex, age and other factors upon some blood parameters in small mammals. J. Mammal. *45*:598, 1965.
Umbreit, W. W., Burris, R. H., and Stauffer, J. F. *Manometric Techniques*. Burgess Publ. Co. Minneapolis. 1957.
Weier, E. T. and Benson, A. A. The molecular organization of chloroplast membranes. Amer. J. Bot. *54*:389, 1967.
White, A., Handler, P. and Smith, E. L. *Principles of Biochemistry*. 3rd edition. McGraw-Hill Book Co. New York. 1968.
Wilson, C. L. and Loomis, W. E. *Botany*. 3rd edtion. Holt, Rinehart, and Winston. New York. 1962.
Zeuthen, E. Oxygen uptake as related to body size in organisms. Quart. Rev. Biol. *28*:1, 1953.
Zuckerkandl, E. The evolution of hemoglobin. Sci. Amer. *212*(5): 110, 1965.

Chapter Eight

Intermediary Metabolism

A. CARBOHYDRATE METABOLISM
B. LIPID METABOLISM
C. PROTEIN METABOLISM
D. NUCLEIC ACIDS METABOLISM
 1. PYRIMIDINES
 2. PURINES
 3. RNA
 4. DNA

The act of photosynthesis is directed toward the capture of radiant energy for the production of carbohydrates by carbon dioxide fixation. In doing so, the act produces hexoses and intermediate compounds. From these series of reactions the major foodstuffs of the cell and organism are formed, *i.e.*, lipids, proteins, and carbohydrates; nucleic acids as well are metabolized by plant cells. These foodstuffs, carbohydrates, lipids, and proteins, are major in that these form the chief dietary intake by supplying nearly all of the caloric and structural requirements of animals whereas the nucleic acids can be synthesized *in vivo*. As a consequence of photosynthesis, the animal cell is dependent of the plant cell for sustenance. With all of the chemical activity occurring in a cell it is remarkable that chemical stability is maintained.

The sum of all of the chemical processes in a cell or organism is termed metabolism. Maintaining control of this metabolism are the stabilizing enzymes. All of the metabolism and its control depends on the enzymes present in the cell. Enzymes bring about transformation in chemical compounds as discussed in Chapter 5. These vital compounds are called metabolites and the sequences or series of their involvement are called metabolic pathways. These metabolic pathways are usually divided into catabolic or anabolic pathways. There are various types of metabolism in which pathways of reactants in the cell can be arranged. These include basal metabolism, electrolyte metabolism, mineral metabolism, respiratory metabolism as well as intermediary metabolism, and others. Intermediary metabolism may be defined as the sum of chemical events from the time foodstuffs enter an organism until waste products leave it. For example, foodstuff → a → b ····· z → waste products in which a long series of chemical changes occur in individual steps which involve intermediate substances. Therefore, the term intermediary metabolism applies to the enzyme-controlled chemical changes of these intermediate substances.

In metabolism, the catabolic routes have clearly defined beginnings leading to diffuse end-products. Essentially, catabolism is the degradation of large organic molecules (usually by oxidative reactions) to simpler molecules with a release of free energy. In anabolism, however, anabolic routes have diffuse beginnings leading to clearly distinguished end-products. An anabolic route is one of biosynthesis in which complex organic molecules are produced (usually by reduction reactions) from simple molecules requiring the expenditure of energy. Rarely, if ever, do anabolic and catabolic routes follow in detail the same pathways. Yet there are links between anabolism and catabolism.

There are three ways in which anabolism is connected to catabolism. First is the carbon sources. The products of catabolism become the substrates for anabolism. Second is the level of energy supply. Metabolic energy is produced in catabolism usually in the form of ATP. Energy is required and ATP is consumed in anabolism. And thirdly, as already mentioned, catabolism is essentially oxidative, but anabolism is essentially reductive. Both NADH and NADPH are produced in catabolism, but anabolism requires and consumes mostly NADPH. It has been suggested that NADPH, therefore, has a special role, in contrast to NADH, in biosynthetic reactions involving reduction reactions. For example, this pyridine nucleotide (NADPH) is specifically utilized in long-chain fatty acid biosynthesis, reduction of glucose to sorbitol, glucuronic acid to L-gluconic acid, dihydrofolic acid to tetrahydrofolic acid, and reductive decarboxylation of pyruvic acid to malic acid by malic enzyme. NADPH is also involved in hydroxylation reactions, unsaturated fatty acid formation, and the conversion of phenylalanine to tyrosine. That the NADPH/NADH ratio is greater than five in many tissues may also indicate a special role for NADPH in biosynthesis.

There is a fourth metabolic pathway, the anaplerotic route. Both anabolic and catabolic routes lead to a removal and drain on the intermediates pooled in the common pathways. Consequently, the intermediates must be replenished. This replenishment is accomplished by anaplerotic sequences filling the depleted pool of intermediates with 1-carbon (CO_2), or 2-carbon fragments (as acetyl CoA).

A. Carbohydrate Metabolism

The importance of carbohydrates to both plants and animals cannot be overemphasized. Hexose formation and pentose regeneration in plants has just been discussed in the previous chapter. In animals, and specifically mammals, the importance of carbohydrates is indicated in that about 60% of the total foodstuffs ingested in the diet are polysaccharide carbohydrates, mostly starches. Of some dietary importance is the disaccharide sucrose; but lactose of course, is important in the infant diet. Monosaccharides are present in the diet only in small quantities.

As carbohydrates pass along the normal gastrointestinal tract, they are mostly converted to the usable form in the cell, glucose, but some fructose and galactose are also end-products. Absorption of glucose from the intestinal lumen occurs across the mucosal cells into the blood capillaries and then to the portal venous blood. The pathways of carbohydrate digestion for mammals in general are shown in Figure 8-1. After passing through the liver, glucose is carried via blood to all cells of the organism.

Although the mechanism of monosaccharide entry into the cell is unknown, apparently the sugar is transported through the membrane by a carrier system. It seems that the glucose transport system of the intestine involves an active transport. Energy used to drive glucose across the membrane arises from the consumption of the energy present in the electrochemical gradient of sodium ions accumulated by the sodium pump of the cell. In other cells, however, glucose entry into cells apparently is by facilitated diffusion. In this system glucose enters and leaves the cell, but it does not accumulate in a greater concentration inside the cell than outside it. Consequently, cellular energy is not consumed either directly or indirectly in a facilitated diffusion system. But in any event, the rate of glucose transport can be increased 10-fold when large amounts of insulin are secreted by the pancreas. In fact, the rate of insulin secretion by the beta cells in the islets of the pancreas determines the rate of utilization of glucose by cells. Since disaccharides and larger polysaccharides are not utilized in cellular metabolism they, along with some glucose, are excreted in the urine.

After glucose enters a cell, it can be used immediately for release of

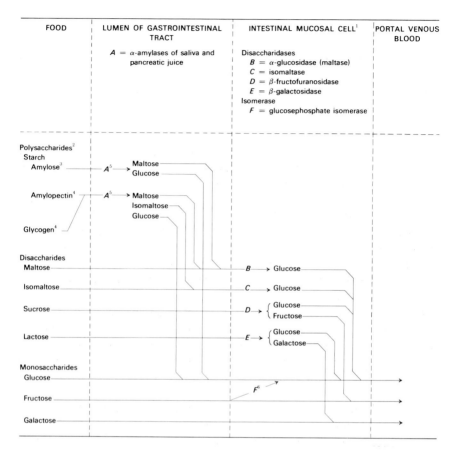

FIG. 8-1. Pathways of carbohydrate digestion: Man and laboratory mammals. [1] Dihexoses and monohexoses are absorbed into the intestinal mucosal cell. Within the microvilli of the intestinal mucosal cell, the dihexoses are split to monohexoses. Apart from the small fraction of hexose metabolized (oxidized) during passage through the intestinal mucosal cell, phosphorylation of hexoses does not occur as a mechanism for absorption of hexoses into the cell or for delivery from the cell into the portal blood. [2] A number of so-called structural polysaccharides occurring in foods are not digestible in the alimentary tract of vertebrates and so pass into the feces essentially unaltered. These include cellulose, lignin, mannan, xylan, pectic acids, alginic acid, and chitin. [3] Amylose is a straight chain polymer of glucose with alpha 1,4 glucosidic linkages. [4] Amylopectin and glycogen are branched chain polymers of glucose with alpha 1,4 linkages in the straight chain portions and alpha 1,6 linkages at the points of branching. [5] α-Amylase hydrolyzes 1,4 glucosidic linkages in chains of glucose containing three or more residues; it does not split maltose. α-Amylase does not split the 1,6 linkages in amylopectin. [6] Some fructose is transformed to glucose in the intestinal mucosal cell and some passes through unchanged. From Altman, P. L. and Dittmer, D. S. 1968. *Metabolism*. Federation of American Societies for Experimental Biology. Bethesda, Md. p. 290. Fig. 47.

energy or stored in the form of glycogen, a large polymer of glucose. Although all cells are capable of storing some glycogen, some cells can store up to 1% of their weight (muscle) and liver cells up to 8% of their weight. Also, fructose and galactose can be converted to glucose by liver cells.

Glycogen may be converted to glucose, and then back to glycogen. The process of glycogen formation from glucose is called glycogenesis. Glucose is first converted to glucose 6-phosphate, then to glucose 1-phosphate, to uridine diphosphate glucose, and finally to glycogen. Glycogen breakdown to reform glucose is termed glycogenolysis; it is not the reversal of the same chemical reactions which form glycogen. In a process of phosphorolysis catalyzed by phosphorylase, each glucose molecule in the glycogen polymer is split away. The phosphorylase enzyme is specific for an α-1,4 glycosidic linkage. Action starts at the nonreducing end of the glycogen chain and cleaves one glucose at a time to form glucose 1-phosphate until an α-1,6 bond is reached. Amylo-$(1,4 \rightarrow 1,6)$-glucosylase, a debranching enzyme, cleaves the α-1,6 bond and produces free glucose, and phosphorylase action can then continue. Glucose 1-phosphate is converted to glucose 6-phosphate by the phosphoglucomutase reaction. Glucose formation is possible in the intestinal mucosa, liver, and kidney tubules which contains glucose 6-phosphatase. This enzyme splits phosphate away from glucose 6-phosphate to form glucose, making it also available for transport out of the cell. (The formation of glucose from noncarbohydrate precursors is called gluconeogenesis). Two hormones, epinephrine and glucagon, act specifically to activate phosphorylase thereby causing rapid glycogenolysis. When the sympathetic nervous system is stimulated the adrenal medulla releases epinephrine. In this way the sympathetic nervous system can function to increase the availability of glucose for rapid metabolism. The alpha cells of the pancreas islets secrete glucagon when blood glucose falls to a low level. It functions primarily in the liver to activate phosphorylase in blood glucose regulation.

Glucose is utilized by the cell in a catabolic process in the cytoplasm called glycolysis in which a series of reactions convert glucose to pyruvic acid. It may be recalled from the discussion in Chapter 6 that pyruvic acid enters the mitochondrion where it is oxidized to carbon dioxide via acetyl CoA and the TCA cycle, with oxidative phosphorylation. Glycolysis is usually referred to as the Embden-

Meyerhof pathway. It is to be noted that this pathway is a series of reactions forming intermediates anaerobically, *i.e.*, no oxygen is consumed in the over-all process in converting glucose to pyruvic acid. In Figure 8-2 is shown the Embden-Meyerhof pathway of glycolysis, and the intermediates formed enzymatically in the conversion of glucose to pyruvic acid.

The Embden-Meyerhof pathway is not only the pathway in muscle glycolysis in which glucose (or glucose units from glycogen) is converted to lactic acid via pyruvic acid, but it is also the pathway of alcoholic fermentation by yeasts in which glucose is converted to ethanol and carbon dioxide. Yet these two pathways differ at the point of pyruvic acid formation. In muscle glycolysis, there is an interaction between NADH and pyruvic acid in the presence of lactic acid dehydrogenase which yields lactic acid and regenerates NAD from NADH. In alcoholic fermentation the pyruvic acid is first decarboxylated to acetaldehyde which is then reduced by NADH to form ethyl alcohol.

Each of the steps in glycolysis is freely reversible. But oxidation will not occur if either NAD or inorganic phosphate is omitted. Phosphate is required for phosphorolytic cleavage and NAD as the acceptor for electrons or hydrogen. NADH can release its hydrogen in two ways. In cells adequately supplied with oxygen, hydrogen can be transferred to the electron transport chain as one pathway (see Fig. 6-1). In the other, regeneration of NAD occurs under anaerobic conditions as pyruvic acid forms lactic acid (see Fig. 8-2).

Lactic acid can form glucose, but not just as a reversal of glycolysis. This is evident from a reaction for glycolysis as

$$\text{Glucose} + 2 \text{ ADP} + 2 \text{ P}_i \rightarrow \text{lactic acid} + 2 \text{ ATP}$$

The reverse process occurs as

$$2 \text{ Lactic acid} + 6 \text{ ATP} \rightarrow \text{glucose} + 6 \text{ ADP} + 6 \text{ P}_i$$

There are alternate pathways for carbohydrate metabolism. One such pathway exists in plants and microorganisms for genesis of dicarboxylic acids. It is called the glyoxylic acid pathway (Fig. 8-9). Isocitric acid lyase catalyzes the formation of succinic acid and glyoxylic acid from isocitric acid. Malic acid synthetase catalyzes the formation of malic acid from a condensation of acetyl CoA with glyoxylic acid. Enzymes of the TCA cycle can then convert succinic acid (or malic acid) to oxaloacetic acid which may be converted to phospho-

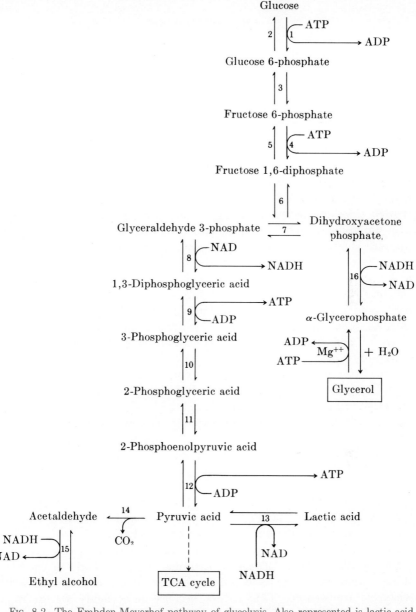

Fig. 8-2. The Embden-Meyerhof pathway of glycolysis. Also represented is lactic acid and ethyl alcohol formation from pyruvic acid. The enzymatic steps are numbered. 1. hexokinase, Mg^{++}; 2. glucose 6-phosphatase, Mg^{++}; 3. phosphoglucose isomerase; 4. phosphofructokinase, Mg^{++}, AMP; 5. diphosphofructose phosphatase, Mg^{++}; 6. aldolase; 7. phosphotriose isomerase; 8. phosphoglyceraldehyde dehydrogenase; 9. phosphoglyceric acid kinase, Mg^{++}; 10. phosphoglyceromutase, Mg^{++}; 11. enolase, Mg^{++} or Mn^{++}; 12. pyruvic acid kinase, Mg^{++}, K^+; 13. lactic acid dehydrogenase; 14. pyruvic acid decarboxylase; 15. alcohol dehydrogenase; 16. glycerophosphate dehydrogenase.

enolpyruvic acid and then to glucose. Isocitric acid lyase and malic acid synthetase are lacking in animal tissues.

In animal tissues, glucose is metabolized by two prominent pathways. One, the Embden-Meyerhof pathway, has just been described above. The other pathway is the phosphogluconate oxidative (PGO) pathway. This designation refers to the first intermediate in the metabolic sequence (6-phosphogluconate), and that the first reaction step is an oxidative one. The PGO pathway is also called the pentose phosphate (PP) pathway because it is a source of pentoses. Still another name for the PGO pathway is the hexose monophosphate shunt (HMS) since the pathway diverges from glycolysis at the level of glucose 6-phosphate. The reactions and intermediates of the PGO pathway are shown in Figure 8-3. Alternate metabolic pathways of the intermediates are also indicated in the figure. The intermediates glucose 6-phosphate, glyceraldehyde 3-phosphate, and fructose 6-phosphate can be utilized in glycolysis. In a reversible reaction, xylulose 5-phosphate and erythrose 4-phosphate form glyceraldehyde 3-phosphate and fructose 6-phosphate. Erythrose 4-phosphate participates in the synthesis of phenylalanine, and ribose 5-phosphate initiates synthesis of nucleotides.

The PGO pathway has a cyclic nature. Initially, glucose 6-phosphate is converted oxidatively to 6-phosphogluconate. After further oxidation ribulose 5-phosphate is formed. Pentose transformations occur until finally fructose 6-phosphate is formed which is readily convertible to glucose 6-phosphate to complete the cycle. The PGO pathway differs particularly from the glycolytic pathway in that oxidation occurs early and CO_2 is produced, but not in glycolysis.

Glucose 6-phosphate is the starting point for the interconversions of a great variety of hexoses. As a result, many different monosaccharides are formed. These monomers form polymers, and the macromolecules are directed toward the biosynthesis of polysaccharides.

Carbohydrates are interrelated through metabolic pathways with other foodstuffs. When glycogen accumulates in the body in excessive storage amounts, there is a net conversion of glucose to fatty acids which is irreversible. This occurs because fatty acids are in rapid equilibrium with acetyl CoA which is formed primarily from carbohydrate through pyruvic acid formation. No net synthesis of carbohydrate occurs since the decarboxylation of pyruvic acid is irreversible in animal cells. In contrast to the fatty acids, glucose transformation to glycerol of lipids is readily reversible. The glycerol moiety of

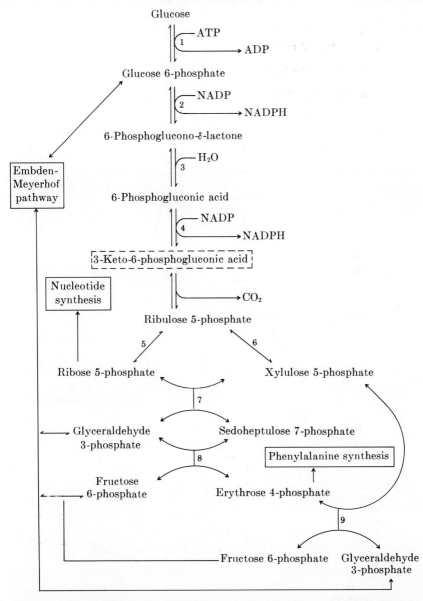

Fig. 8-3. The phosphogluconate oxidative pathway. Alternative metabolic routes for intermediates are indicated. The enzymatic steps are numbered. 1. hexokinase, Mg^{++}; 2. glucose 6-phosphate dehydrogenase; 3. gluconolactonase, Mg^{++}; 4. phosphogluconic acid dehydrogenase, Mg^{++} or Mn^{++}; 5. phosphopentose isomerase; 6. phosphopentose epimerase; 7. transketolase, TPP, Mg^{++}; 8. transaldolase; 9. transketolase.

lipids is directly convertible to trioses and then to hexoses. These conversions may be represented as

glucose ⇌ pyruvic acid → acetyl CoA ⇌ fatty acids

glucose ⇌ glycerol

In Figure 8-4 a scheme is shown for the metabolic interrelations of carbohydrates, glycerol and fatty acids, and some amino acids. Glucogenic amino acids, such as alanine, aspartic acid, and glutamic acid can

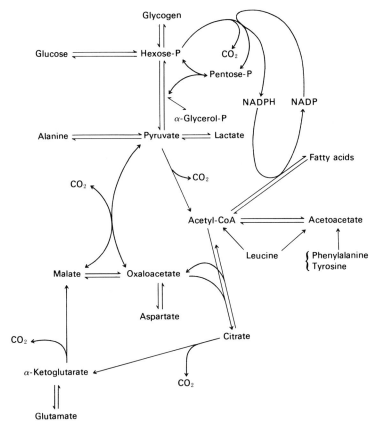

FIG. 8-4. Metabolic interrelations of carbohydrates with other foodstuffs. From Cantarow, A. and Schepartz, B. 1967. *Biochemistry*. 4th edition. W. B. Saunders Co. Philadelphia. p. 447. Fig. 18-17.

produce new molecules of pyruvic acid. Others, such as leucine, phenylalanine, and tyrosine can enter the TCA cycle. From such interrelated metabolic pathways, it can be seen that a deprivation of carbohydrates leads to an increase of the catabolism of lipids and proteins.

B. Lipid Metabolism

Lipids are stored in cells in the form of triglycerides. As discussed in Chapter 3, triglycerides are fats which are esters of glycerol and fatty acids. Triglycerides are neutral fats and are very insoluble. Of all the common dietary foodsutffs, lipids have the highest caloric value because of the characteristic of long-chained fatty acids. Attached to the carboxyl group of the fatty acid is the hydrocarbon radical (R—COOH) where R is $CH_3(CH_2)_n$ for a saturated fatty acid, but if R has one or more double bonds the fatty acid is unsaturated. Upon oxidation of the highly reducible hydrocarbon radical there is a large increase in the negative free energy $(-\Delta F')$.

The lipid molecule is the chief storage form of available energy in animal cells. The average metabolic energy derived from fat is 9.3 Cal/gm but only 4.1 Cal/gm from carbohydrate and protein. When the caloric value of foodstuffs ingested exceeds their rate of utilization, excess food is stored as fat. A variation of depot fat occurs among different species, but in any one individual of a species the depot lipid composition is rather uniform.

The major products of lipid digestion are formed by pancreatic lipase and bile salts in the lumen of the upper small intestine. The products include free fatty acids, glycerol, monoglycerides, and some diglycerides and triglycerides. Micelles are formed which are complexes of bile salts, fatty acids, and monoglycerides. Present evidence indicates that the intact micelle does not enter columnar epithelium, but absorption of micellar contents does occur with the exception of bile salts which absorb in the ileum. This mixture of fatty acids and monoglycerides is absorbed by the columnar epithelial cells of the mucosa as indicated in Figure 8-5 which shows pathways of lipid digestion. Furthermore, water-soluble glycerol and fatty acids containing less than 10 carbon atoms cross the epithelial cells and are absorbed into the portal blood. Long-chained fatty acids absorb and form mostly into chylomicrons made in the columnar cell. The chylomicrons consist of a core of triglycerides resembling dietary fat in fatty acid com-

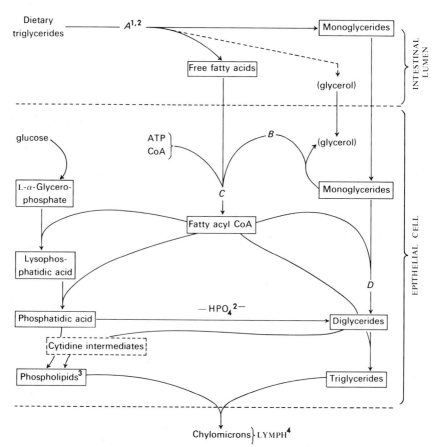

FIG. 8-5. Pathways of lipid digestion: man and laboratory mammals. [1] Pancreatic lipase acts preferentially on ester linkages at the terminal or 1 position of glycerol. Thus the major products of digestion are fatty acids and monoglycerides. [2] Bile salts in their conjugated form participate in at least three reactions during fat digestion and absorption: (i) as a co-factor for pancreatic lipase; (ii) to form micelles containing monoglyceride and fatty acid, as well as other lipids (these micelles are probably the form in which lipid is absorbed into the cell); (iii) as a cofactor for thiokinase in the intestinal mucosal cell. [3] Absorbed fatty acids go mainly into the triglycerides of chylomicrons, but small amounts are synthesized into cholesterol esters and phospholipids which also are constituents of chylomicrons. [4] Fatty acids with chain lengths shorter than 10 carbon atoms are absorbed mainly into the portal blood, those with longer chain lengths mainly into the lymph. From Altman, P. L. and Dittmer, D. S. 1968. *Metabolism*. Federation of American Societies for Experimental Biology. Bethesda, Md. p. 289. Fig. 46.

position. In the particle core is all of the cholesterol. An envelope can be separated from the core. The envelope consists mostly of phospholipid with phosphotidyl choline as the primary phospholipid fraction. Also in the outer layer is a small amount of protein and some free cholesterol. Electron micrographs of chylomicrons do not reveal evidence of any membrane, or structure in the core material.

Although pinocytosis has been shown to be active in a pathway of fat absorption, recent evidence supports the view that no appreciable uptake of fat results from pinocytosis. The uptake of fat is apparently by simple diffusion into the absorptive cells. The products of triglyceride hydrolysis enter the microvilli and apical portion of the absorptive cell (Fig. 8-6). The triglyceride products enter the smooth endoplasmic reticulum, located strategically in the apical portion, and enzymes in the membranes, or cavities, reform triglycerides. Presumably, chylomicrons are resynthesized in the luminal part of the columnar cell in smooth endoplasmic reticulum, having a size about 0.5 μ. The smooth and rough endoplasmic reticulum exhibit a clear continuity with each other. Apparently the rough endoplasmic reticulum can change over to smooth endoplasmic reticulum during fat uptake. This changeover supports the view that the rough endoplasmic reticulum ribosomes are active in making enzymes for this pathway of fat synthesis. It may be, too, that the rough endoplasmic reticulum makes and stores the protein in the outer envelope of the chylomicron to make it available during triglyceride resynthesis (Fig. 8-7).

As fat absorption continues, smooth endoplasmic vesicles form and in some way move laterally across the cell.

The contents of the vesicles are released by a reverse pinocytosis into extracellular spaces of the lamina propria. The chylomicrons have been seen to move between the cells into the lacteals. From the lacteals, the chylomicrons are transported by chyle via the lymph vessels to the thoracic duct where they enter the blood circulatory system. Most of the fat present in the chyle is in chylomicrons. The function of the Golgi complex is unclear in fat absorption. It may act for the transport of fat, or as a storage depot.

In the postabsorptive state in man, *i.e.*, no chylomicrons in blood, the normal blood plasma contains about 700 mg % of total lipid; about 25% of the total lipid is composed of triglycerides. There are also about 180 mg % of cholesterol; about two-thirds is esterified with fatty acids and one-third appears as free cholesterol. The phospholipids are present in a concentration of about 160 mg %. The remain-

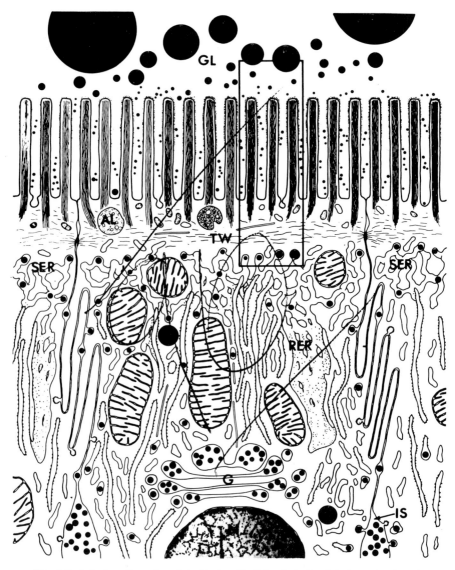

Fig. 8-6. Apical portion of an intestinal epithelial cell during fat absorption based on electron micrographs. The lower portion of the enclosed area shows several stages in formation of a smooth endoplasmic vesicle containing a fat droplet. This area is enlarged in Figure 8-7 where the significant events in the initial phases of fat absorption are shown. From Porter, K. R. 1969. Independence of fat absorption and pinocytosis. Fed. Proc. 28: 35. Fig. 3.

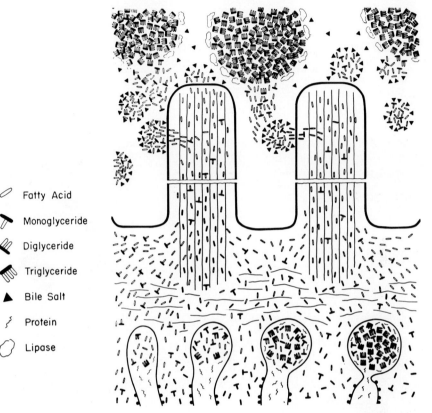

FIG. 8-7. Biochemical events in (a) the digestion of triglycerides in the intestinal lumen, (b) the selective absorption by intestinal epithelial cells of monoglycerides and free fatty acids from micelles, and (c) the synthesis and segregation of triglycerides within the cavities of the smooth endoplasmic reticulum. Refer to Figure 8-6 for the morphological counterparts of this information. From Porter, K. R. 1969. Independence of fat absorption and pinocytosis. Fed. Proc. 28:35. Fig. 4.

---

der of the lipids exists as lipoproteins. The first step in the utilization of fats by cells involves their hydrolysis by lipases to yield three molecules of fatty acid and one of glycerol. The glycerol is glycogenic. It can enter the Embden-Meyerhof pathway via formation of α-glycerophosphate (Fig. 8-2) by action of ATP.

Animal and plant cells degrade fatty acids through a principal pathway called β-oxidation. In this sequence of reactions, the even-chain fatty acids are shortened two units at a time to the 2-carbon fragment, acetyl CoA. Interestingly enough, nearly all of the acids in the naturally occurring fats have an even number of carbon atoms. The acetyl CoA formed by the degradation of fatty acids mixes with acetyl CoA arising from the oxidative decarboxylation of pyruvic acid and many amino acid degradations. The reduced coenzymes, NADH and $FADH_2$, are the other major products formed in the catabolism of fatty acids. The β-oxidation of fatty acids occurs exclusively in the mitochondria, and this is the locale of the enzymes catalyzing fatty acid degradation. There are five enzymes responsible for β-oxidation. These enzymes and the type reactions they catalyze are:

1. Fatty acid thiokinase

fatty acid + CoA + ATP $\underset{}{\overset{Mg^{++}}{\rightleftarrows}}$ fatty acyl CoA + AMP + pyrophosphate

2. Acyl dehydrogenase (contains FAD)

saturated fatty acyl CoA + FAD $\rightarrow$ α,β-unsaturated fatty acyl CoA + $FADH_2$

3. Enoyl hydrase

α,β-unsaturated fatty acyl CoA + H_2O $\rightleftarrows$ β-hydroxy fatty acyl CoA

4. Hydroxyacyl dehydrogenase

β-hydroxy fatty acyl CoA + NAD $\rightleftarrows$ β-keto fatty acyl CoA + NADH + H^+

5. Ketoacyl thiolase

β-keto fatty acyl CoA + CoA $\rightarrow$ fatty acyl CoA + acetyl CoA

The cycle for these reactions of β-oxidation of fatty acids is shown in Figure 8-8. It can be seen in the figure that complete oxidation of a fatty acid with an even number of carbons ultimately is degraded to acetyl CoA. But in degradation of a fatty acid with an odd number of carbons, acetyl CoA and one equivalent of propionyl CoA is produced. Propionyl CoA can undergo a series of reactions to form succinyl CoA, an intermediate of the TCA cycle.

In animal cells carbohydrates can be converted to fats, but fats are not converted back to carbohydrates. In plant cells, however, fat deposits can be rapidly converted to sucrose and other complex sugars and into proteins. In Figure 8-9 the cyclic reaction series which convert fats to carbohydrates is shown. This cycle is called the glyoxylate

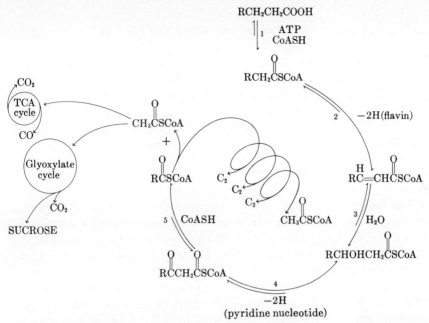

FIG. 8-8. The classical β-oxidation sequence by which fatty acids are degraded to acetyl CoA. The enzymatic steps are numbered 1-5: 1. fatty acid thiokinases; 2. acyl dehydrogenases; 3. enoyl hydrase; 4. β-hydroxyacyl dehydrogenase; and 5. β-ketoacyl thiolase. From Stumpf, P. K. 1965. Lipid metabolism. In *Plant Biochemistry*. Bonner, J. and Varner, J. E. (editors). Academic Press. New York. p. 335. Fig. 3.

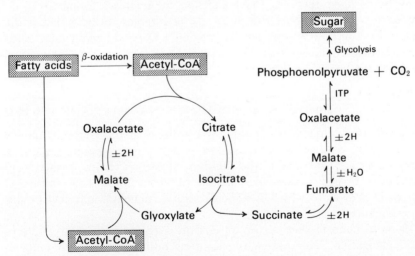

FIG. 8-9. Glyoxylate cycle: conversion of fatty acids to sugars. This cycle occurs in plant cells, but not animal cells. From Conn, E. C. and Stumpf, P. K. 1966. *Outlines of Biochemistry*. 2nd edition. John C. Wiley & Sons. New York. p. 293. Fig. 13-3.

cycle. It functions to convert acetyl CoA, produced in β-oxidation of fatty acids, to sucrose via the glycolytic pathway.

Acetyl CoA is the immediate starting point for synthesis of fatty acids. When the degradative steps of β-oxidation of fatty acids were finally worked out, it was thought that the major steps were reversible for lipogenesis. This, however, turned out not to be true. The catabolism of fatty acids is restricted to mitochondria, but independent anabolic pathways are found in the cytoplasm as well as in mitochondria.

The extramitochondrial system catalyzes the *de novo* synthesis of fatty acids by an enzyme complex. The complex converts acetyl CoA to malonyl CoA to palmitic acid. The enzyme complex is tightly bound in animals, but dissociates readily in plants. In the initial step acetyl CoA is converted to malonyl CoA. This reaction is one of carbon dioxide fixation which is catalyzed by acetyl CoA carboxylase. ATP provides the energy for the carbon—carbon linkage; the reaction also requires Mn^{++}, and biotin is essential since it serves as a carrier for carbon dioxide. In the over-all conversion, one molecule of acetyl CoA condenses with seven molecules of malonyl CoA in a series of reactions. Involved in the anabolic series is decarboxylation, dehydration, and the equivalent of double reduction forming palmitic acid. The pathway is favored by the high NADPH/NADP ratio in the cytoplasm. It may be recalled that the PGO pathway for pentoses formation is an important source of NADPH. It is also of intereest to note that if the acetyl CoA primer is replaced by propionyl CoA odd-chain fatty acids are synthesized rather than the even-chained fatty acids. The over-all reaction for cytoplasm lipogenesis is represented as

Acetyl CoA + 7 malonyl CoA + 14 NADPH + 14 $H^+ \rightarrow$
    palmitic Acid + 7 CO_2 + 8 CoA + 14 NADP + 6 H_2O

In cytoplasmic lipogenesis, one of the proteins of the enzyme complex discussed above has been designated as acyl carrier protein (ACP). It substitutes for CoA during the build-up of carbon chains in fatty acid synthesis. It has a molecular weight of about 9100. One sulfhydryl group is present to which is attached acetyl, malonyl, and acyl groups in covalent linkage.

In mitochondria there exists an enzyme system which catalyzes fatty acid elongation. The system functions by adding acetyl CoA units to preformed fatty acids, and utilizes both NADH and NADPH. The system is active with C_{12}, C_{14}, and C_{16} chain length fatty acids and its products are mostly fatty acids of chain lengths C_{18}, C_{20}, C_{22}, and C_{24}. The pathway requires four steps for chain elongation. The first three steps are a reverse of the three last reactions described above for β-oxidation of fatty acids. The fourth step requires enoyl reductase and NADPH instead of the acyl CoA dehydrogenase containing FAD. The overall reaction is represented as

RCH_2COCoA + acetyl CoA + NADH + NADPH + 2 H$^+ \rightarrow$
$R(CH_2)_3COCoA$ + NAD + NADP + CoA

An enzyme system has been isolated from microsomes which catalyzes elongation of fatty acid acetyl CoA derivatives. In this system, fatty acid elongation utilizes both malonyl CoA and NADPH.

Questions have risen concerning the transport of acetyl CoA in and out of mitochondria. Acetyl CoA has its site of origin in mitochondria primarily as a result of glycolysis, and β-oxidation of fatty acids. But acetyl CoA is mostly utilized in the cytoplasm for lipogenesis. And thioesters of CoA do not traverse mitochondrial membranes readily. Two acetyl-carriers are now recognized, carnitine and citrate. Carnitine is capable of reversible transacetylation with CoA both inside and outside the mitochondria. Carnitine stimulates fatty acid oxidation in mitochondria; it may be involved also in transporting fatty acid into the mitochondria. Inside the mitochondrion, citrate synthetase can catalyze the condensation of acetyl CoA and oxaloacetate to form citrate; outside the mitochondrial citrate lyase and ATP can cleave citrate to the same compounds.

Interconversions between saturated fatty acids also occur in mechanisms apart from lipogenesis. In these interconversions, saturated fatty acids with different chain lengths are formed by addition or removal of 2-carbon units. For example,

$$\text{palmitic Acid} \underset{-C_2}{\overset{+C_2}{\rightleftharpoons}} \text{stearic acid.}$$

With the exception of blood plasma, free fatty acids occur sparsely in tissues. Since it has been mentioned that fat is stored, however, in the form of triglycerides it is important to note their synthesis. Triglycerides are synthesized in liver cells and adipose cells in the following manner. A precursor of triglycerides, as well as various phospholipids, is α-glycerophosphoric acid. As indicated in Figure 8-2, this compound is formed from free glycerol by phosphorylation with ATP and in the presence of glycerokinase, or from reduction of dihydroxyacetone. A phosphatase hydrolyzes the α-glycerophosphoric acid to a 1,2-diglyceride. The 1,2-diglyceride can then react with acyl CoA and form a neutral triglyceride. Apparently triglyceride formation can follow the same route in intestinal mucosa cells, but another reaction also is present. In this synthesis, acyl CoA can react with a monoglyceride to form a diglyceride and CoA.

The rate of degradation and synthesis of fatty acids is under the control of both dietary and hormonal factors. Isotopic studies show that liver, adipose tissue, and intestinal mucosa has the highest rate, but the lowest is in muscle, skin, and nervous system.

C. Protein Metabolism

A discussion of protein metabolism necessarily includes a consideration of the metabolism of amino acids in relation to their formation by protein degradation in catabolic pathways, their interactions, and their polymerization in anabolic pathways to form proteins. Protein is the major source of dietary amino acids. The ingested protein is hydrolyzed in the gastrointestinal tract by enzymatic action. Some of the amino acids from protein degradations in the gut cannot be synthesized by mammals, but others can in metabolic pathways in the cell. Whether dietary or metabolized, amino acids serve as precursors for protein synthesis in the cell.

Although not definitive, present evidence indicates that amino acids are absorbed in the upper part of the small intestine in man. This evidence is based on nitrogen balance studies, and isotope experiments with ^{131}I-labeled albumin and tritium-labeled albumin as absorption markers. There appears to be one or more carrier mechanisms by which amino acids are actively transported through the cell membrane. One of these carrier systems involves pyridoxal phosphate, a vitamin B_6 derivative, which is required in the transport of some amino acids through cell membranes.

The pathways of protein digestion for man and some laboratory animals are shown in Figure 8-10. Pepsin is the proteinase of the gastric juice. It is derived from pepsinogen which is the zymogen precursor secreted by the chief cells of the gastric mucosa. Pepsinogen is converted to pepsin by the acidity of the gastric juice and the autocatalytic action of pepsin. Pepsin, as well as trypsin and chymotrypsin in pancreatic juice, is an endopeptidase. These enzymes hydrolyze peptide bonds in the interior of peptide chains as well as terminal bonds.

The acinar cells of the exocrine pancreas produce the enzymes of

INTERMEDIARY METABOLISM

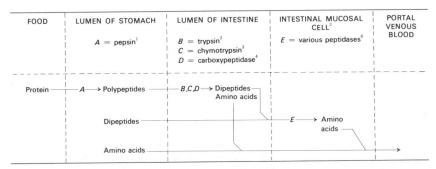

FIG. 8-10. Pathways of protein digestion: man and laboratory mammals. [1] Pepsin hydrolyzes many types of peptide bonds but splits most rapidly those in which an aromatic amino acid provides the amino group. [2] Trypsin hydrolyzes peptide bonds to which L-arginine, or L-lysine, contributes the carbonyl group. [3] Chymotrypsin hydrolyzes many types of peptide bonds, but splits most rapidly those in which an aromatic amino acid contributes the carbonyl group. [4] Carboxypeptidase does not exhibit absolute specificity with respect to the terminal amino acid forming the bond being split; it acts most rapidly on those linkages in which aromatic amino acids are in the terminal position. The terminal amino acid must have a free carboxyl group. [5] Amino acids and dipeptides enter the intestinal mucosal cells. Amino acids pass through unaltered (with a few exceptions, such as transamination of glutamic acid), and dipeptides are split to amino acids in the microvilli of the cell where the peptidases are localized. [6] Only a few of the intestinal mucosal peptidases have been characterized. The best known is leucine aminopeptidase. From Altman, P. L. and Dittmer, D. S. 1968. *Metabolism.* Federation of American Societies for Experimental Biology. Bethesda, Md. p. 291. Fig. 48.

---

the pancreatic juice. Ribosomes attached to the endoplasmic reticulum synthesize zymogen which is transferred to the Golgi complex. Here the zymogen is concentrated and stored as zymogen granules in the Golgi complex which lies in the apical portion of the cell. Membrane-enclosed vesicles from the Golgi complex containing the zymogen granules coalesce with the cell membrane and a reversed pinocytosis results in the release of zymogen to the lumen of the acinar gland. The lumina are continuous with the duct system to the small intestine where the zymogen precursors are converted to the active enzymes. For example, enterokinase secreted by the mucosal cells of the duodenum converts trypsinogen to trypsin, and trypsin in turn acts on procarboxypeptidase to convert it to active carboxypeptidase. The action of trypsin, chymotrypsin, and carboxypeptidase on polypeptides and dipeptides in the lumen of the small intestine is indicated in Figure 8-10 also. There is evidence that the microvilli of the columnar cells of the mucosa of the small intestine contain various peptidases. These peptidases hydrolyze dipeptides absorbed by the microvilli to amino acids.

The amino acids from the mucosal cells enter the portal venous blood. They are quickly carried to the other cells of the body, and do not ordinarily build up in large concentrations in the blood. Upon entry into body cells, amino acids are almost immediately conjugated by enzymes into protein or other complexes so that a large storage of amino acids *per se* does not occur. Cathepsins, however, can convert intracellular proteins quickly into amino acids for transport out of the cell.

Although the plasma of the blood transports amino acids to various body cells, they are not the only proteins present. In fact, the major types of plasma proteins are albumin, globulin, and fibrinogen. Each of these serves a different function. Albumin provides the colloidal osmotic pressure which prevent plasma loss in the capillaries. Globulins are responsible for the natural and acquired immunities. Fibrinogen aids in forming blood clots by polymerizing into long fibrin threads during leaks from the blood circulatory system. Nearly all of the albumin and fibrinogen, and about three-fourths of the globulins are formed in liver cells.

Cells vary widely in their capacity to synthesize amino acids. Many microorganisms and higher plants can synthesize all of the amino acids found in proteins. This is possible because sulfate and nitrate undergo a change in valency before being incorporated into organic form. As shown in the sulfate and nitrate cycles in Figure 8-11, this fact allows sulfate and nitrate to serve as electron acceptors, and after being reduced, to be assimilated into cellular organic compounds, *i.e.*, amino acids. There are three major reactions in which ammonia is fixed; namely, glutamic acid, glutamine, and carbamyl phosphate. Inorganic sulfur is fixed into organic linkage as cysteine.

Deamination reactions are mostly carried out by liver cells. Deamination of one amino acid provides the nitrogen in the amidogen for the synthesis of other amino acids. The amino acids are deaminated to produce NH_3 and the corresponding α-keto acid. The NH_3 may be converted in extrahepatic cells to glutamine in which form the NH_3 is stored or transported. Or glutamine may be converted to urea in liver cells which is transported to the kidney, and excreted in the urine. The α-keto acid is oxidized via the TCA cycle to carbon dioxide, water, and energy.

Glutamine synthetase catalyzed the formation of glutamine as

$$\text{glutamic acid} + ATP + NH_3 \xrightarrow{Mg^{++}} \text{glutamine} + ADP + P_i.$$

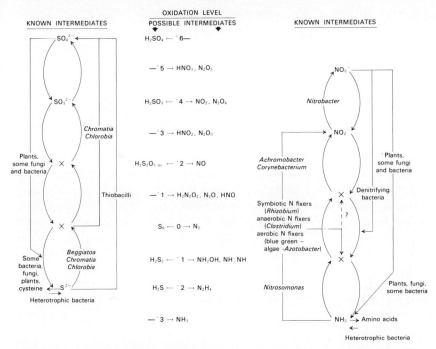

FIG. 8-11. A diagrammatic representation of the sulfur and nitrogen cycles. From Bandurski, R. S. 1965. Biological reduction of sulfate and nitrate. In *Plant Biochemistry*. Bonner, J. and Varner, J. E. (editors). Academic Press. New York. p. 468. Fig. 1.

Glutamine is a nontoxic substance which readily transports across cell membranes. In the liver cell, glutaminase splits glutamine back to glutamic acid. The NH_3 produced in this reaction may be converted to urea which, as indicated below, is a cyclic process:

$$\text{carbamyl phosphate} + ADP + P_i \xleftarrow{\text{acetyl glutamate} \atop Mg^{++}} NH_3 + H_2O + CO_2 + ATP$$

$$+$$

$$\text{ornithine} \xrightarrow{\text{ornithine transcarbamylase}} \text{citrulline} + P_i$$

$$+$$

$$PP_i + AMP + \text{arginino-succinic acid} \xleftarrow{\text{argininosuccinic acid synthetase; } Mg^{++}} \text{aspartic acid} + ATP$$

$$\xrightarrow{\text{argininosuccinase}} \text{arginine} + \text{fumaric acid}$$

$$+$$

$$H_2O$$

$$\xrightarrow{\text{arginase}} \text{urea}$$

Arginine is required for protein synthesis and is common to all animal cells. Arginase makes urea synthesis possible for disposal, ultimately, of nitrogenous wastes.

Deamination may be accomplished by three different mechanisms: (1) nonoxidative deamination: (2) oxidative deamination: and (3) transamination. Nonoxidative deamination is catalyzed by enzymes which remove water or hydrogen sulfide from the amino acid. These enzymes are hydrases or desulfhydrases. An imino acid is produced which hydrolyzes spontaneously to form NH_3 and an α-keto acid. Pyridoxal phosphate is required as a cofactor. Oxidative deamination is carried out by dehydrogenases and flavin enzymes (containing FAD as a prosthetic group) known as amino acid oxidases. Glutamic acid dehydrogenase is widely distributed in animals, plants, and bacteria. It is the most important example of the dehydrogenases. Glutamic acid dehydrogenase catalyzes the following reaction

glutamic acid + NAD (NADP) + $H_2O \rightleftharpoons$
α-ketoglutaric acid + NH_3 + H^+ + NADH (NADPH)

The reaction occurs with either of the pyridine nucleotides present. One important aspect of the reaction is that it provides a bridge between α-ketoglutaric acid, an intermediate of the TCA cycle, and glutamic acid. Another important aspect of the reaction is that it is one of the few reactions whereby NH_3 can be incorporated enzymatically into organic compounds. Transamination reactions are so general that nearly every amino acid participates. Transamination is a process in which an amino group is transferred from one amino acid to the carbon skeletal of another amino acid. The transfer reactions are catalyzed by transaminases, and require either pyridoxal phosphate or pyridoxamine phosphate as cofactors. Invariably, the reactants most abundantly involved are glutamic acid and α-ketoglutaric acid. For example,

glutamic acid + oxaloacetic acid $\rightleftharpoons$ α-ketoglutaric acid + aspartic acid

or

glutamic acid + pyruvic acid $\rightleftharpoons$ α-ketoglutaric acid + alanine.

Isotopic studies with ^{15}N have shown that amino nitrogen is constantly shifting from one amino acid to another, primarily by transamination. Furthermore, there is evidence that amino acids are continually being produced by protein catabolism, and continually used in protein anabolism.

INTERMEDIARY METABOLISM 249

Besides deamination and transamination, decarboxylation is a third type of generalized enzymatic reaction of amino acids. Amino acid decarboxylases are distributed widely throughout living cells. Mostly, these enzymes catalyze the α-decarboxylation of L-amino acids. These reactions lead to the formation of such important metabolites as alanine, dopamine, lysine, γ-aminobutyric acid, and histamine.

The metabolism of individual amino acids is very complex. In contrast to the carbohydrates and lipids, all of the individual amino acids do not follow one cycle or chain of metabolic reactions or pathways. Brief mention has been made, however, of common pathways for the amidogen portion of amino acids, and the enzymatic reactions common to their metabolism. As such, there are various alternative metabolic pathways available to the carbon skeletons of the deaminated amino acids. In Figure 8-12, the direct interconversions among amino acids are shown in a metabolic schema illustrating the metabolic pools shared by two or more amino acids. Several points are worth mentioning. Amino acids are more closely related to carbohydrates than lipids (ketone bodies); leucine is the only ketogenic amino acid. The glucogenic amino acids cluster about pyruvate or α-ketoglutarate. Depending on the test employed, other amino acids are either ketogenic or glucogenic, or neutral. Reference should be made to a textbook of biochemistry for the multitudinous detailed metabolic pathways of individual amino acids. However, glycine is a good example of amino acid metabolism.

Porphyrin synthesis is mentioned here inasmuch as it relates to an individual amino acid, glycine. Porphyrin structure is basic to several molecular configurations in the cell. Chlorophyll is a magnesium porphyrin, and hemoglobin, myoglobin, cytochromes, catalase, and peroxidase are all iron porphyrin proteins. Consequently, the porphyrins are functionally related to systems in the cell which involve photosynthesis, oxygen transport, electron transport, and catalysis.

All of the carbon and nitrogen atoms in the porphyrin ring (shown in Chapter 3) are derived from glycine and succinyl CoA (see Fig. 6-3), the active form of succinic acid. Synthesis of the porphyrin structure may be represented as occurring in the following steps showing the enzymes and the reactions catalyzed in heme synthesis:

1. δ-Aminolevulinic acid synthetase

$$\text{glycine} + \text{succinyl CoA} \xrightarrow{\text{pyridoxal phosphate}} \delta\text{-aminolevulinic acid}$$

2. δ-Aminolevulinic acid dehydrase

$$2\ \delta\text{-aminolevulinic acid} \rightarrow \text{porphobilinogen}$$

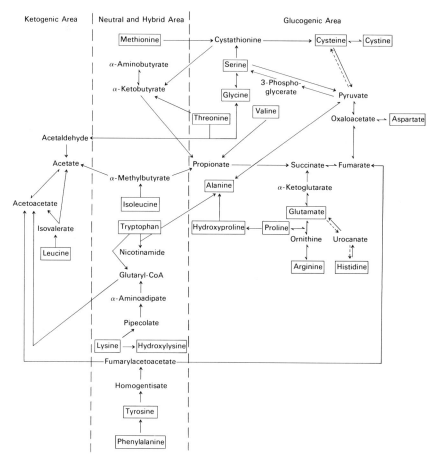

FIG. 8-12. Metabolic interrelations of the amino acids. From Cantarow, A. and Schepartz, B. 1967. *Biochemistry*. 4th edition. W. B. Saunders Co. Philadelphia. p. 576. Fig. 21-7.

3. Porphobilinogen deaminase

    porphobilinogen → polypyrryl methane

4. Isomerase

    polypyrryl methane → uroporphyrinogen III
                         (tetrapyrrole)

5. Uroporphyrinogen decarboxylase

    uroporphyrinogen III → coproporphyrinogen III

6. Coproporphyrinogen oxidase

    coproporphyrinogen III → protoporphyrin III (9)

7. Heme synthetase

    protoporphyrin III (9) $\xrightarrow{Fe^{++}}$ heme

Other enzymatic actions are believed to convert protoporphyrin III (9) to chlorophyll. Although little is known of the synthesis of catalases and peroxidases, it is known that hemoglobin is synthesized in reticulocytes during erythrocyte maturation, and that cytochrome c is synthesized by mitochondria of liver and heart cells.

Protein synthesis involves the formation of a complex molecule from amino acids in an exact sequence each time the molecule is produced. This biosynthetic sequence involves a coding system whereby a specific amino acid is inserted in the formative chain at a specific site. Consequently, the mechanism of protein synthesis and the way the coding system operates will be discussed in the next chapter in the part devoted to ribosomes and protein synthesis. Proteins not only serve as a source of the amino acids required for *de novo* protein synthesis, but proteins also serve as a source of carbon skeletons for the synthesis of purines and pyrimidines as well as other nitrogenous compounds.

D. Nucleic Acids Metabolism

Nucleic acids are polynucleotides, *i.e.*, they consist of a mixture of nucleotides. Upon hydrolysis, nucleotides yield nucleosides and phosphoric acid. And nucleosides can be further hydrolyzed to form pyrimidines and purines and a pentose sugar either ribose or deoxyribose. As discussed in Chapter 3, ribonucleic acid (RNA) and deoxyribonucleic acid (DNA) are the two most common nucleic acids in the cell. The biosynthesis of nucleic acids involve the formation of pyrimidines and purines from precursor molecules, their phosphorylation to nucleotides, and the polymerization of the nucleotides to form the polynucleotides, the nucleic acids RNA and DNA.

The pyrimidines and purines can be synthesized in animal cells and are not required in the diet. The nucleoproteins ingested, however, are cleaved in the stomach by gastric acidity which does not affect the nucleic acids. Digestion of the nucleic acids occur in the upper part of the small intestine, mostly in the duodenum. The acinus of the exocrine pancreas secretes nucleases into the pancreatic juice. Pancreatic ribonucleases hydrolyze only ribonucleic acids, and pancreatic deoxyribonucleases hydrolyze only deoxyribonucleic acid, presumably to oligonucleotides. Nucleases and diesterases are believed to be secreted by mucosal cells to digest oligonucleotides to nucleotides. Intestinal nucleotidases hydrolyze nucleotides to nucleosides which are probably absorbed in the mucosal cells. Cellular nucleosidases then hydrolyze nucleosides to the pentose sugar and pyrimidines and purines.

1. Pyrimidines

The initial step in pyrimidine biosynthesis is the formation of carbamyl phosphate. This compound has been shown earlier in the chapter to be an important intermediate in urea formation. Carbamyl phosphate is formed from one of a few major reactions in which ammonia is fixed. The N-1 of the ring structure of the pyrimidine nucleus is derived from ammonia and C-2 from carbon dioxide, both by way of carbamyl phosphate. Aspartic acid contributes N-3 and C-4, C-5, and C-6 to the ring. As shown in the schema for pyrimidine biosynthesis of Figure 8-13 carbamyl phosphate condenses initially with aspartic acid. A series of reactions follow which result in the formation of nucleotide triphosphates via orotidylic acid (orotidine 5'-phosphate). RNA polymerase converts certain of these nucleotide triphosphates into RNA-bound uracil and cytosine. DNA polymerase converts other nucleotide triphosphates into DNA-bound cytosine, methylcytosine, and thymine.

In the degradation of pyrimidines presumably cytosine is deaminated to form uracil, and deamination of methylcytosine yields thymine. After reduction and hydrolytic reactions, the end-products of uracil and thymine metabolism are β-amino acids. β-Alanine from uracil undergoes transamination, oxidation, and decarboxylation reactions to form acetate which can enter the TCA cycle. The β-aminoisobutyric acid from thymine can form succinyl CoA, an intermediate of the TCA cycle, via reactions producing methylmalonate and methylmalonyl CoA.

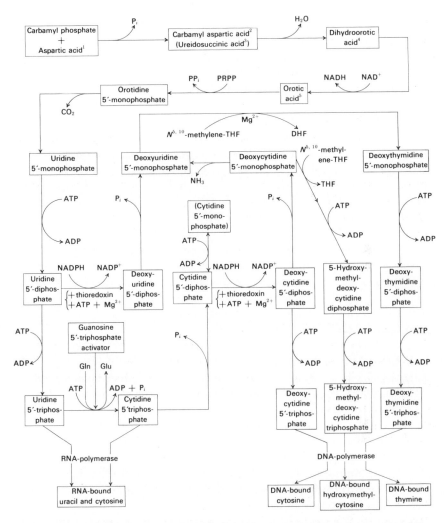

FIG. 8-13. Pathways of polynucleotide biosynthesis: pyrimidines. [1] Aspartate. [2] Carbamyl aspartate. [3] Ureidosuccinate. [4] Dihydroorotate. [5] Orotate. The abbreviations used are ADP = adenosine 5′-diphosphate; ATP = adenosine 5′-triphosphate; DHF = dihydrofolic acid coenzyme; THF = tetrahydrofolic acid coenzyme; Gln = L-glutamine; Glu = glutamic acid (glutamate); NAD^+ and NADH = nicotinamide adenine dinucleotide (oxidized and reduced forms, respectively); $NADP^+$ and NADPH = nicotinamide adenine dinucleotide phosphate (oxidized and reduced forms, respectively); P_i = inorganic orthophosphate; PP_i = inorganic pyrophosphate; PRPP = 5-phosphoribosyl-1-pyrophosphate. From Altman, P. L. and Dittmer, D. S. 1968. *Metabolism*. Federation of American Societies for Experimental Biology. Bethesda, Md. p. 454. Fig. 88.

2. Purines

In purine biosynthesis, the purine carbon skeleton is synthesized from small fragments. The route seems to be generally the same in animals and plants. Isotope studies indicate that the purine nucleus is derived from five different precursors. In the numbered purine rings (see page 101), N-1 is derived from the amino group of aspartate, N-3 and N-9 from the amide nitrogen of glutamine, C-2 and C-8 from formate or the 1-carbon unit from various compounds, C-6 from carbon dioxide, and the sequence C-4-C-5-N-7 from glycine. As the ring structures for pyrimidine and purine are indicated in Chapter 3, the purine ring structure is a pyrimidine ring fused to an imidazole ring. Actually, the purine ring system is formed on C-1 of ribose 5-phosphate via initial coupling with N-9. In the biosynthesis of the purine ribonucleotides, the initial step is the pyrophosphorylation of ribose 5-phosphate by ATP which yields 5-phosphoribosyl-1-pyrophosphate (PRPP). A successive series of reactions as shown in Figure 8-14 leads directly to the formation of purine ribonucleotides via inosinic acid (inosine 5'-monophosphate). In this sequence, neither nucleosides nor free purines appear as intermediates, but the actual purine precursors of RNA and DNA are the nucleoside triphosphates.

In the catabolism of purines, the parent nucleic acids are initially hydrolyzed by polynucleotidases and other phosphodiesterases to mononucleotides. These are then hydrolyzed by phosphomonesterases to nucleosides. Guanosine undergoes phosphorolysis by a nucleoside phosphorylase to form free guanine. Presumably, adenosine is acted upon by adenosine deaminase to form inosine. This hydrolytic deamination reaction is followed by a phosphorolytic reaction in which inosine is converted to the free base, hypoxanthine, by a nucleoside phosphorylase. Xanthine oxidase can convert hypoxanthine to xanthine which is also formed from guanine by guanine deaminase. Xanthine is an intermediate common to the catabolism of both adenine and guanine. Xanthine is then oxidized by xanthine oxidase to uric acid. Uric acid is the major end-product of purine catabolism in primates, birds, snakes, and lizards. When the levels of uric acid are elevated, monosodium urate are deposited in cartilage as tophi causing gout. Renal calculi are also formed by uric acid deposits. But in animals having uricase, uric acid is converted to allantoin. This is the end-product of purine metabolism secreted by mammals other then primates, and turtles. In some fishes, allantoin is converted by allantoinase to allantoic acid for elimi-

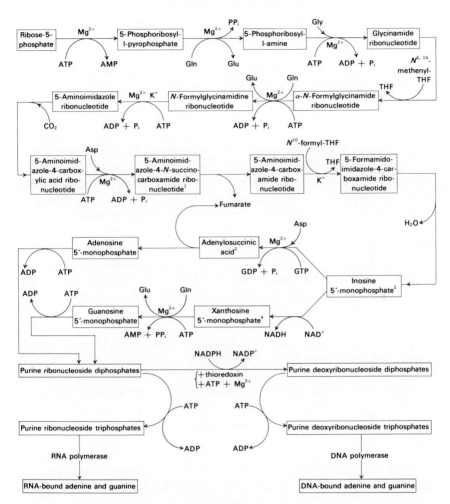

Fig. 8-14. Pathways of polynucleotide biosynthesis: purines. [1]Succinylaminoimidazole carboxamide ribonucleotide. [2]Inosinic acid or inosinate. [3]Adenylosuccinate. [4]Xanthylic acid. The abbreviations used are: ADP = adenosine 5'-diphosphate; AMP = adenosine 5'-monophosphate: ATP = adenosine 5'-triphosphate; Asp = aspartic acid (aspartate); Gln = glutamine; Glu = glutamic acid (glutamate); Gly = glycine; GDP = guanosine 5'-diphosphate; GTP = guanosine 5'-triphosphate; NAD^+ and NADH = nicotinamide dinucleotide (oxidized and reduced forms, respectively); NAP^+ and NADPH = nicotinamide adenine dinucleotide phosphate (oxidized and reduced forms, respectively); P_i = inorganic orthophosphate; PP_i = inorganic pyrophosphate; THF = tetrahydrofolic acid coenzyme. From Altman, P. L. and Dittmer, D. S. 1968. *Metabolism*. Federation of American Societies for Experimental Biology. Bethesda, Md. p. 453. Fig. 88.

nation. Most fishes, however, and amphibians secrete urea formed from allantoic acid by allantoicase. And some invertebrates even secrete ammonia. Apparently, the enzymes allantoicase, allantoinase, urease, and uricase have been lost during vertebrate evolution resulting in less complete degradation of purines in higher animals.

Purine synthesis is considerably more complicated than pyrimidine biosynthesis and differs in two respects. One is that the starting molecule on which the build-up of the purine ring occurs, 5-phosphoribosyl-1-pyrophosphate (PRPP), is an intermediate in pyrimidine biosynthesis. The other difference is that the purine ring is synthesized on ribose 5-phosphate, but in pyrimidine synthesis the pyrimidine ring is formed before coupling to ribose 5-phosphate. It is of interest also that both inosinic acid and orotidylic acid, although the initial nucleotides formed in purine and pyrimidine syntheses respectively, are not major constituents of nucleic acids.

A schema showing the principal pathways for the major foodstuffs of the cell is presented in Figure 8-15. It may be noted in the scheme that some pathways are more closely related to each other than are other pathways. For example, amino acids arising from protein degradation are closely aligned in synthesis to the TCA cycle, acetyl CoA, and pyruvic acid. Note also that lipids are related metabolically to glycolysis, acetyl CoA, and cholesterol formation. Carbohydrates, through ribulose 5-phosphate, give rise to nucleic acids. Carbohydrates are also a primary source for the TCA cycle through pyruvic acid and acetyl CoA. And porphyrins arise from glycine combining with succinyl CoA of the TCA cycle. Throughout all of these interrelated pathways, one may visualize a concept of energy flow extending from the entrapment of radiant energy in light reactions and carbon fixation and energy transfer in the dark reactions of photosynthesis, with the consequent degradation (and energy conserving) reactions associated with intermediary metabolism, to the TCA cycle and electron transport system and the ultimate formation of carbon dioxide and water.

3. RNA

Function in living cells is directly related to structure of nucleic acids and proteins. As mentioned earlier, nucleic acids are important because they include RNA and DNA. It is the DNA molecule in which genetic messages are encoded. And it is the RNA which aids in transmitting and transcribing the messages into thousands of different proteins in a living cell.

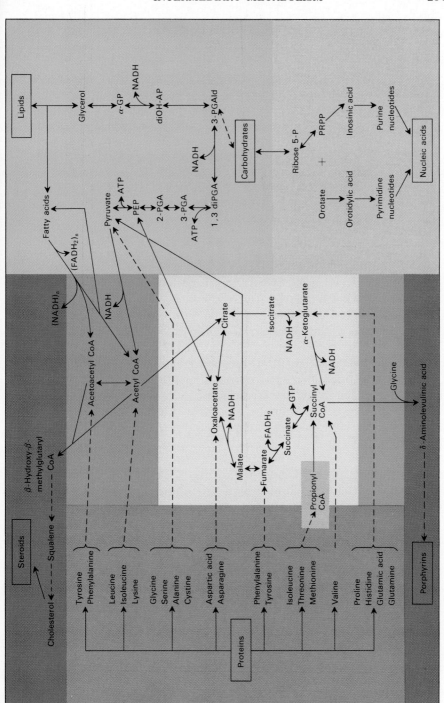

Fig. 8-15. A schema showing an integration of metabolic pathways (modified from lecture notes of Dr. E. S. Higgins, Medical College of Virginia)

There are several different ribonucleic acids. One is messenger RNA which transcribes the genetic message for each protein from its original site in the nucleus encoded in DNA to the ribosome in the cytoplasm. Ribosomal RNA is a part of the ribosome. Transfer RNA's, each one specific for one amino acid, presumably transfer amino acids to the ribosomal site where the message is transcribed into a growing polypeptide chain to form, ultimately, proteins. The structure for alanine tRNA was reported in 1965. This tRNA was shown to contain 77 nucleotides.

Recently Ottensmeyer of Toronto has applied dark field optic techniques to electron microscopy. His electron micrographs confirm the cloverleaf shape of tRNA. He also has obtained similar direct evidence that DNA exhibits a double helix pattern, and that the ribonuclease molecule is globular with a small cleft in the protein.

4. DNA

In 1869 Miescher discovered DNA while working in Hoppe-Seyler's laboratory. Miescher called the substance nuclein since it was isolated from nuclei. He isolated nuclei from white blood cells of pus with dilute hydrochloric acid. He cleaned the nuclei with a hydrochloric acid extract made from a pigs stomach. The nuclein was then obtained with dilute alkali. Miescher reported this work in 1871. In 1874, he reported an improved procedure for preparing nuclein from fish sperm nuclei. Zacharis in 1881 identified nuclein with chromatin. In 1882 Sachs stated that nucleins of sperm and egg are different and in 1884 Hertwig claimed that nuclein is responsible for the transmission of hereditary characteristics! Altmann introduced the term nucleic acid to replace nuclein in 1899. By 1941, Caspersson and Brachet, independently, related that nucleic acids were connected to protein synthesis. A remarkable experiment by Avery, MacLeod, and McCarty in 1944 showed that bacteria could be genetically transformed by DNA. And from LeBlond's laboratory in 1948 came radioautographs showing that newly formed DNA was found in tissues where cell divisions were numerous. By 1949 it was recognized that the quantity of DNA per set of chromosomes is constant in different cell types of any organism. In 1950, Chargaff reported that the numbers of adenine and thymine groups were always equal as were the numbers of cytosine and guanine. Hershey and Chase demonstrated in 1952 that the bacteriophage DNA carried the viral genetic information from parent to progeny. And finally, in 1953, Watson and Crick proposed that the DNA molecule consisted of two polypeptide chains jointed together by base pairs, and the bases were joined in pairs by hydrogen bonding. By this time x-ray diffraction studies by Wilkins indicated a double-helix structure for DNA. Now a completely synthetic DNA has been made. In 1967 Kornberg and Sinsheimer reported that, starting with natural DNA as a template, the synthetic DNA has the full biological activity of the native DNA.

Recently it has been found that DNA of higher organisms reassociate more rapidly than predicted. Separated complementary strands of purified DNA recognize each other and specifically reassociate. The conclusion is that DNA has certain segments which are repeated hundreds of thousands of times. A survey of many species of both invertebrates and vertebrates indicated that repeated sequences in DNA occur widely and are probably universal in higher organisms.

Suggested Reading—Chapter 8

Altman, P. L and Dittmer, D. S. *Metabolism*. Federation of American Societies for Experimental Biology. Bethesda, Md. 1968.
Atkinson, D. E. Biological feedback control at the molecular level. Science *150:*851. 1965.
Avery, O. T., MacLeod, C. M., and McCarty, J. Studies on the chemical nature of the substance inducing transformation of pneumococcal type III. J. Exp. Med. *79:*137, 1944.
Bandurski, R. S. Biological reduction of sulfate and nitrate. In *Plant Biochemistry*. Bonner, J. and Varner, J. E. (editors) Academic Press. New York. 1965.
Bonner, J. and Varner, J. E. *Plant Biochemistry*. Academic Press. New York. 1965.
Booth, C. C. Sites of absorption in the small intestine. Fed. Proc. *26:*1583, 1967.
Britten, R. J. and Kohne, D. E. Repeated sequences in DNA. Science *161:*529, 1968.
Cantarow, A. and Schepartz, B. *Biochemistry*. 4th edition. W. B. Saunders Co. Philadelphia. 1967.
Cardell, R. R., Jr., Badenhausen, S., and Porter, K. R. Intestinal triglyceride absorption in the rat. J. Cell. Biol. *34:*123, 1967.
Chargaff, E. Chemical specificity of nucleic acids and mechanism of their enzymatic degradation. Experientia *6:*201, 1950.
Conn, E. C. and Stumpf, P. K. *Outlines of Biochemistry*. 2nd edition. John C. Wiley & Sons. New York. 1966.
Cornforth, J. W. Biosynthesis of fatty acids and cholesterol considered as chemical processes. J. Lipid Res. *1:*3, 1959.
Dawkins, M. J. R. and Hull, D. The production of heat by fat. Sci. Amer. *213(2):*62, 1965.
Dawson, R. M. C. and Rhoades, D. N. (editors). *Metabolism and Physiological Significance of Lipids*. John C. Wiley & Sons. New York. 1964.
Dietschy, J. M. Effects of bile salts on intermediate metabolism of the intestinal mucosa. Fed. Proc. *26:*1589, 1967.
Green, D. E. The synthesis of fat. Sci. Amer. *202(2):*46, 1960.
Greenberg, D. M. (editor). *Metabolic Pathways*. 3rd edition. Academic Press. New York. 1966.
Guyton, A. C. *Textbook of Medical Physiology*. 3rd edition. W. B. Saunders Co. Philadelphia. 1966.
Hershey, A. D. and Chase, M. Independent functions of viral protein and nucleic acid in growth of bacteriophage. J. Gen. Physiol. *36:*39, 1952.
Holley, R. W., Marquisee, M., Merrill, S. H., Penswick, J. R., and Zamir, A. Structure of a ribonucleic acid. Science *147:*1462, 1965.
Ito, S. Structure and function of the glycocalyx. Fed. Proc. *28:*12, 1969.

Kleiner, I. S. and Orten, J. M. *Biochemistry*. 7th edition. C. V. Mosby Co. St. Louis. 1966.
Korn, E. D. Current concepts of membrane structure and function. Fed. Proc. *28:*6, 1969.
Kornberg, A. The synthesis of DNA. Sci. Amer. *219(4):*64, 1968.
Kornberg, A. and Sinsheimer, R. L. Enzymatic synthesis of DNA. XXIV. Synthesis of infectious phage ϕX174 DNA. Proc. Nat. Acad. Sci. USA *58:*2321, 1967.
Krebs, H. A. and Kornberg, H. L. A survey of the energy transformation in living matter. Ergebn. Physiol. *49:*212, 1957.
Lacy, D. and Taylor, A. B. Fat absorption of epithelial cells of the small intestine of the rat. Amer. J. Anat. *110:*155, 1962.
LeBlond, C. P., Stevens, C. E., and Borgoroch, R. Histological localization of newly formed desoxyribonucleic acid. Science *108:*531, 1948.
Lowenstein, J. M. Reductions and oxidations in mammalian biosyntheses. J. Theor. Biol. *1:*98, 1961.
Lynen, F. and Ochoa, S. Enzymes of fatty acid metabolism. Biochim. Biophys. Acta *12:*299, 1953.
Mahler, H. R. and Cordes, E. H. *Biological Chemistry*. Harper & Row. New York. 1966.
Majerus, P. W. Acyl carrier protein. VII. The primary structure of the substrate-binding site. J. Biol. Chem. *240:*4723, 1965.
Meister, A. *Biochemistry of Amino Acids*. Academic Press. New York. 1965.
Mirsky, A. E. The discovery of DNA. Sci. Amer. *218(6):*78, 1968.
Palay, S. L., and Karlin, L. J. An electron microscopic study of the intestinal villus. II. The pathway of fat absorption. J. Biophys. Biochem. Cytol. *5:*373, 1959.
Porter, K. R. Independence of fat absorption and pinocytosis. Fed. Proc. *28:*35, 1969.
Pugh, E. L. and Wakil, S. J. Studies on the mechanism of fatty acid synthesis. XIV. The prosthetic group of acyl carrier protein and the mode of its attachment to the protein. J. Biol. Chem. *240:*4727, 1965.
Senior, J. R. Intestinal absorption of fats. J. Lipid Res. *5:*495, 1964.
Smellie, R. M. S. The biosynthesis and function of nucleic acids. In *The Biological Basis of Medicine*. Bittar, E. E. and Bittar, N. (editors). Academic Press. New York. 1968.
Stein, W. D. The transport of sugars. Brit. Med. Bull. *24:*146, 1968.
Stumpf, P. K. Lipid metabolism. In *Plant Biochemistry*. Bonner, J. and Varner, J. E. (editors). Academic Press. New York. 1965.
Tubbs, P. K. Membranes and fatty acid metabolism. Brit. Med. Bull. *24:*158, 1968.
Wakil, S. J. The mechanism of fatty acid synthesis. Proc. Nat. Acad. Sci. USA *52:*106, 1964.
Watson, J. D. and Crick, F. H. C. Molecular structure of nucleic acids. A structure for deoxyribose nucleic acid. Nature (London) *171:*737, 1953.
Weiss, J. M. The role of the Golgi complex in fat absorption as studied with the electron microscope with observations on the cytology of the duodenal absorptive cells. J. Exp. Med. *102:*775, 1955.
White, A., Handler, P., and Smith, E. L. *Principles of Biochemistry*. 4th edition. McGraw-Hill Book Co. New York. 1968.
Wilkins, M. H. F. Molecular configuration of nucleic acids. Science *140:*941, 1963.
Zilversmit, D. B. Formation and transport of chylomicrons. Fed. Proc. *26:*1599, 1967.

Section **III.**
Nucleocytoplasmic Relations

Chapter Nine

Theory of the Gene

A. CHROMOSOME MORPHOLOGY

B. DNA REPLICATION AND CELL DIVISION

C. RIBOSOMES AND PROTEIN SYNTHESIS

D. GENETIC CODE

E. ENZYME REGULATION

Although Miescher isolated nucleic acid as a nucleoprotein complex from pus cell nuclei in 1871 which he called nuclein, it was not until 1874 that he isolated pure nucleic acid from the DNA-protamine complex in salmon sperm nuclei. He called this more purified nuclein by the term protamine without recognizing its protein nature. Five years later chemists had isolated purine bodies. In 1884 Kossel recognized that histones and protamines were associated with nucleic acid. He also found that the histones were basic proteins. Ten years later cytosine was identified, and in 1909 uracil was isolated. By 1930 the sugar moiety of nucleic acid was identified as 2-deoxyribose. Since then the type of pentose in nucleic acids has been used to classify the compounds as ribonucleic acids.

A. Chromosome Morphology

DNA was localized as a constituent of the cell nucleus in 1924. By means of the Feulgen reaction, DNA was shown to be a part of the chromosomes. And in 1940, Brachet identified RNA as a component of cytoplasmic basophilia. DNA occurs in the cell as a DNA-protein complex called chromatin, or often nowadays referred to as nucleohistone. During interphase, with the aid of electron micrographs, condensed portions of the chromosomal material (heterochromatin) can be distinguished from the less dense regions (euchromatin). The chromatin in the interphase nucleus represents an aggregation of chromosomes. Since the chromatin fiber is about 200–300 A in diameter, it is not seen in light microscopes techniques. It is the heterochromatin which is seen in aggregates. In 1943 chromatin fibers were morphologically identified as chromosomes, and found to be Feulgen positive. And in 1948 chromosomes were isolated for *in vitro* study by Mirsky and Ris.

During mitosis chromatin condenses to become visible as chromosomes. Chemical analyses consistently show that chromatin is made up of DNA, RNA, histones, and nonhistone proteins; in sperm nuclei of some fishes and birds histones are replaced by the more basic proteins, protamines. The nonhistone proteins are acidic.

The DNA molecule appears to be the most stable one present in the chromosome. Labeling experiments with proteins, RNA, and DNA show that DNA is the only macromolecule capable of persisting through many cell generations. Evidence indicates, too, that RNA is not a primary structural unit, but may be bound at the surface of the chromosome.

Some of the difficulty in studying structure and organization of chromatin is in separating pure DNA from proteins. It appears as though the DNA is tightly packed, and the histones seem to lie in the large and small grooves of the DNA molecule. An example that the DNA is tightly packed is the metaphase chromosome of the golden hamster. Here the continuous strand of DNA has a molecular weight of 3×10^9 and is about 1.6 mm long; this makes the DNA molecule about 160 times longer than the chromosome. The DNA present per chromosome set is constant for a particular species. The DNA content of diploid somatic cells is exactly twice that of haploid cells.

There is a close correlation between DNA content and chromosome size. It has been found that about 31 cm of double helix DNA equals a pikogram (1 pg = 10^{-12} gm). In man, a diploid cell contains about 5.6 pg of DNA corresponding to about 174 cm of DNA. In comparison, a giant salivary gland cell from *Drosophila* contains about 293 pg of DNA which amounts to some 91 m in length. The largest dipteran salivary gland chromosome is about 4 μ in diameter by 400 μ in length as compared with the largest human chromosome which has a diameter of about 0.6 μ and a length of 10 μ. In fact, if the total end-to-end length of human chromosomes were measured, they would be about 220 μ long. Although the haploid chromosome number has been established as 23 in man, more than half of other animals in which the haploid num-

THEORY OF THE GENE

ber is known have less than 23 chromosomes. The haploid number may be as low as 2 in a roundworm, or as high as 120 in a fern or 127 in a crustacean.

An interesting and useful aspect of chromosome morphology has been the identification of sex chromatin, *i.e.*, the heteropyknotic X chromosome. Sex chromatin is sometimes called the Barr body. The drumstick phenomenon identifiable in about 3% of the neutrophils of the human female apparently is the equivalent to sex chromatin. The drumstick is peculiar to the polymorphonuclear leukocyte nucleus. Sex chromatin is usually located against the inner surface of the nuclear membrane, and is positive to the Feulgen reaction. Sex chromatin can be identified in oral mucosa smears (Fig. 9-1). Cells are stained in cresyl echt violet, differentiated in ethyl alcohol, cleared in xylene, and mounted in neu-

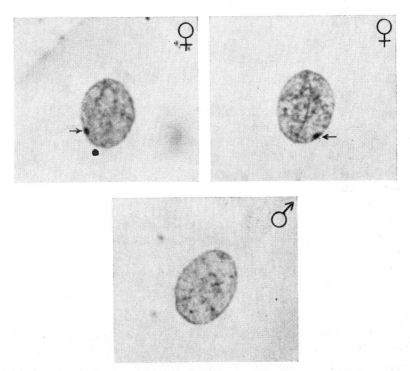

FIG. 9-1. Epithelial cells, stained with cresyl echt violet, in smears from the oral mucous membrane (2000 ×). Left, adult female; right, female infant, aged 2 days; lower, adult male. From Moore, K. L. and Barr, M. L. 1955. Smears from the oral mucosa in the detection of chromosal sex. Lancet 2:57. Figs. 1–3.

tral balsam or picolyte. When gonadal dysgenesis is suspect, the sex chromatin test reveals whether or not sex reversal has obtained. Sex chromatin is about 1 μ in diameter.

Aneuploidy is any chromosomal aberration characterized by a deviation from the total number. The normal diploid chromosome number in man is 46. A karyotype of the normal chromosomal picture for both male and female is shown in Figure 9-2. The karyotype of a person with the Turner syndrome showing 45 chromosomes is illustrated in Figure 9-3. These individuals are sex chromatin negative (XO) with incomplete female genitalia. The Turner syndrome occurs once in about 5000 female births. The Klinefelter syndrome occurs more frequently, appearing about once in 500 male births. In this anomalie 47 chromosomes are present (Fig. 9-4). People with Klinefelter syndrome have poorly developed external male genitalia, and are sex chromatin positive (XXY). From these examples it is seen that the sex chromatin or Barr body results from the two XX chromosomes. The number of Barr bodies is always one less than the number of X chromosomes.

Interest in the histones stems from their role in genetic regulation. Histones and DNA occur about equally in the nucleus; histones are about 32% of the nuclear mass, DNA about 30%. Histones do not appear to turnover which makes them stable structures. They contain a high proportion of lysine and arginine, but lack tryptophan. Histone synthesis seems to occur at the time of, or slightly before, DNA synthesis. Not only is DNA linked to histones (or protamines), but also nonhistones may be linked to DNA. However, the specific binding between DNA and the basic and acidic proteins is essentially unknown.

The great majority of DNA (about 80%) is nonfunctional; the other 20% is used as template for RNA synthesis. Fractionation studies show that heterochromatin is inert, but that euchromatin is active in RNA synthesis. An active site for RNA synthesis has been demonstrated in the puffs and rings of insect chromosomes as shown in Figure 9-5. As shown in Figure 9-6 RNA synthesis is suppressed by actinomycin D. RNA synthesis occurs at the boundary site between clumped and diffuse chromatin.

Although the term histone is a general one and includes a variety of compounds three classes of histones have been identified by column chromatography. The F1 is a lysine-rich histone, F2 is slightly lysine-rich, and F3 histone is arginine-rich. On the basis of end-group analysis, F2a histone possesses an N-terminal acetylalanine, and F2b has an N-terminal proline. The F2a and F2b subfractions are fairly homogeneous preparations, but F1 and F2 each seem to contain four or five com-

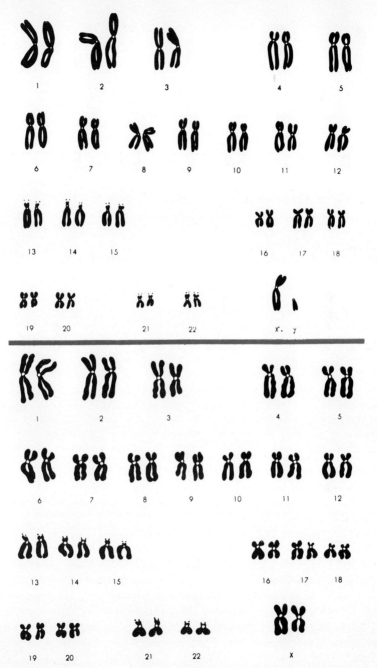

FIG. 9-2. Upper, the mitotic metaphase chromosomes of a somatic cell of a male, arranged in a karyotype. Lower, the mitotic metaphase chromosomes of a somatic cell of a female, arranged in a karyotype. From McKusick, V. A. 1964. *Human Genetics*. Prentice-Hall. Englewood Cliffs, N. J. p. 8. Figs. 2.1 and 2.2.

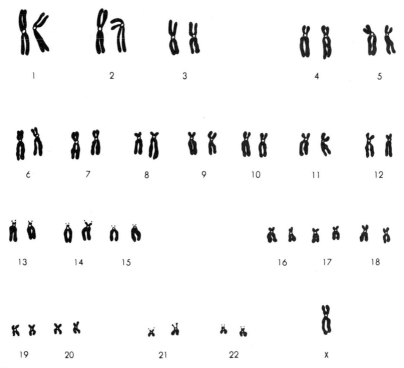

Fig. 9-3. The Turner syndrome. The karyotype shows 45 chromosomes with an XO sex-chromosome constitution which is chromatin-negative. The features of a person with this syndrome are female external genitalia, short stature, webbed neck, low-set ears and typical facies, broad shield-like chest with widely spaced nipples and undeveloped breasts, small uterus, and ovaries represented only by fibrous streaks. In some cases coarctation of the aorta leads to severe hypertension in the upper part of the body. From McKusick, V. A. 1964. *Human Genetics*. Prentice-Hall. Englewood Cliffs, N. J. p. 3. Fig. 2.6.

---

pounds. The number of histones are less than 10. They are alike in different cell types of the same specimen, and show little variation from one species to another.

That histones inhibit RNA synthesis in the nucleus, *i.e.*, inhibit gene action by suppressing DNA, was first experimentally demonstrated in 1961 by Allfrey. He has studied the changes in chromosomal proteins at times of gene activation. The F2a1 histone seems best to work with in acetylation studies. With carbon-14 acetyl CoA as the direct acetyl donor, it has been found that acetylation increases RNA synthesis in isolated nuclei of calf thymus. When histone acetylation is removed, RNA synthesis decreases. Although histone acetylation is but one event

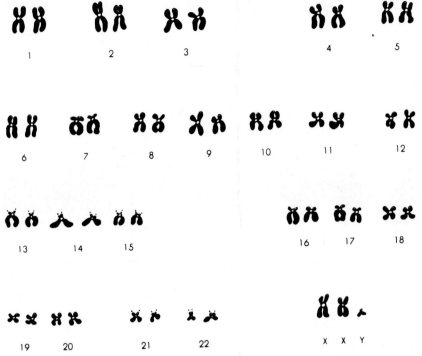

Fig. 9-4. The Klinefelter syndrome. The karyotype shows 47 chromosomes with an XXY sex-chromosome constitution which is chromatin-positive. The features of a person with this syndrome are male external genitalia but the testes are consistently very small and body hair is sparse. Most individuals have gynecomastia (female-like breast development), and tend to be long-legged. From McKusick, V. A. 1964. *Human Genetics*. Prentice-Hall. Englewood Cliffs, N. J. p. 4. Fig. 2.7.

---

in gene activation, the site of acetylation is the site of RNA synthesis in the nucleus. It appears that the amino terminal of ϵ-N-acetyl lysine is attached to the DNA molecule. A model of the F2a1 histone constructed on the DNA molecule model indicates that the attachment points appear most likely at N-7 of guanine. All histones probably fit in the small and large grooves of the DNA molecule.

Phosphorylation of nuclear proteins is also correlated with function, *i.e.*, RNA synthesis, but methylation is not correlated with function. The process of phosphorylation is ATP-dependent, and proceeds independently of protein synthesis. Phosphorylation is more pronounced in the region of diffuse chromatin. Once phosphate is incorporated, phosphorylation is readily reversible. It may be that phosphorylation changes the state of the fibrils in heterochromatin by an uncoiling

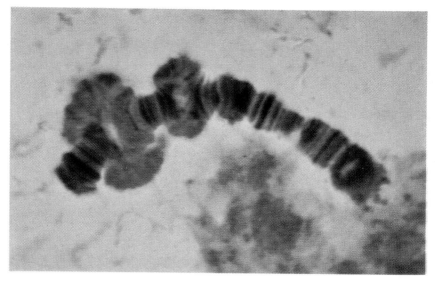

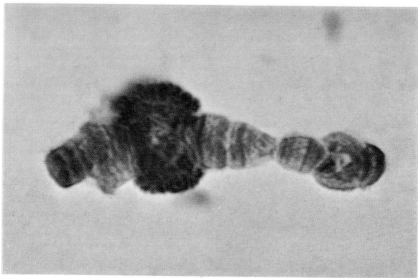

FIG. 9-5. Upper, chromosome puffs are the protuberances on the left-hand portion of the giant chromosome. Very large puffs, two of which are seen, are called Balbiani rings. Protein has been stained green, DNA brown. Lower, products of puffs, RNA, is reddish-violet when dyed with toluidine blue. Here the DNA is blue. In this figure are two different specimens of the giant chromosome IV from the salivary gland of the midge *Chironomus tentans*. (1600 ×). From Beermann, W. and Cleaver, U. 1964. Chromosome puffs. Sci. Amer. 210(4):51.

THEORY OF THE GENE 271

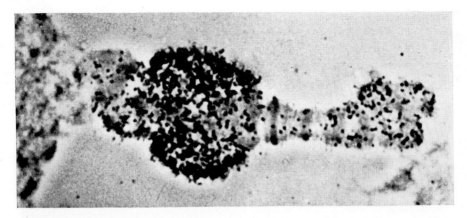

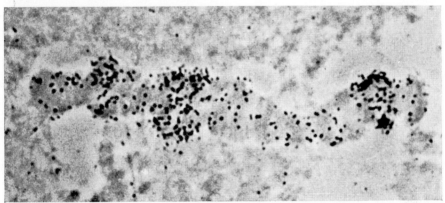

FIG. 9-6. Inhibition of puffing and of RNA synthesis is accomplished by treatment with the antibiotic actinomycin D. At top an autoradiogram of a chromosome IV of *Chironomus tentans* shows the incorporation of much radioactive uridine (black spots), which takes place during the production of RNA. Another chromosome (bottom) that had been puffing shows puff regression and little radioactivity after ½ hour of treatment with minute amounts of actinomycin D which inhibits RNA synthesis by DNA. From Beermann, W. and Clever, U. 1964. Chromosome puffs. Sci. Amer. *210(4)*:53.

---

mechanism, and dephosphorylation may lead to a tight coiling of the DNA-histone complex.

Chromosomal proteins have been studied in higher plants and animals because proteins associated with DNA occur there in large amounts. But chromosomal proteins binding to DNA have been identified in bacteria. They fulfill the repressor function postulated by Jacob and Monod in 1961 which is discussed later in section E of this chapter. Histones may be regarded as repressors to function as gene regulators from another point. Chromatin stripped of its protein is a more effective primer for DNA-dependent RNA synthesis than native chromatin. But in regard to nonhistone proteins concerning their repressor, or expressor, capabilities not enough information is available to make a general statement concerning their function. However, there is some evidence to indicate that most chromosomal puff RNA is associated with nonhistone protein in the form of ribonucleoprotein particles. The acidic proteins may then be involved in binding with RNA to facilitate its removal from the DNA template. Histones have been found to have an effect on embryonic development. Three fractions of calf thymus histone, F1, F2a, and F3, exhibit an independent effect on chick embryo causing an arrested development at the primitive streak or early head process stage.

RNA synthesis is also related to cellular differentiation in embryos. It is known that tRNA is synthesized in the nucleus, and seems dependent on DNA templates as does mRNA, but rRNA is derived from nucleolar DNA. The nucleolar RNA first becomes labeled at the side of the nucleolus in contact with the chromosome. The chromosomal DNA, at the point of nucleolar attachment, contains the cistrons which prescribe the structure of rRNA. In fact, it has been shown in HeLa cells of metaphase isolations that DNA complementary to rRNA is confined to smaller chromosomes including those carrying a nucleolar organizer, but DNA complementary to mRNA is distributed among chromosomes of various sizes. In embryonic development, some mRNA may be preformed in the ovum, but most of the RNA synthesized during cleavage is of the mRNA type. Near the end of the cleavage stages, tRNA synthesis occurs. Ribosomal RNA synthesis begins at gastrulation and increases as embryonic development proceeds.

Visualization of the process of RNA synthesis has now been accomplished with the aid of electron microscopy. The developing oocyte of the amphibian, *Triturus viridescens*, has several hundred extrachromosomal nucleoli within each nucleus formed by the chromosomal organizer. Typically, each extrachromosomal nucleolus exhibits a fibrous core region and a granular cortex. The core contains only DNA, and both the core and cortical regions contain RNA and protein. Each nucleolar core is composed of a thin axial fiber about 100–300 A in diameter coated with a matrix substance. Deoxyribonuclease breaks the axial fiber, and treatment with trypsin indicates that the core axis is a double helix DNA molecule about 30 A in diameter. Matrix segments occur intermittantly along the core axis. Each matrix segment consists of some 100 thin fibrils connected at one end to the core axis. The matrix fibrils can be removed from the core axis by treatment with ribonuclease, trypsin or pepsin. Autoradiography with tritiated nucleotides shows RNA synthesis only over the matrix segments. As shown in Figures 9-7 and 9-8, the strand must be DNA, and at intermittant points along its core axis a matrix segment covers the DNA gene coding for rRNA precursor molecules of the matrix segment.

The gene may be defined in terms of function and DNA structure. A gene is a functional region of the DNA molecule which codes for different polypeptides. It has been proposed by Benzer that the unit of func-

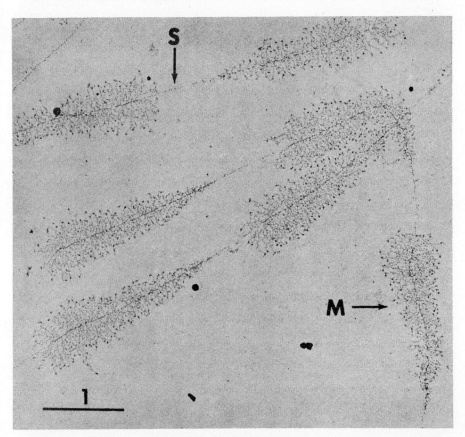

FIG. 9-7. Portion of a nucleolar core isolated from *Triturus viridescens* oocyte showing matrix units (M) separated by matrix-free segments (S) of the core axis. The axial fiber can be broken by treatment with deoxyribonuclease, whereas the matrix fibrils can be removed by ribonuclease, trypsin, or pepsin. Scale 1 μ. From Miller, O. L. Jr. and Beatty, B. R. 1969. Visualization of nucleolar genes. Science *164:*956. Fig. 2.

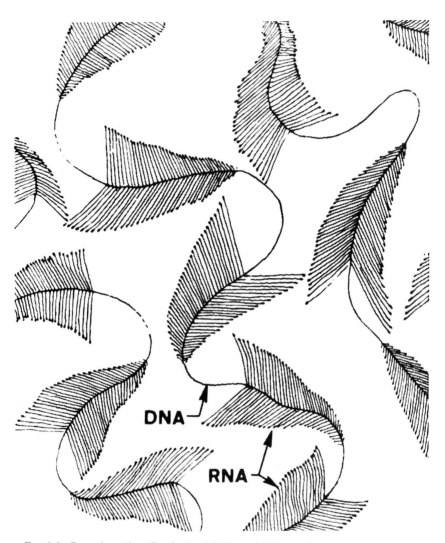

Fig. 9-8. Genes in action: Synthesis of RNA on DNA strand. From Oak Ridge Associated Universities Newsletter. July 1, 1969. ORNL biologists photograph functioning genes. p. 5.

tion be called a cistron to distinguish it from a unit of mutation, the muton, and a unit of recombination, the recom. A muton or recom may vary from one to several DNA nucleotide pairs. Both the muton and recom may contain fewer nucleotide pairs than a cistron. In fact, it has been estimated that a gene may be a linear arrangement of about 1500 base pairs.

B. DNA Replication and Cell Division

Replication of DNA requires an unwinding of the two polynucleotide chains which form the double helix of the DNA molecule. At the present time, it is believed that both unwinding and replication of the two chains can be accomplished through rotation. During the process of cell division, chromosomes are characteristically coiled and appear in variously condensed forms. As discussed above in Section A, it is presumed that the DNA molecule is a singular structure forming a continuous axial fiber or backbone of an entire chromosome in higher organisms. But autoradiographs of the bacterial chromosome indicate the presence of one continuous DNA double helix forming a circle. This circular strand of a bacterial chromosome has been verified morphologically in electron micrographs. When the process of cell division is occurring, the chromosome is in its most condensed, and coiled, phase. Therefore, it would seem that DNA replication would be less likely to occur. Alternately, when cell division is not occurring, *i.e.*, during interphase, the chromatin fiber is characteristically elongated, and less condensed. This interphase condition would seem to facilitate DNA replication as the double helix rotates and unwinds.

Tritiated thymidine has been used to show how DNA is organized in the individual chromosome. Analysis of autoradiographs show that both daughter chromosomes resulting from duplication appear equally and uniformly labeled in the presence of labeled thymidine. The label appears in only one of the daughter chromosomes after a duplication in the absence of the labeled DNA precursor. A model which accounts for these results strongly suggests that a single interphase chromosome contains two DNA units before its replication, and these two units correspond to the two polynucleotide chains of the Watson-Crick double helix DNA molecule.

Not all DNA, however, has two polynucleotide chains. The DNA synthesized in 1967 by Kornberg and Sinsheimer mentioned in the previous chapter, is single-stranded. The single-strand DNA is in the ϕX174 virus which preys on the colon bacillus, *Escherichia coli*, as a bacteriophage. Other viruses, however, have single-stranded RNA as well as double-stranded RNA and DNA in their genetic component.

Mitosis is the usual process of cell reproduction involving both karyokinesis and cytokinesis. Karyokinesis is a process of indirect nuclear division and cytokinesis is a division of the cytoplasm. If karyokinesis occurs without being followed by cytokinesis, the process results in the formation of a multinucleated cell. Continued and repeated karyokinesis results in a multinucleated mass of protoplasm called a coenocyte; a slime mold is an example. A coenocyte differs from a syncytium. A syncytium is a multinucleated mass of protoplasm which results from a fusion of cells. An example of a syncytium is a striated muscle cell.

Five phases are usually recognized in describing mitosis. These are (1) prophase, (2) prometaphase, (3) metaphase, (4) anaphase, and (5) telophase. These five phases are represented in Figure 9-9. Although these phases are artificial in terms of stop-go mechanisms since mitosis is a continuous process, they are convenient markers for the time flow of events happening in a mitotic reproductive cycle. But the time spent in each phase differs in cells. Interphase exists between mitotic periods.

A most precise marker in the sequence of events in mitosis is the metaphase climax. At this point the mitotic apparatus has been formed and chromosome movement is about to occur. An idealized mitotic apparatus is schematized in Figure 9-10 at a metaphase climax. Here the chromosomes are aligned on the equatorial plate. A spindle fiber, believed to be a microtubule, is attached to the chromosome at a particular point called a centromere (kinetochore). Centriolar polarity has been established at the asters. As shown in Figure 9-11 certain timing events occur before and after the metaphase climax. Metaphase chromosomes contain less water than any other nuclear constituent. Interestingly enough, from one cell division to another, chromosomes tend to occupy the same characteristic position on the metaphase plate.

From the very beginning of a typical prophase, individual chromosomes are separate from one another. There is no spireme which is continuous as described by early cell biologists. Each chromosome is doubled, *i.e.*, each chromosome consists of two chromatids closely approximating each other throughout their length. Chromosomes in some cells become visible in the same position they occupied at the end of the preceding cell division. The chromosomes condense and exhibit centromeres. A centromere corresponds to a constriction of the chromosome. The nucleolus becomes less stainable at the chromosomal portion

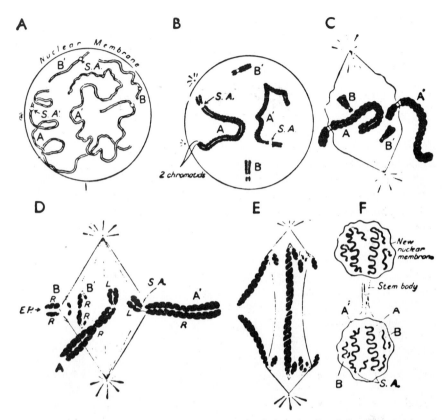

Fig. 9-9. Diagrams of the main stages of mitosis. Only two pairs of chromosomes A and A', B and B' are shown. Both of these have subterminal centromeres. Mitotic stages are A and B = prophase; C = prometaphase; D = metaphase; E = anaphase; F = telophase going into interphase. S. A. = centromeres, E. P. = equatorial plane of the spindle, R and L are regions of the chromosomes which are spiralized in a right- or left-handed direction at metaphase. At early prophase the relic spirals are clearly seen. From White, M. J. D. 1961. *The Chromosomes*. John Wiley & Sons. New York. p. 9. Fig. 2.

---

of the nucleolar organizer, and finally disappears. Centrioles separate and move poleward as the nuclear envelope disappears.

In prometaphase, polarity is established at aster formation. The achromatic spindle is formed, and a spindle fiber attaches to a centromere of a chromosome. The centromere is constant for an individual chromosome, and persists throughout the mitotic cycle. Its position, however,

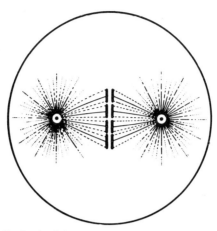

Fig. 9-10. The idealized mitotic apparatus at metaphse climax. From Mazia, D. 1961. Mitosis and the physiology of cell division. In *The Cell*. Vol. 3. Brachet, J. and A. E. Mirsky (editors). Academic Press, New York. p. 96. Fig. 1.

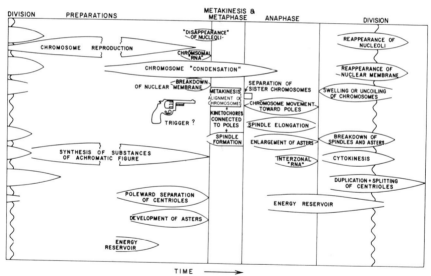

Fig. 9-11. A generalized plan of the time flow of events in mitosis. The time span for a given process represents the period during which it *may* take place; the actual interval will vary in different kinds of cells. The diagrammatic representation of each process converges to a point at the earliest time when it is known to begin or the latest time at which it may be completed. The representation is open-ended in cases where the time of initiation or termination of the process cannot be fixed. From Mazia, D. 1961. Mitosis and the physiology of cell division. In *The Cell*. Vol. 3. Brachet, J. and A. E. Mirsky (editors). Academic Press. New York. p. 97. Fig. 2.

on the chromosome may be median, submedian, or terminal. Centromeres divide, and each chromatid, now a single chromosome, moves to align itself on the equatorial plate of metaphase.

Chromosome movement begins following the metaphase climax described above. Asters enlarge, and the chromosomes move toward the poles although they never reach them; typically, plant cells do not have centrioles. Although the point is controversial, it seems as though the chromosomes move poleward in response to spindle contraction. As the spindle elongates, its middle region forms the stem body which exhibits longitudinal striations. The stem body may persist to the formation of two new cells. The anaphase begins with chromosome movement.

Telophase is a reversal of prophase activity. Activities associated with telophase includes the reappearance of the nuclear envelope and nucleolus. The chromosomes decondense and uncoil so that they become elongated; they also become more hydrated. There is a breakdown of the spindle and asters. Cytokinesis occurs. In animal cells, cytokinesis typically begins, and continues, as a peripheral constriction which results in the formation of the two new cells. But typically in plant cells, cytokinesis begins when a cell plate forms at the central equator to initiate the division of cytoplasm. Each new cell begins a period of interphase during which cell synthesis and growth occurs. The centriole duplicates and forms the centrosome.

It is presumed that actual chromosome replication takes place during interphase since the usual chromosome appears single in late telophase, but double (two chromatids) in early prophase. There is also a general conclusion for somatic cells that DNA replication occurs during interphase. A mass of evidence favors this view.

Following mitosis, there is an increase in the total mass of the cell and a synthesis of DNA during interphase. The general regularity is such that a DNA cycle has been formulated. It may be represented as

$$\cdots > G_1 \to S \to G_2 \to \text{Mitosis} \to G_1 \cdots >$$

In this scheme, G_1 represents a period or gap following a completion of the mitotic reproductive cycle in which no DNA synthesis takes place. The S period is the one in which DNA synthesis occurs, and the DNA content of the nucleus is exactly doubled. A second gap period is represented by G_2 where no DNA synthesis takes place. Many cells in an organism, however, can remain in an interphase for varying periods of time before completing a division cycle. A sufficient amount of DNA polymerase and the necessary precursors are prerequisites for DNA synthesis. As shown in Figure 9-12, DNA replication is preceded by a substantial rise in the concentration of deoxyribonucleotides and an increase in DNA polymerase activity. DNA synthesis in the S period occurs first in euchromatin, followed by DNA replication in heterochromatin.

Many studies point to the fact that one DNA molecule has an identity with one chromosome in animals, plants, and microorganisms. Among these studies are those which show that a single genetic event causes a specific amino acid substitution. One example is the replacement of the glutamic acid in the 6-position on the two β-chains of human hemoglobin A by valine which results in hemoglobin S, sickle-cell anemia. Other studies clearly show that both DNA and a chromosome displays semiconservative replication and linear genetic recombination. Irradiation by x-rays causes fragmentation of both structures, and mutations caused by ultraviolet light-rays.

Three mechanisms of replication are possible for the double helix DNA. Studies with electron microscopy and autoradiography have been offered for the most likely method of replication. The mechanisms studied involved (1) two polynucleotide chains, (2) one polynucleotide chain, or (3) nucleotide base pairs. The first is the conservative scheme which has it that the original double helix molecule remains intact and a new molecule is formed. The second is the semiconservative scheme which was proposed when the structure for DNA was suggested in 1953.

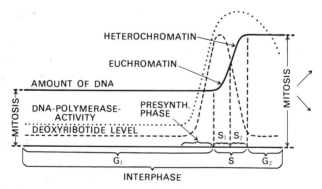

FIG. 9-12. Illustration of the time course of DNA synthesis (S phase) preceded by synthesis of DNA polymerase and DNA precursors within the interphase. During the S period, first the DNA of euchromatin is replicated (S_1), followed by the DNA of heterochromatin (S_2). From Harbers, E., Domagk, G. F., and Müller, W. 1968. *Introduction to Nucleic Acids.* Reinhold Book Corp. New York. p. 182. Fig. 87.

(The fascinating story of the discovery of the DNA double helix has now been related. The discovery of the Watson-Crick double helix represents an elegant case of scientific sleuthing in the literature, thinking, and master-model building.) In semiconservative replication, the process proceeds by separation of the two chains without breakage, and each polynucleotide chain serves as a template for the formation of a complementary new chain. The new DNA molecule would then be composed of one parent chain and complementary new chain. The usual mode of DNA replication favors the semiconservative scheme based on studies from chloroplasts, bacterial cells and mammalian somatic cells. The third possibility of replication is the nonconservative scheme. Here parental chains are broken so that nucleotide bases are distributed in both strands of a new molecule.

Studies concerned with meiosis have indicated that DNA replication occurs during premeiotic interphase, and that DNA replication is semiconservative. Meiosis is a process of cell division in which gametes are formed. This process of gametogenesis is accomplished by two successive nuclear divisions, but it is important to note that chromosomes replicate only in the first karyokinesis. As a result the number of chromosomes (2 N where N = 1 set) initially present in the predestined somatic cell at the onset of meiosis is reduced to the chromosome number (N) by being distributed to four gametocytes at the end of the second meiotic division. When gametes fuse in fertilization a haploid set (N) of chromosomes from each gamete combine to restore the diploid number (2 N) in the zygote. In Figure 9-13 is shown a sequence of stages in the life cycle of a sexually reproducing animal correlated with changes in the amount of DNA per cell. From the figure it can be seen that DNA replicates only once during a meiotic cycle.

As in the process of mitosis, meiotic cell division is a continuous process where time intervals show characteristic patterns. Thus, it is convenient to name the stages as markers in the meiotic process. The four stages in the first nuclear division are called prophase I, metaphase I, anaphase I, and telophase I. Prophase I is subdivided into five separate substages for descriptive convenience. These substages are leptonema, zygonema, pachynema, diplonema, and diakinesis; they apply only to the first karyokinesis, not to the second karyokinesis. A schematic diagram of meiosis is shown in Figure 9-14. Reference should be made to the figure frequently as an aid in the following discussion.

At the beginning of meiosis, the nucleus of the somatic cell starts to swell. Swelling results from an influx of water coming from the cytoplasm. Nuclear volume increases about threefold. This increase in nu-

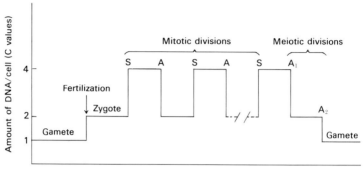

FIG. 9-13. Sequences of stages in the life cycle of a sexually reproducing animal correlated with the changes in the amount of DNA per cell. S = period of synthesis in interphase; A = mitotic anaphase; A_1 = first meiotic anaphase; A_2 = second meiotic anaphase. From Swanson, C. P., Mertz, T., and Young, W. J. 1967. *Cytogenetics*. Prentice-Hall. Englewood Cliffs, N. J. p. 48. Fig. 3.1.

---

clear volume is followed by a gradual modification in nuclear structure. The chromosomes condense by contraction, coiling, and loss of water. The spiraling of the chromosomes resembles spirals of the mitotic prophase, but the chromosomes are longer and thinner. The leptonema chromosomes appear double, *i.e.*, a bivalent chromosome consisting of two chromatids which appose each other closely throughout their entire length. Whether chromosome replication has followed DNA replication during interphase and is just now becoming visible, or whether chromosome replication takes place in leptonema has not been clearly settled. The important point, however, is that the chromosome is composed of two chromatids and that DNA of the nucleus has doubled. (If one views each chromatid as a chromosome, however, the chromosome complement has also doubled to 4 N.) Chromomeres appear along the length of each bivalent chromosome at constant positions.

During zygonema homologous bivalent chromosomes (one derived from each parent) undergo an active pairing or synapsis along their entire length. Chromosomes continue to coil and thicken.

Thus, pachynema exhibits two homologous paired chromosomes each of which is composed of two chromatids. Therefore, two bivalents (or two dyads) comprise the pachynema tetrad pictured in Figure 9-14. In pachynema, the nucleoli are particularly evident. Each one is attached to certain chromosomes at the nucleolar organizer.

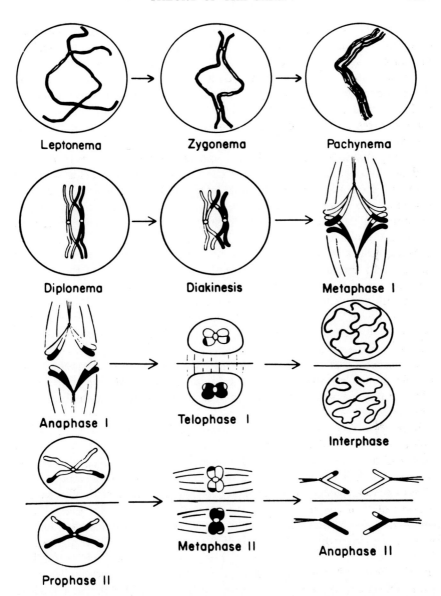

FIG. 9-14. Schematic diagram of meiotic behavior of a bivalent chromosome with localized centromeres. Pairing only is indicated in leptonema whereas chiasmata may begin in pachynema. From Rhoads, M. M. 1961. Meiosis. In *The Cell*. Vol. 3. Brachet, J. and Mirsky, A. E. (editors). Academic Press. New York. p. 13. Fig. 1.

Diplonema is recognized by a separation of the paired homologous chromosomes, but one chromatid of one bivalent chromosome remains in contact with a chromatid of the other bivalent chromosome. The point of contact between chromatids is called a chiasma. Chiasmata are the points of genetic cross-over. It is of interest that in the human female, oocytes are already formed by the fifth month of prenatal life; these oocytes all arrest in the diplonema, and remain so until ovulation occurs over an approximate period between 12 and 50 years.

During diakinesis, the bivalent chromosomes become more condensed and evenly distributed throughout the nucleus. The nucleolus becomes detached from its associated chromosome and disappears. The nuclear envelope also undergoes dissolution. As it appears, the bivalent chromosomes move toward the equatorial plate to become established on the forming bipolar spindle.

Metaphase I is characterized by the complete formation of a bipolar spindle and the alignment of the bivalent chromosomes on the equatorial plate at the metaphase climax. Each bivalent chromosome now has two undivided centromeres which are attached to the spindle fiber.

In anaphase I, each bivalent chromosome (or dyad) moves toward a pole since two chromomeres of each bivalent chromosome are undivided. This of course results in the poleward movement of a bivalent chromosome composed of two chromatids. As chromosome movement continues the chiasmata lose their adhering influence causing a flaring apart of the two chromatids which are then held together only by two undivided centromeres.

Usually in meiosis at the end of the first karyokinesis a regular telophase I forms, followed by an interphase. In this case the two new cells each will have an N number of chromosomes since the two chromomeres of each bivalent chromosome did not divide. On the other hand, if one views each chromatid as a chromosome then each new cell will have the 2 N number of chromosomes the same as the parent cell at the onset of meiosis. If the former view is held, then the chromosomes have undergone a reduction division from diploid to haploid at the end of the first karyokinesis. If the latter view is considered, however, then reduction division from 2 N to N will occur at the end of the second nuclear division. The more prevalent view seems to be to consider the first nuclear division as the reduction division, and the second nuclear division as a type of regular mitosis since the centromeres divide here.

Each interphase cell now has the same DNA content as that of a somatic cell, and no doubling of DNA takes place between the first and second meiotic divisions. Usually the interphase is of a short duration.

The chromosomes do not become extended as in a typical interphase. A single large nucleolus usually does not form.

The second meiotic division superficially resembles an ordinary mitotic cycle. The stages are prophase II, metaphase II, anaphase II, and telophase II. In prophase II the chromatids shorten some, a bipolar spindle forms to which the centromeres attach, and the nuclear envelope disappears. The centromeric regions become aligned on the equatorial plate at the metaphase climax. Now the centromeres divide and each chromatid (or monad), and more specifically called at this point the chromosome, moves in anaphase II toward the poles. Following this poleward movement telophase II ensues. In telophase II, the cell parts become reestablished and two new cells are formed, the gametes. Each gamete contains an N number of chromosomes and one-half the DNA content of a somatic cell. The first synthesis activity shown by the zygote after fertilization is DNA replication.

C. Ribosomes and Protein Synthesis

In most cells, the synthesis of proteins from amino acids is a continuous anabolic process. The catabolic process whereby proteins are broken down to amino acids is also continuous, but protein synthesis and breakdown occur by different chemical processes. Protein synthesis involves aligning a large number of amino acids, which typically may vary from about 100 to 500 or more, linking them together in a specific sequence by formation of peptide bonds between adjacent amino acids, and the production of the characteristic polypeptide chain forming the complete protein.

The central dogma in protein synthesis has been described as

$$\text{DNA} \xleftarrow{\text{replication}} \text{DNA} \xrightarrow{\text{transcription}} \text{RNA} \xrightarrow{\text{translation}} \text{PROTEIN}$$

in which the arrows indicate the direction of transfer of the genetic material. DNA replicates itself on complementary strands, but DNA is not the direct template on which proteins are constructed. Instead, the DNA contains the basic instructions for the synthesis of the hundreds of different proteins in a cell which are transcribed to another molecule, RNA. In turn, the message in the RNA molecule is translated into a protein molecule.

The scheme of protein synthesis is as follows. The DNA molecule is composed of nucleotide chains. Each nucleotide contains a base side group. The sequence of bases in DNA specify what the sequence is to be of amino acids which will comprise the protein. The DNA bases are (A) adenine, (G) guanine, (T) thymine, and (C) cytosine. Three bases constitute a triplet, or codon, which specifies a particular amino acid. There would be 64 codons if the four bases were taken three at a time in various sequences. These 64 codons are more than enough to code for the 20-odd amino acids found in nature. In fact, some triplets code for more than one amino acid. The genetic message encoded in DNA is transcribed into messenger RNA. The bases found in the single-stranded mRNA molecule are the same as in the DNA molecule except thymine which is replaced by uracil. Direct evidence was obtained by Hall and Spiegelman in 1961 that the mRNA molecule is formed on one DNA template strand. The process is called transcription. The mRNA molecule formed is complementary to DNA in the same sense that the two DNA strands are complementary to each other. A coupling process operates during the transcription in which U couples to A, and G couples to C. During transcription, then the four DNA bases A, G, T, and C give rise to U, C, A, and G, respectively, on mRNA. In fact, it is RNA polymerase which links together the monomeric precursors that are the nucleotide triphosphates, UTP, CTP, ATP, and GTP. The coded information transcribed on the mRNA molecule is now carried from the nucleus to the ribosomes in the cytoplasm.

Once attached to the ribosomes, the message is translated into a protein with the aid, somehow, of ribosomal RNA, and tRNA. The tRNA brings amino acids to the sites on the ribosome where protein synthesis occurs. It has been found that each tRNA is specific for a particular amino acid. The tRNA attachment to an amino acid is catalyzed by an enzyme. The tRNA has an anticodon present which binds to the codon on the mRNA. Thus, several ribosomes of a polyribosome moves along the mRNA molecule reading each codon, acting to direct the placement of each amino acid on the protein chain as it is delivered by the tRNA to the ribosomal site. The amino acid is joined to the chain by a peptide bond formed by enzymes. In keeping with this abbreviated discussion of protein synthesis, the central dogma may be restated

DNA makes RNA makes PROTEIN.

The sequence of these events leading to protein synthesis is unidirectional, and as such, the cellular tRNA, rRNA, and mRNA never serve as templates for replication of DNA. Nuclear DNA also contains cistrons for forming tRNA and rRNA which prescribe their complementary structures as in the case of mRNA. The synthesis of rRNA, however, is known to occur in the nucleolus whereas tRNA and mRNA are synthesized extranucleolarly in the nucleoplasm.

Most of the information known today about protein synthesis has been obtained from cell-free systems. In particular, most of the information has derived from *E. coli*. Ribosomes from bacterial cells are readily isolated by differential ultracentrifugation. They can be got out of the cell in solution free from other cellular components. Their kind and amounts depend on the Mg^{++} concentration used in the isolation medium. In a concentration medium of Mg^{++} less than 0.01 M, *E. coli* ribosomes have a sedimentation constant of 70 S. In lower concentrations of Mg^{++}, a 30 S and a 50 S component is present in solution. And a 30 S and 50 S subunit can combine to give a 70 S complete ribosome as indicated in electron micrographs and sedimentation studies. The 30 S particle is composed of some 20 different proteins and one species of rRNA, the 16 S subunit. About 30 different proteins have been found in the 50 S particle. It contains two species of rRNA, the subunits 5 S and 23 S.

Protein synthesis is carried out on an aggregate of ribosomes varying usually from 3 to 10 ribosomes. This aggregate is called a polyribosome, and it is held together by a strand of mRNA.

Each day the relation between ribosomes and protein synthesis has become, and is becoming, increasingly more complex. Specifically, ribosomes are now known to bind mRNA, amino acyl-tRNA, peptidyl-tRNA, GTP, and at least six different protein factors. The sites for chain initiation, amino acid polymerization, and chain termination are all located on ribosomes. These ribosomal sites are called the peptide tRNA site and the amino acid tRNA site. Each site apparently overlaps the junction of the 30 S and 50 S particles. Some evidence, however, indicates that the peptide tRNA site is more related to the 30 S particle than the 50 S. On the other hand, the amino acid tRNA site seems more related to the 50 S than to the 30 S particle. But in order for chain initiation and chain elongation to occur, activated amino acids must be combined with tRNA which transfers the amino acid to the ribosome site. Amino acids combine with tRNA in the following manner.

In the first reaction,

$$\text{Amino acid} + \text{ATP} \rightleftharpoons \text{amino acyl-AMP-enzyme} + PP_i.$$

Amino acyl synthetases activate amino acids in the presence of Mg^{++} in which the carboxyl group is attached to the phosphate of an adenylic acid group (AMP) by high-energy bonding. Presumably, there is a specific amino acyl synthetase for each amino acid in the cell.

In the second reaction,

amino acyl-AMP-enzyme + tRNA ⇌
amino acyl-tRNA + AMP + enzyme.

The amino acyl-AMP remains tightly bound to the enzyme until the intermediate collides with its specific tRNA molecule. The amino acyl synthetase transfers the amino acid to the terminal adenylic acid residue (at the —CCA end) of the tRNA. All tRNA molecules contain the base sequence (—CCA) at their terminal which holds the amino acid. The amino acid attachment is by its carboxyl group to a 3′-hydroxyl group of the ribose of the terminal adenylic acid residue of the tRNA. When the amino acid bonds to the tRNA, the amino acyl synthetase and the AMP are split off. The amino acyl-tRNA can now transfer its amino acid to the ribosomal site.

The initiation of protein synthesis in *E. coli* involves the formation of a 70 S complex. The *in vivo* pool of ribosomes exists mainly as 30 S and 50 S particles. Three initiating factors bring together the elements of the initiating complex at the peptide tRNA ribosomal site. The initiating complex consists of 30 S-formyl methionyl-tRNA, guanosine triphosphate (GTP), and mRNA. A 50 S particle joins the initiation complex to form a 70 S complex. When the AUG codon is in the front end of mRNA, a modification of the methionine molecule is specified in which a formyl group (CHO) replaces a hydrogen atom in the amino group ($-NH_2$) to form a formyl methionine and prevent the formation of a peptide bond. The formylation of the α-amino group of methionine occurs after the amino acid has been esterified by tRNA.

Formation of the protein chain starts with formyl methionine in the first position of the chain followed by alanine and serine. If the AUG codon is internal in mRNA, regular methionine is synthesized. The codon GUG also codes for formyl methionine if the codon is at the front end of mRNA and initiates protein synthesis, but internally GUG codes for valine.

After formation, the 70 S complex has a high capacity to bind amino acyl-tRNA at the amino acid ribosomal site, as specified by the second codon of mRNA, followed by peptide bonding. The specific triplet anticodon on the tRNA of the amino acyl-tRNA can bind with the codon on the mRNA.

A model for the central dogma of replication, and initial translation, is illustrated in Figure 9-15 showing the 70 S complex formed by the three initiating factors. Very recent evidence indicates that the formyl

THE CENTRAL DOGMA

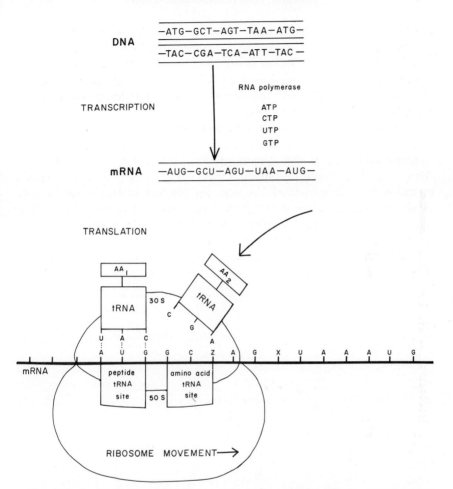

FIG. 9-15. In this figure and in Figures 9-16–9-18 are diagrams illustrating a possible schema of protein synthesis. *Transcription:* The mRNA molecule formed is complementary to one strand of the DNA template. The bases in the mRNA molecule are the same as in DNA with the exception of thymine which is replaced by uracil in mRNA. The genetic message encoded in mRNA is carried from the nucleus to the cytoplasm. The mRNA joins with a 30 S ribosome particle and formyl methionyl-tRNA to form an initiation complex, aided by GTP and initiating factors. A 50 S ribosome particle joins the initiation complex to form the 70 S complex where the translation to a specific protein occurs. *Translation:* The first amino acid (AA_1) of the protein chain locates on the peptide tRNA site. The amino acid tRNA site is activated to receive the incoming second amino acid (AA_2). AA_1 = formylated methionine; AA_2 = alanine; AUG = codon for formyl methionine when in front position on mRNA; GCZ = codon for alanine where Z can be U, C, A, or G; here Z is taken as U for binding with the anticodon CGA; X = either base U or C; here X is U for anticondon binding; UAA = codon for chain termination.

methionyl-tRNA locates initially at the peptide tRNA site. Note in the illustration that the AUG codon for formyl methionine (AA₁) is indicated on the mRNA strand as being bound to the anticodon triplet UAC located on the formyl methionyl-tRNA at the peptide tRNA site. At the amino acid tRNA site, the amino acyl-tRNA is indicated as bringing in the next amino acid (AA₂).

Chain elongation by amino acid polymerization in which the amino acids are joined together by formation of peptide bonds involves two enzymes and GTP in the cytoplasm. Transferase I kicks off the tRNA from formyl methionine and flips the AA₁ to the amino acyl-tRNA bound at the amino acid tRNA site; the enzyme snaps the methionine to the second amino acid (AA₂) by formation of a peptide bond. The peptide bond is formed between the carboxyl group of formyl methionine and the amino group of AA₂. The peptide tRNA site is opened by this action. The model for this first transfer action is shown in Figure 9-16.

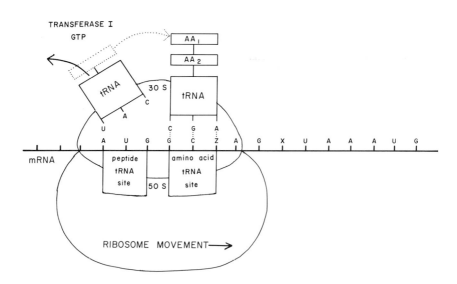

Fig. 9-16. The translation continues from Figure 9-15. Transferase I and GTP moves the formyl methionine (AA₁) from the peptide tRNA site, freeing it in the process, to form a peptide bond with tRNA-AA₂ at the amino acid site. Letters have the same meaning as shown in Figure 9-15.

THEORY OF THE GENE

A model for transferase II (translocase) activity is indicated in Figure 9-17. Here the tRNA-AA_2-AA_1 translocates from the amino acid tRNA site to the peptide tRNA site. Associated with this translocating activity by transferase II is relative movement between the ribosome and mRNA as the next codon triplet moves into a new position. The mRNA is translated in the 5'- to 3'-nucleotide direction. This translocating activity thus opens the amino acid tRNA site for the third amino acid (AA_3) coming in as tRNA-AA_3.

A repetitive action continues for chain elongation. Each amino acid added follows the cycle of amino acyl-tRNA binding, formation of a peptide bond by transferase I, and translocation by transferase II. Figure 9-18 illustrates the continuing schema for chain elongation. Transferase I kicks off the tRNA from the tRNA-AA_2-AA_1 and flips the amino acids from the peptide tRNA site to the amino acid tRNA site. In the

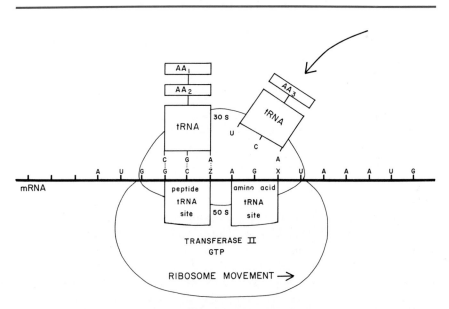

FIG. 9-17. Translation of the message into the protein chain continues (from Fig. 9-16). Transferase II and GTP effect a translocation of the tRNA-AA_2-AA_1 from the amino acid tRNA site to the peptide tRNA site as relative motion occurs between the ribosome and mRNA, and the next codon moves to the free amino acid tRNA site. The incoming tRNA-AA_3 binds at the amino acid tRNA site. The letters in this illustration have the same meaning as those shown in Figures 9-15 and 9-16. AA_3 = serine.

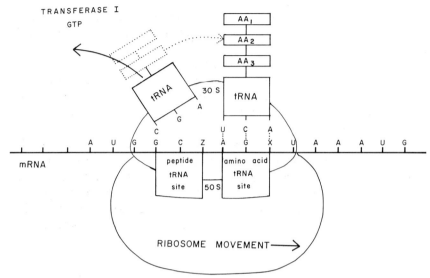

Fig. 9-18. Translation of the message into the protein chain continues (from Fig. 9-17). The transferring action by transferase I and GTP is similar to that shown in Figure 9-16. The tRNA is kicked off the peptide tRNA site, freeing the site, and the AA_2-AA_1 is transferred to the amino acid tRNA site where it forms a peptide bond with AA_3 to form tRNA-AA_3-AA_2-AA_1. Depending on the succeeding codon on the mRNA, the chain can be terminated or the polypeptide chain can continue to grow. Alternate action by transferase I in peptide bonding and transferase II in translocation, aided by GTP, forms the polypeptide chain. Letters have the same meaning as shown in Figures. 9-15–9-17.

---

transfer process a peptide bond forms between the third amino acid (AA_3) and the peptide chain being transferred to form tRNA-AA_3-AA_2-AA_1 at the amino acid tRNA site. The peptide bond is formed between the carboxyl terminal amino acyl residue and the α-amino group of the AA_3 in the tRNA-AA_3. Transferase II activity effects relative movement between the ribosome and mRNA, and the tRNA-AA_3-AA_2-AA_1 translocates to the peptide tRNA site; an amino acyl-tRNA brings in the fourth amino acid to the amino acid tRNA site. As the cycle is repeated, the polypeptide chain is elongated one amino acid at a time.

The chain termination codon at the end of the cistron will cause a dissociation of the 70 S complex. Apparently the termination is triggered when the codon signal reaches the amino acid tRNA site. The nonsense codons, UAA, UAG, and UGA serve as terminator signals. The terminators UAA and UAG have been named ochre and amber, respectively, but have nothing to do with color. Once dissociation occurs, the 30 S particles, either on the mRNA or close to it, are free to recognize the initiator AUG codon of the next cistron.

D. Genetic Code

The genetic code for protein synthesis is contained in the base sequence of DNA. The base sequences from one DNA strand are transferred to a complementary strand of ribosomal RNA, or messenger RNA, or transfer RNA. The mRNA is unstable and is synthesized continuously. It makes up about 5% of all of the RNA in the cell; tRNA and rRNA comprise the rest of the cell RNA, both of which are stable.

Ribosomal RNA and ribosomal proteins combine to form ribosomes, but little information is available concerning rRNA. Transfer RNA's carry specifically activated amino acids to sites on the ribosome for protein synthesis. Messenger RNA carries the transcribed message to the ribosome where, attached to the mRNA, the ribosome translates the message encoded in the mRNA to facilitate protein synthesis. The anticodon of the tRNA appears to bind with the codon consisting of a trinucleotide in the mRNA. In more than a dozen tRNA's, it has been determined that each molecule varies from about 75 to 85 nucleotides. The primary sequence of an alanine tRNA molecule is shown in Figure 9-19. The sequence can be arranged in a hypothetical clover-leaf model which may be the configuration of tRNA. An arrangement of this kind satisfies base-pairing requirements. At one end of the tRNA model is the coupling sequence (—CCA) and at the other end of the molecule is the presumptive anticodon (—IGC). The region G-T-ψ-C-G may be common to all transfer RNA molecules (ψ = ribosyluracil = pseudouridine; I = ionsine).

The genetic code is a code for amino acids. Specifically, it is concerned with what codons specify what amino acids. Or to view it in another sense, what is the code built on the four bases in DNA which is transcribed to mRNA that in turn dictates the sequence of amino acids in a forming polypeptide chain? The answer has obtained mostly from cytochemical studies in cell-free systems.

Cell-free systems have been the instruments for studying key reactions related to protein synthesis. As a result, a significant chronology

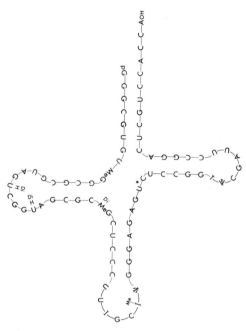

FIG. 9-19. Upper, base sequence in the structure of an alanine tRNA. Lower, probable conformation of the alanine tRNA. From Holley, R. W. et al. 1965. Structure of a ribonucleic acid. Science 147:1464. Fig. 2.

---

developed between the genetic code and cell-free systems. Studies in 1941 by Claude on homogenized liver cells indicated a fraction which contained submicroscopic granules with a high RNA content. He referred in 1943 to the submicroscopic particles as microsomes. *In vivo* experiments in 1950 by Hultin demonstrated that labeled glycine was incorporated in the liver microsome fraction. In 1952 Siekevitz demonstrated labeled alanine incorporation *in vitro* in the RNA microsome fraction of liver establishing cell-free system studies in protein synthesis. By 1952 Porter had described the endoplasmic reticulum and Palade the ribosome in 1953, respectively, from electron micrographs. In 1956 Palade and Siekevitz identified the submicroscopic particles, the microsomes, in the high RNA liver fractions as being composed of endoplasmic reticulum and ribosomes.

It was also in 1956 when Hoagland's laboratory reported that separate enzymes catalyzed the activation of different amino acids. The next year Hoagland's laboratory also reported tRNA, and that it combined with amino acids before protein synthesis occurred. Then in the year following, Tissières and Watson isolated, cytochemically, ribosomes from *E. coli*. They identified the 70 S ribosome unit and the two 30 S and 50 S subunits. In 1960, Huxley and Zubay used electron microscopy to verify morphologically that the 70 S ribosome appeared to be a two-part unit; it was clefted into a smalller cap, the 30 S particle, which fitted onto a larger dome, the 50 S particle. And in 1961, Jacob and Monad, as well as Crick, postulated the presence and function of mRNA in protein synthesis which was demonstrated by Brenner, Jacob, and Meselson in the same year.

As early as 1954, the year following the description of the Watson-Crick double helix for DNA, Gamow suggested that a minimum code would be three nucleotides for one amino acid which would give $4 \times 4 \times 4 = 64$ possible triplets. Only two nucleotides would give $2 \times 2 \times 2 = 16$ possibilities which would be too few for the 20 amino acids. Experimental evidence for the code continued to accumulate until finally in 1961 Nirenberg and Matthaei reported their stable template system for cell-free protein synthesis. They prepared a cell-free system from *E. coli* containing washed ribosomes, tRNA, and a supernatant fraction. When only one type of template, uridylic acid, was added to the system, it was found that the triplet UUU coded for phenylalanine!

Following this direct experimental analysis, polyribonucleotides containing different base compositions were used as messengers under similar conditions. Information was obtained in this manner about the composition of the codons and their correlation with coded amino acids in the laboratories of Nirenberg, Ochoa, and Khorana. But the sequence of bases within the triplets were yet to be determined. Then in 1964 Nirenberg and Leder reported that mRNA dictates the binding of specific amino acyl tRNA's to the ribosome independent of protein synthesis. In a cell-free system containing ribosomes, Mg^{++}, amino acyl tRNA's, and poly-U, ^{14}C-labeled phenyl alanyl-tRNA was found bound specifically and irreversibly to the ribosomes. This effect was found by filtering the ribosomes from the soluble amino acyl-tRNA's at the end of the reaction, and measuring their radioactivity. This direct experimental evidence established the concept of the triplet code.

The arrangement of the triplets in the template are recognized to be arranged sequentially as

thereby permitting 64 different combinations (4^3).

A degeneracy was found in the code by Bernfield and Nirenberg in 1965. Degeneracy means that there is more than one codon, or triplet, for an amino acid. For example, the code triplet for phenylalanine is not only UUU, but also UUC. In general, the various codons differ only in the last base. Exceptions, however, are arginine and serine which have two groups of triplets differing either in the first, or in the first and second bases. All amino acids are coded for by more than one triplet except methionine and tryptophan. A pattern of degeneracy in which C = U for the 3′-nucleotide has also been found in which A = G in the 3′-nucleotide. Although the code is read in three nucleotides, specific information seems to be only in the first and second nucleotide. Crick has put forth a speculation called the wobble hypothesis to explain how the third base in the codon may vary. As an example, inosine in the anticodon of tRNA may pair with U, C, or A. Inosine closely resembles G and would be expected to pair with C. Thus, in anticodon-codon pairing the third base in the codon would vary, or wobble. The 64 codons presently known in mRNA corresponding to their amino acids are shown in Table 9.1. Two nonsense codons, UAG and UAA, are shown in the table which apparently are chain terminators. By 1968 Caskey and other workers in Nirenberg's laboratory reported that UGA is also a terminator codon. Both AUG and GUG are chain initiators when in the first position on the mRNA strand.

Although some cells differ in specificity of RNA codon translation, many studies have indicated that the genetic code may be universal. Marshall, Caskey, and Nirenberg in 1967 showed that the code is essentially universal, but tRNA may differ from one organism to another in relative response to some codons. They found that bacterial, amphibian, and mammalian amino acyl-tRNA's use almost the same genetic language as indicated in Table 9-2. In the table are listed 50 nucleotide sequences of RNA codons recognized by amino acyl-tRNA from *E. coli*, and *Xenopus laevis* and guinea pig livers. The sequences were determined by stimulating the binding of amino acyl-tRNA to *E. coli* ribosomes with trinucleotide codons.

TABLE 9-1
mRNA Codons

1st Base	2nd Base				3rd Base
	U	C	A	G	
U	<u>Phe</u>	<u>Ser</u>	<u>Tyr</u>	<u>Cys</u>	U
	<u>Phe</u>	Ser	<u>Tyr</u>	Cys	C
	<u>Leu</u>	<u>Ser</u>	(CT)	(CT)	A
	<u>Leu</u>	Ser	(CT)	Try	G
C	<u>Leu</u>	Pro	His	Arg	U
	<u>Leu</u>	<u>Pro</u>	His	Arg	C
	<u>Leu</u>	Pro	Gln	Arg	A
	Leu	Pro	Gln	Arg	G
A	Ileu	<u>Thr</u>	Asn	<u>Ser</u>	U
	<u>Ileu</u>	Thr	Asn	Ser	C
	<u>Ileu</u>	<u>Thr</u>	<u>Lys</u>	Arg	A
	<u>Met</u> (CI)	Thr	<u>Lys</u>	Arg	G
G	<u>Val</u>	Ala	Asp	Gly	U
	<u>Val</u>	Ala	<u>Asp</u>	Gly	C
	<u>Val</u>	Ala	<u>Glu</u>	Gly	A
	<u>Val</u>	Ala	<u>Glu</u>	Gly	G

The first base is the initial nucleotide of a triplet bearing a 5'-OH or 5'-phosphate; the third nucleotide of the triplet bears a 3'-phosphate connecting to the next triplet. (CI) is the chain-initiating codon for the NH_2-end of a peptide chain. (CT) is the chain-terminating codon for the COOH-end of a peptide chain. The key is as follows: the assignments *not underlined* are on the basis of binding experiments only; *doubly underlined* assignments are from binding data and have been confirmed by cell-free polypeptide synthesis with completely defined polymers; *singly underlined* assignments are deduced from incorporation experiments with defined polymers but have essentially no binding. Modified from Morgan, A. R., Wells, R. D., and H. G. Khorana. 1966. Studies on polynucleotides. Proc. Nat. Acad. Sci. USA *56:* 1905. Table 7 and from White, A., Handler, P., and Smith, E. L. 1968. *Principles of Biochemistry.* 4th edition. McGraw-Hill. New York. p. 675. Table 29.3.

No doubt but that the genetic code is at least 3 billion years old. Fossil records of bacteria are known to be at least 3.1 billion years old. The code was probably fixed by the time complex bacteria evolved. Vertebrates appeared about 510 million years ago, and amphibians about 355 million years ago. Mammals arose about 181 million years ago. Thus, the universality of the code could be related to organismal evolution.

TABLE 9-2
Nucleotide Sequences of RNA Codons

1st Base	2nd Base				3rd Base
	U	C	A	G	
U	PHE	SER	TYR	Cys	U
	PHE	SER	TYR	Cys	C
	(leu, phe?)	SER	term?	[Cys]	A
	leu, [F-MET]	[SER]	term?	TRP	G
C	leu	PRO	HIS	ARG	U
	leu	PRO	HIS	ARG	C
	leu	PRO	gln	ARG	A
	leu	PRO	gln	[ARG]	G
A	ILE	THR	asn	[SER]	U
	ILE	THR	asn	[SER]	C
	[ILE]	THR	LYS	[ARG]	A
	MET, [F-MET?]	THR	[LYS]	[ARG]	G
G	VAL	ALA	ASP	GLY	U
	VAL	ALA	ASP	GLY	C
	VAL	ALA	GLU	gly	A
	VAL [F-MET?]	[ALA]	GLU	gly	G

Nucleotide sequences of RNA codons recognized by AA-tRNA from bacteria, amphiban liver, and mammalian liver were determined by stimulating, with trinucleotide codons, the binding of AA-tRNA to *Escherichia coli* ribosomes. The key is as follows: *Boxed areas*, relative response of AA-tRNA from one organism to degenerate trinucleotides differs from that of another organism; *capital letters*, AA-tRNA from guinea pig liver, *Xenopus laevis* liver, and *E. coli* assayed with trinucleotides; *lower case letters*, AA-tRNA only from *E. coli* assayed with the exception of Cys-codons which were assayed with guinea pig liver and *E. coli* AA-tRNA. From Marshall, R. E., Caskey, C. T., and Nirenberg, M. 1967. Fine structure of RNA codewords recognized by bacterial, amphibian, and mammalian transfer RNA. Science 155: 824. Table 3.

E. Enzyme Regulation

The general pathway in which genetic information in the DNA molecule is replicated and translated into amino acid sequences of protein is better understood than the mechanisms regulating gene expression. Insight into genetic regulatory mechanisms in the synthesis of protein has derived mostly from studies in microorganisms concerned with induction and repression experiments. For example, enzyme synthesis can be induced in cells. If yeast cells are grown for several generations in a medium containing lactose, fermentation starts early. Within an hour a high concentration of lactase begins to hydrolyze lactose to glucose and galactose. But yeast cells grown in a medium not containing lactose are devoid of the enzyme lactase. If these same cells are grown in a lactose medium, no fermentation occurs initially. After about 14 hours, however, the cells begin to ferment the sugar. The presence of the lactose induces the cells to form lactase. Lactase, therefore, is said to be an inducible enzyme.

The induction effect by inducer substrates has been particularly investigated in the lactose system of *E. coli*. Not only lactose, but other β-galactosides are metabolized in *E. coli* cells by β-galactosidase. A formation of this inducible enzyme represents a *de novo* synthesis of the protein. Induction begins rapidly after the β-galactoside inducer is added to the chemically defined medium. At 37°C β-galactosidase synthesis occurs within 3 minutes. When the inducer is removed formation of the enzyme ceases in about 5 minutes. Some substrates are good inducers, others are not. Extensive studies have shown that the genetic information for enzyme synthesis is already present, and adding an inducer only increases the production of the enzyme.

Enzyme synthesis can be repressed as well as induced. One way repression inhibits synthesis of protein is by negative feedback. In a negative feedback (end-product) system the presence of the end-product inhibits one of its own precursors by inhibiting the action of the first enzyme as in the pathway

$$A \xrightarrow{\text{a enzyme}} B \xrightarrow{\text{b enzyme}} C \xrightarrow{\text{c enzyme}} \text{End-Product}$$
$$\text{INHIBITION}$$

In another example, when *E. coli* cells are grown in an abundance of tryptophan, the cells stop producing the amino acid by inhibiting the enzyme tryptophan synthetase.

An analysis of the genetic basis for induction and repression was presented by Jacob and Monod in 1961. Their postulate contained the view that both induction and repression was primarily negative in operation by inhibiting, or releasing an inhibition of, protein synthesis. Jacob and Monod proposed that the regulation of protein synthesis resulted from an interaction of three different kinds of genes. Two of these were structural and operator genes; they made up the operon. The third type of gene was called a regulator gene. Their concept of the operon has been a most useful one in the study of enzyme regulation.

The operon is the unit of genetic transcription. The DNA structural genes determine the amino acid sequence of a protein by transcription of the nucleotide sequence via mRNA. The transcription of the nucleotide sequence by structural genes is controlled by an operator gene. It is on an adjacent site of the DNA molecule and may control several structural genes. A diagram of an operon unit and its relation to protein synthesis is illustrated in Figure 9-20 in relation to both transcription and translation.

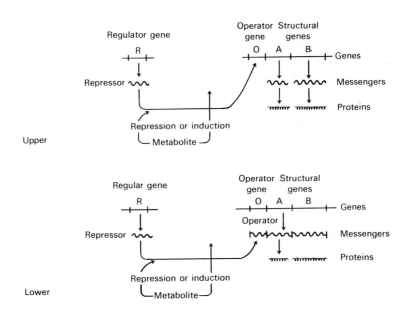

Fig. 9-20. An operon unit is illustrated at the levels of transcription and translation in which protein synthesis is regulated. The operon consists of an operator and structural genes. Upper, the regulator gene produces a repressor molecule which acts on the operator gene to repress transcription, or induce it. If a repressor combines with a co-repressor metabolite, the repressor is activated to affinity for the operator gene to prevent mRNA transcription. In the induction of enzyme synthesis, a repressor combines with an inducer metabolite which renders the repressor to inactivation so that it has no effect on the operator gene, and the operon functions to transcribe mRNA. Lower, at the level of translation, the repressor acts to prevent the ribosome-mRNA complex to function for synthesis of a specific protein. From Jacob, F. and Monad, J. 1961. Genetic regulatory mechanisms in the synthesis of proteins. J. Molec. Biol. 3:314. Fig. 6.

The regulator gene synthesizes a repressor which can block the function of an operator gene. The repressor molecules can interact with small molecules. These small molecules are specific metabolites termed effectors. Effectors are either inducers or co-repressors. In enzyme induction, the repressor interacts with the operator gene and thus blocks the activity of the operon. When the repressor combines with an inducer, however, the combination has no affinity for the operator, and the operon is activated to function. In enzyme repression, the repressor has no affinity for the operator gene. Upon combining with a co-repressor, though, the combined repressor has an affinity for the operator gene thereby blocking the activity of the operon. When the co-repressor is eliminated, the functional activity of the operon is restored.

Precise levels of enzymes in metabolic pathways are maintained by feedback control, induction, and repression. These levels in turn regulate the proper concentrations of substrates. Inducible enzyme systems operate in the degradation of exogenous substrates in sequential catabolic processes. Repressible enzyme systems operate in sequential anabolic processes in which amino acids and nucleotides are synthesized. The inducers are the substrates of the metabolic sequences, and the co-repressors are the products of metabolism.

The synthesis of enzymes in the cell may, or may not, be affected by either inducers or co-repressors as indicated in Figure 9-21. If a regulatory gene mutates so that its specific repressor is not synthesized, or if an operator gene is deleted by a point mutation, an uncontrolled, or constitutive, synthesis of enzymes occurs. In such case the structural genes apparently code for enzymes by some other mechanism.

The repressor molecule seems to be an allosteric protein which binds specifically to the operator region of the DNA blocking its transcription to mRNA. Some evidence exists that there is a promotor gene between the operator and structural genes. The promotor gene may be the initial point of mRNA synthesis. In any event, DNA transcription into mRNA appears to be at the operator end of the operon. It also seems that translation of mRNA into protein begins at the operator end of the message.

Whether interaction of the repressor and operator occurs at the level of DNA or RNA is difficult to decide in all cases. Reference to Figure 9-20 again shows the schema proposed for regulation of protein synthesis at the level of translation. At the translation level, there may be some regulation by tRNA for the histidine operon. Histidyl-tRNA is more effective in repressing the histidine operon than is free histidine. A similar finding has been shown for the valine operon.

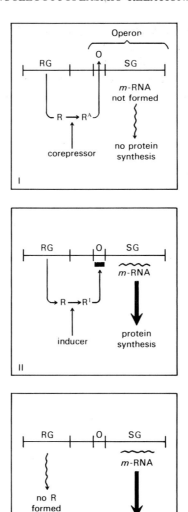

Fig. 9-21. Possible mechanisms for (I) repression, (II) induction, and (III) constitutive enzymes, where RG is not functional. RG is regulatory gene; O, operator; SG, structural gene; R, repressor; R^A, activated repressor; R^I, inactivated repressor. From Conn, E. C. and Stumpf, P. K. 1966. *Outlines of Biochemistry*. 2nd edition. John Wiley & Sons. New York. p. 409. Fig. 21.

Although the genetic regulatory mechanisms are still somewhat obscure, it seems that the concepts of regulation of enzyme synthesis depends on repressors being activated, or inactivated, by inducers or co-repressors. As mentioned earlier, these concepts have obtained primarily in bacterial systems, but some data are accumulating that the concepts apply in systems of both animals and plants. Having a single, defined unit of genetic transcription would facilitate greatly many experiments on regulation and control mechanisms of enzyme synthesis. Such a unit now exists. Late in 1969 Shapiro and his coworkers published the first picture of an isolated gene as shown in Fig. 9-22. They purified chemically the *lac* operon DNA and photographed it using electron microscopy. The promoter-operator region was estimated to have an upper limit of 410 bases.

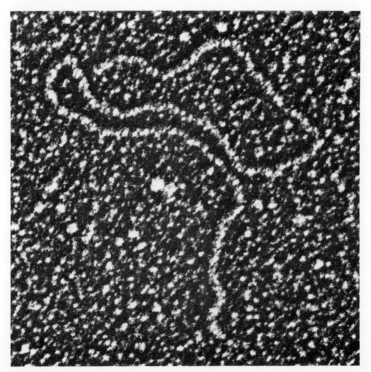

FIG. 9-22. Electron microscope photo of a single pure *lac* duplex. The calculated length of the gene is 1.4 microns. The magnification was not determined. Photograph was taken and kindly furnished by Dr. Lorne MacHattie, Harvard University. From Shapiro, J., MacHattie, L., Eron, L., Ihler, G., Ippen, K., and Beckwith, J. 1969. Isolation of pure *lac* operon DNA. Nature (London). 224:768. Plate 5.

Suggested Reading—Chapter 9

Abuelo, J. G. The human chromosome. J. Cell Biol. *41:*73, 1969.
Allfrey, V. G. Observations on the mechanism and control of protein synthesis in the cell nucleus. In *Functional Biochemistry of Cell Structures.* Lindberg, O. (editor). Pergamon Press. New York. 1961.
Allfrey, V. G. Changes in chromosomal proteins at times of gene activation. Second Sarotoga Conference on Molecular Biology and Pathology, Skidmore College. Sarotoga Springs, New York. August, 1969. (Unpublished).
Ames, B. N. and Martin, R. G. Biochemical aspects of genetics: the operon. Ann. Rev. Biochem. *33:*235, 1964.
Anderson, J. S., Bretscher, M. S., Clark, B. F. C., and Marcker, K. A. A GTP requirement for binding initiator tRNA to ribosomes. Nature (London) *215:*490, 1967.
Barghoorn, E. and Schnopf, J. Microorganisms three billion years old from the Precambrian of South Africa. Science *152:*758, 1967.
Barnicot, N. A. and Huxley, H. E. Electron microscope observations on mitotic chromosomes. Quart. J. Micr. Sci. *106:*197, 1965.
Barr, M. L. The significance of the sex chromatin. Int. Rev. Ctyol. *19:*35, 1966.
Barr, M. L. and Bertram, E. G. A morphological distinction between neurons of the male and female, and the behaviour of the nucleolar satellite during accelerated nucleoprotein synthesis. Nature (London) *163:*676, 1949.
Beermann, W. and Clever, U. Chromosome puffs. Sci. Amer. *210(4):*50, 1964.
Birnstiel, M. The nucleolus in cell metabolism. Ann. Rev. Plant Physiol. *18:*25, 1967.
Brachet, J. La détection histochimique des acides pentosenucléiques. C. R. Soc. Biol. (Paris) *133:*88, 1940.
Brachet, J. Nucleocytoplasmic interactions in unicellular organisms. In *The Cell.* Vol. 2. Brachet, J. and Mirsky, A. E. (editors). Academic Press, New York. 1961.
Brachet, J. Quelques aspects moléculaires de la cytologie et de l'embryologie. Biol. Rev. *43:*1, 1968.
Brenner, S., Barnett, L., Katz, E. R., and Crick, F. H. C. UGA: A third nonsense triplet in the genetic code. Nature (London) *213:*449, 1967.
Brenner, S., Jacob, F., and Meselson, M. An unstable intermediate carrying information from genes to ribosomes for protein synthesis. Nature, (London) *190:*576, 1961.
Briggs, R. and King, T. J. Nucleocytoplasmic interactions in eggs and embryos. In *The Cell.* Vol. 1. Brachet, J. and Mirsky, A. E. (editors). Academic Press. New York. 1959.
Buck, R. C. The central spindle and the cleavage furrow. In *The Cell in Mitosis.* Levine, L. (editor). Academic Press. New York. 1963.
Cairns, J. The form and duplication of DNA. Endeavour *22:*141, 1963.
Cairns, J. The bacterial chromosome and its manner of replication as seen by autoradiography. J. Molec. Biol. *6:*208, 1963.
Cairns, J. The bacterial chromosome. Sci. Amer. *214(1):*36, 1966.
Caspersson, T. and Schultz, J. Ribonucleic acids in both nuclei and cytoplasm and the function of the nucleolus. Proc. Nat. Acad. Sci. USA *26:*507, 1940.
Changeux, J. P. The control of biochemical reactions. Sci. Amer. *212(4):*36, 1965.
Clark, B. F. C. and Marcker, K. A. How proteins start. Sci. Amer. *218(1):*36, 1968.
Claude, A. Particulate components of cytoplasm. Sympos. Quant. Biol. *9:*263, 1941.
Claude, A. The constitution of protoplasm. Science *97:*451, 1943.
Claude, A. and Potter, T. S. Isolation of chromatin threads from the resting nucleus of leukemic cells. J. Exp. Med. *77:*345, 1943.
Cohen, N. R. The control of protein biosynthesis. Biol. Rev. *41:*503, 1966.
Commoner, B. Failure of the Watson-Crick theory as a chemical explanation of inheritance. Nature (London) *220:*334, 1968.

Conn, E. C. and Stumpf, P. K. *Outlines of Biochemistry*. 2nd edition. John Wiley & Sons. New York. 1966.
Costello, D. P. On the orientation of centrioles in dividing cells and its significance. A new contribution to spindle mechanisms. Biol. Bull. *120:*285, 1961.
Crick, F. H. C. Genetic code. Science *139:*461, 1963.
Crick, F. H. C. The genetic code: III. Sci. Amer. *215(4):*55, 1966.
Dales, S. A study of the fine structure of mammalian somatic chromosomes. Exp. Cell. Res. *19:*577, 1960.
Davidson, W. M. and Smith, D. R. A morphological sex difference in the polymorphonuclear neutrophil leucocytes. Brit. Med. J. *1:*6, 1954.
de Reuck, A. V. S. (editor). *Histones*. CIBA Foundation Study Group No. 24. Little, Brown and Co. Boston. 1966.
Droz, B. and LeBlond, C. P. Axonal migration of proteins in the central nervous system and peripheral nerves as shown by radioautography. J. Comp. Neurol. *121:*325, 1963.
DuPraw, E. J. *Cell and Molecular Biology*. Academic Press. New York. 1968.
Eigsti, O. J. and Dustin, P. *Colchicine in Agriculture, Medicine, Biology, and Chemistry*. Iowa State College Press. Ames. 1955.
Feldherr, C. M. The nuclear annuli as pathways for nucleocytoplasmic exchanges. J. Cell Biol. *14:*65, 1962.
Feulgen, R. and Rossenbeck, H. Mikroskopisch-chemischer Nachweis einer Nucleinsäure vom Typus der Thymonucleinsäure und die daruf beruhende elektive Färbung von Zellkernen in mikroskopischen Pröparaten. Z. Physiol. Chem. *135:*203, 1924.
Frenster, J. H. Mechanisms of repression and de-repression within interphase chromatin. In Vitro *1:*78, 1966.
Gahan, P. B. and Chayen, J. Cytoplasmic deoxyribonucleic acid. Int. Rev. Cytol. *18:*223, 1965.
Geiduschek, E. P. and Haselkorn, R. Messenger RNA. Ann. Rev. Biochem. *38:*647, 1969.
Glinos, A. D. Environmental feedback control of cell division. Ann. N. Y. Acad. Sci. *90:*592, 1960.
Goldstein, L. Nucleocytoplasmic relationships. In *Cytology and Cell Physiology*. 3rd ed. Bourne, G. (editor). Academic Press. New York. 1964.
Goldstein, L. and Prescott, D. M. Proteins in nucleocytoplasmic interactions. II. Turnover and changes in nuclear protein distribution with time and growth. J. Cell Biol. *36:*53, 1968.
Gross, P. R. (editor). Second conference on the mechanisms of cell division. Ann. N. Y. Acad. Sci. *90:*345, 1960.
Grundmann, E. *General Cytology*. The Williams & Wilkins Co. Baltimore. 1966.
Gurdon, J. B. and Woodland, H. R. The cytoplasmic control of nuclear activity in animal development. Biol. Rev. *43:*233, 1968.
Hadek, R. The structure of the mammalian egg. Int. Rev. Cytol. *18:*29, 1965.
Hall, B. D. and Spiegelman, S. Sequence complementarity of T2-DNA and T2-specific RNA. Proc. Nat. Acad. Sci. USA *47:*137, 1961.
Harbers, E., Domagk, G. F., and Müller, W. *Introduction to Nucleic Acids*. Reinhold Book Corp. New York. 1968.
Harris, H. *Nucleus and Cytoplasm*. Oxford University Press. London. 1968.
Henneguy, L. F. Sur les rapports des cils vibratiles avec les centrisomes. Arch. Anat. Micr. Morph. Exp. *1:*481, 1897.
Heller, C. G. and Clermont, Y. Spermatogenesis in man: An estimate of its duration. Science *140:*184, 1963.
Hoagland, M. B. Nucleic acids and proteins. Sci. Amer. *201(6):*55, 1959.
Hoagland, M. B., Keller, E. B., and Zamecnik, P. C. Enzymatic carboxyl activation of amino acids. J. Biol. Chem. *218:*345, 1956.

Hoagland, M. B., Zamecnik, P. C., and Stephenson, M. L. Intermediate reactions in protein biosynthesis. Biochim. Biophys. Acta 24:215, 1957.
Holley, R. W. The nucleotide sequence of a nucleic acid. Sci. Amer. 214(2):30, 1966.
Holley, R. W., Apgar, J., Everett, G., Madison, J., Marquisee, M., Merrill, S., Penswick, J., and Zamir, A. Structure of a ribonucleic acid. Science 147:1462, 1965.
Horowitz, N. H. and Metzenberg, R. L. Biochemical aspects of genetics. Ann. Rev. Biochem. 34:527, 1965.
Huberman, J. A. and Attardi, G. Studies of fractional HeLa cell metaphase chromosomes. J. Molec. Biol. 29:487, 1967.
Huettner, A. F. Continuity of the centrioles in *Drosophila melanogaster*. Z. Zellforsch. 19:119, 1933.
Hughes, A. *The Mitotic Cycle: The Cytoplasm and Nucleus During Interphase and Mitosis.* Academic Press. New York. 1952.
Hultin, T. Incorporation of ^{15}N-labeled glycine into liver fractions of newly hatched chicks. Exp. Cell Res. 1:376, 1950.
Hurwitz, J. and August, J. T. The role of DNA in RNA synthesis. Progr. Nucleic Acid Res. 1:59, 1963.
Hurwitz, J. and Furth, J. J. Messenger RNA. Sci. Amer. 206(2):41, 1962.
Huskins, C. L. Nuclear reproduction. Int. Rev. Cytol. 1:9, 1952.
Huxley, H. E. and Zubay, G. Electron microscope observations of the structure of microsomal particles from *Escherichia coli*. J. Molec. Biol. 2:10, 1960.
IUPAC-IUB Combined Commission on Biological Nomenclature. Abbreviations and symbols for chemical names of special interest in biological chemistry. J. Biol. Chem. 241:527, 1966.
Jacob, F. and Monod, J. Genetic regulatory mechanisms in the synthesis of proteins. J. Molec. Biol. 3:318, 1961.
Johns, E. W., Phillips, D. M. P., Simson, P., and Butler, J. A. V. Improved fractionations of arginine-rich histones from calf thymus. Biochem. J. 77:631, 1960.
Kaye, J. S. Changes in the fine structure of nuclei during spermatogenesis. J. Morph. 103:311, 1958.
Khorana, H. G. Polynucleotide synthesis and the genetic code. Fed. Proc. 24:1473, 1965.
Lacy, J. C. and Pruitt, K. M. Origin of the genetic code. Nature (London). 223:799, 1969.
Lafontaine, J. G. and Chouinard, L. A. A correlated light and electron microscope study of the nuclear material during mitosis in *Vicia faba*. J. Cell Biol. 17:167, 1963.
LeBlond, C. P. Time dimension in histology. Amer. J. Anat. 116:1, 1965.
LeBlond, C. P. and Amano, M. Synthetic activity in the nucleolus as compared to that in the rest of the cell. J. Histochem. Cytochem. 10:162, 1962.
Lengyel, P. and Söll, D. Mechanism of protein biosynthesis. Bact. Rev. 33:264, 1969.
Levitt, M. Detailed molecular model for transfer ribonucleic acid. Nature (London). 224:759, 1969.
Levine, L. (editor). *The Cell in Mitosis.* Academic Press. New York. 1963.
Lewis, W. H. The relation of the viscosity changes of protoplasm to ameboid locomotion and cell division. In *Structure of Protoplasm.* Seifriz, W. (editor). Iowa State University Press. Ames. 1942.
Liau, M. C., Hnilica, L. S., and Hurlbert, R. B. Regulation of RNA synthesis in isolated nucleoli by histones and nuclear proteins. Proc. Nat. Acad. Sci. USA 53:626, 1965.
Locke, M. *Role of Chromosomes in Development.* 23rd Symposium on the Study of Growth. Academic Press. New York. 1965.
McElroy, W. D. and Glass, B. (editors). *The Chemical Basis of Heredity.* John Hopkins Press. Baltimore. 1957.
McIntosh, J. R., Hepler, P. K., and Van Wie, D. G. Model for mitosis. Nature (London). 224:659, 1969.
McKusick, V. A. *Human Genetics.* Prentice-Hall. Englewood Cliffs, N. J. 1964.

Makino, W. *An Atlas of the Chromosome Numbers in Animals*. Iowa State College Press. Ames. 1951.
Mans, R. J. Protein synthesis in higher plants. Ann. Rev. Plant Physiol. *18:*127, 1967.
Marshall, R. E., Caskey, C. T., and Nirenberg, M. Fine structure of RNA codewords recognized by bacterial, amphibian, and mammalian transfer RNA. Science *155:*820, 1967.
Mazia, D. Mitosis and the physiology of cell division. In *The Cell*. Vol. 2. Brachet, J. and Mirsky, A. E. (editors). Academic Press. New York. 1961.
Mazia, D. How cells divide. Sci. Amer. *205(3):*100, 1961.
Meselson, M. and Stahl, F. W. The replication of DNA in *Escherichia coli*. Proc. Nat. Acad. Sci. USA *44:*671, 1958.
Miller, O. L., Jr. and Beatty, B. R. Visualization of nucleolar genes. Science *164:*955, 1969.
Mirsky, A. E. and Osawa, S. The interphase nucleus. In *The Cell*. Vol. 2. Brachet, J. and Mirsky, A. E. (editors). Academic Press. New York. 1961.
Mirsky, A. E. and Ris, H. Isolated chromosomes. J. Gen. Physiol. *31:*1, 1948.
Mirsky, A. E. and Ris, H. The chemical composition of isolated chromosomes. J. Gen. Physiol. *31:*7, 1948.
Mitchell, J. S. (editor). *The Cell Nucleus*. Academic Press. New York. 1960.
Monod, J., Changeux, J. P., and Jacob, F. Allosteric proteins and cellular control systems. J. Molec. Biol. *6:*306, 1963.
Moore, K. L. and Barr, M. L. Smears from the oral mucosa in the detection of chromosol sex. Lancet *2:*57, 1955.
Morgan, A. R., Wells, R. D., and Khorana, H. G. Studies on polynucleotides. LIX. Further codon assignments from amino acid incorporations directed by ribopolynucleotides containing repeating trinucleotide sequences. Proc. Nat. Acad. Sci. USA *56:*1899, 1966.
Moses, M. J. Breakdown and reformation of the nuclear envelope at cell division. In *Fourth International Conference on Electron Microscopy*. Vol. 2. p. 230. Springer Verlag. Berlin. 1960.
Moses, M. J. The nucleus and chromosomes. A cytological perspective. In *Cytology and Cell Physiology*. 3rd edition. Bourne, G. H. (editor). Academic Press. New York. 1964.
Needham J. *Biochemistry and Morphogenesis*. Cambridge University Press. London. 1942.
Newton, A. Effect of nonsense mutations on translation of the lactose operon of *Escherichia coli*. Sympos. Quant. Biol. *31:*181, 1966.
Nicklas, R. B. Chromosome micromanipulation. II. Induced reorientation and the experimental control of segregation in mitosis. Chromosoma *21:*17, 1967.
Nirenberg, M. W. The genetic code: II. Sci. Amer. *208(3):*80, 1963.
Nirenberg, M., Caskey, T., Marshall, R., Brimacombe, R., Kellogg, D., Doctor, B., Hatfield, D., Levin, J., Rottman, F., Pestka, S., Wilcox, M., and Anderson, F. The RNA code and protein synthesis. Sympos. Quant. Biol. *31:*11, 1966.
Nirenberg, M. W. and Leder, P. The effect of trinucleotides upon the binding of sRNA to ribosomes. Science *145:*1399, 1964.
Nirenberg, M., Leder, P., Bernfield, M., Brimacomb, R., Trupin, J., Rottman, F., and O'Neal, C. RNA codewords and protein synthesis. VII. On the general nature of the RNA code. Proc. Nat. Acad. Sci. USA *53:*1161, 1965.
Nirenberg, M. W. and Matthaei, J. H. The dependence of cell-free protein synthesis in *E. coli* upon naturally occurring synthetic polynucleotides. Proc. Nat. Acad. Sci. USA *47:*1588, 1961.
Nomura, M. Ribosomes. Sci. Amer. *221(4):*28, 1969.
Oak Ridge Associated Universities Newsletter. ORNL Biologists Photograph Functioning Genes. July 1, 1969. p. 4. 1969.

Ochoa, S. The chemical basis of heredity—the genetic code. Bull. N. Y. Acad. Med. *40:*387, 1964.
Osgood, E., Jenkins, D., Brooks, R., and Lawson, R. Electron micrographic studies of the expanded and uncoiled chromosomes from human leukocytes. Ann. N. Y. Acad. Sci. *113:*717, 1964.
Palade, G. E. and Siekevitz, P. Liver microsomes. J. Biophys. Biochem. Cytol. *2:*171, 1956.
Pelc, S. R. Labelling of DNA and cell division in so called non-dividing tissues. J. Cell Biol. *22:*21, 1964.
Person, C. and Suzuki, D. T. Chromosome structure—a model based on DNA replication. Canad. J. Genet. Cytol. *10:*627, 1968.
Phillips, G. R. Primary structure of transfer RNA. Nature (London). *223:*374, 1969.
Porter, K. R. and Machado, R. D. Studies on the endoplasmic reticulum. IV. Its form and distribution during mitosis in cells of onion root tip. J. Biophys. Biochem. Cytol. *7:*167, 1960.
Prescott, D. M. Relation between cell growth and cell division. Exp. Cell Res. *9:*328, 1955.
Prescott, D. M. Nuclear function and nuclear-cytoplasmic interactions. Ann. Rev. Physiol. *22:*17, 1960.
Prescott, D. M. Cellular sites of RNA synthesis. Progr. Nucleic Acid Res. *3:*35, 1964.
Prescott, D. M. Comments on cell life cycle. Nat. Cancer Inst. Monogr. *14:*55, 1964.
Rhoads, M. M. Meiosis. In *The Cell.* Vol. 3. Brachet, J. and Mirsky, A. E. (editors). Academic Press. New York. 1961.
Rich, A., Warner, J. R., and Goodman, H. M. The structure and function of polyribosomes. Sympos. Quant. Biol. *28:*269, 1963.
Rich, A. Polyribosomes. Sci. Amer. *209(6):*44, 1963.
Richardson, J. P. RNA polymerase and the control of RNA synthesis. Progr. Nucleic Acid Res. *9:*75, 1969.
Robbins, E. and Gonatas, N. K. The ultrastructures of a mammalian cell during the mitotic cycle. J. Cell Biol. *21:*429, 1964.
Roels, H. "Metabolic" DNA: A cytochemical study. Int. Rev. Cytol. *19:*1, 1966.
Shapiro, J., MacHattie, L., Eron, L., Ihler, G., Ippen, K., and Beckwith, J. Isolation of pure *lac* operon DNA. Nature (London). *224:*768, 1969.
Siekevitz, P. Uptake of radioactive alanine *in vitro* into the proteins of rat liver fractions. J. Biol. Chem. *195:*549, 1952.
Siminovitch, L. The chemical basis of heredity in viruses and cells. Canad. Med. Ass. J. *86:*1137, 1962.
Sinsheimer, R. L. The nucleic acids of the bacterial viruses. In *The Nucleic Acids.* Vol. 3. p. 187. Chargaff, E. (editor). Academic Press. New York. 1960.
Sinsheimer, R. L. Single stranded DNA. Sci. Amer. *207(1):*109, 1962.
Sirlin, J. L. The cell nucleus as the centre of RNA synthesis. In *The Biological Basis of Medicine.* Bittar, E. E. and Bittar, N. (editors). Academic Press. New York. 1968.
Sonneborn, T. M. Beyond the gene. Amer. Sci. *37:*33, 1949.
Spiegelman, S. Hybrid nucleic acids. Sci. Amer. *210(5):*48, 1964.
Srivastava, P. N., Adams, C. E., and Hartree, E. F. Enzymatic action of acromal preparations on the rabbit ovum *in vitro.* J. Reprod. Fertil. *10:*61, 1965.
Stanners, C. P. and Till, J. E. DNA synthesis in individual L-strain mouse cells. Biochim. Biophys. Acta *37:*406, 1960.
Stellwagen, R. H. and Cole, R. D. Chromosomal proteins. Ann. Rev. Biochem. *38:*712, 1969.
Stent, G. S. The operon: on its third anniversary. Science. *144:*816, 1964.
Stern, H. Function and reproduction of chromosomes. Physiol. Rev. *42:*271, 1962.
Stockinger, L. and Cirelli, E. Eine bisher unbekannte Art der Zentriolenvermehrung. Z. Zellforsch. *68:*733, 1965.

Stretion, A. O. W., Kaplan. S., and Brenner, S. Nonsense codons. Sympos. Quant. Biol. *31:*173, 1966.
Swanson, C. P., Mertz, T., and Young, W. J. *Cytogenetics*. Prentice-Hall. Englewood Cliffs, N. J. 1967.
Swift, H. S. Molecular morphology of the chromosome. In Vitro *1:*26, 1966.
Symposium. The genetic code. Sympos. Quant. Biol. *31:*762, 1966.
Symposium. Chromosome mechanics at the molecular level. Sponsored by ORNL Biological Division Oak Ridge, Tenn. J. Cell Physiol. *70(Suppl.):* 230 pp., 1967.
Szollosi, D. Centrioles, centriolar satellites and spindle fibers. Anat. Rec. *148:*343, 1964.
Taylor, J. H. (editor). *Molecular Genetics*. Part II. Academic Press. New York. 1967.
Taylor, J. H. Chromosome reproduction. Int. Rev. Cytol. *13:*39, 1962.
Taylor, J. H., Woods, P. S. and Hughes, W. S. The organization and duplication of chromosomes as revealed by autoradiographic studies using tritium-labeled thymidine. Proc. Nat. Acad. Sci. USA *43:*122, 1957.
Tjio, J. H. and Puck, T. T. The somatic chromosomes of man. Proc. Nat. Acad. Sci. USA. *44:*1229, 1958.
Tissières, A. and Watson, J. D. Ribonucleoprotein particles from *Escherichia coli*. Nature (London) *182:*778, 1958.
Tomkins, G. M. and Ames, B. N. The operon concept in bacteria and in higher organisms. Nat. Cancer Inst. Monogr. *27:*221, 1967.
Varricchio. 70S ribosomes are active. Nature (London) *233:*1364, 1969.
Warner, J. R., Rich, A. and Hall, C. E. Electron microscope studies of ribosomal clusters synthesizing hemoglobin. Science *138:*1399, 1962.
Watson, J. D. Involvement of RNA in the synthesis of proteins: The orderly interaction of three classes of RNA controls the assembly of amino acids into proteins. Science *140:*17, 1963.
Watson, J. D. *Molecular Biology of the Gene*. W. A. Benjamin. New York. 1965.
Watson, J. D. *The Double Helix*. Antheneum. New York. 1968.
Watson, J. D. and Crick, F. H. C. A structure for DNA. Nature (London) *171:*737, 1953.
Watson, J. D. and Crick, F. H. C. Genetic implications of the structure of deoxyribonucleic acid. Nature (London) *171:*964, 1953.
Weber, R. Ultrastructural changes in regressing tail muscles of *Xenopus* larvae at metamorphosis. J. Cell Biol. *22:*481, 1964.
Webster, R. E., Engelhardt, D. L. and Zinder, N. D. *In vitro* protein synthesis: Chain initiation. Proc. Nat. Acad. Sci. USA *55:*155, 1966.
Weiss, J. M. The ergastoplasm; its fine structure and relation to protein synthesis as studied with the electron microscope in the pancreas of the Swiss albino mouse. J. Exp. Med. *98:*607, 1953.
Welshons, W. J. Analysis of a gene in Drosophila. Science *150:*1122, 1965.
Went, H. A. The behavior of centrioles and the structure and formation of the achromatic figure. Protoplasmatologia *6:*1, 1966.
White, A., Handler, P. and Smith, E. L. *Principles of Biochemistry*. 4th edition. McGraw-Hill Book Co. New York. 1968.
White, M. J. D. *The Chromosomes*. John Wiley & Sons. New York. 1961.
Wischnitzer, S. The ultrastructure of the nucleus and nucleocytoplasmic relations. Int. Rev. Cytol. *10:*137, 1960.
Wittmann, H. G. and Schuster, H. (editors). *Molecular Genetics*. Springer-Verlag. New York. 1968.
Woese, C. R., Dugre, D. H., Dugre, S. A., Kondo, M. and Saxinger, W. C. On the fundamental nature and evolution of the genetic code. Sympos. Quant. Biol. *31:*723, 1966.
Yanofsky, C. Gene structure and protein structure. Sci. Amer. *210(5):*80, 1967.

Chapter Ten

Cell Radiation Biology

A. PHYSICAL ASPECTS OF RADIATION
1. HISTORICAL
2. IONIZING RADIATIONS
3. MEASUREMENTS OF RADIOACTIVITY
4. THEORIES OF ACTION

B. RADIOSENSITIVITY
1. INFLUENCE OF ENVIRONMENTAL FACTORS
2. CELLULAR CONDITIONS

C. RADIATION EFFECTS ON THE CELL
1. DNA
2. CHROMOSOME STRUCTURE
3. MITOTIC APPARATUS
4. INFLUENCING FACTORS

D. GROSS EFFECTS OF RADIATION
1. VIRUSES
2. MICROORGANISMS
3. PLANTS
4. ANIMALS
5. MAMMALIAN SYSTEMS

A. Physical Aspects of Radiation

1. Historical

Among the many advances in the field of radiation, there were two significant dates of historic importance. One of these dates was November 8, 1895. On that day Röentgen discovered x-rays. Working with a cathode-ray tube enclosed in a black box in a dark room, he observed a shadow of the bones of his hand on a fluorescent screen coated with barium platinocyanide. The other date was December 2, 1942. On a squash court under the stadium of the University of Chicago the first man-made, self-sustaining nuclear chain reaction was started. Fermi gave the order to remove the last foot and one-half of the neutron absorbing cadmium strip in the lattice pile consisting of graphite and uranium to start the chain reaction. Surprisingly, the intensity of the reaction was easily controlled by moving the cadmium strip in and out of the pile. These results proved that not only could nuclear fission occur in a chain reaction, but also that the reaction could be controlled. The act climaxed a race between Nazi Germany and the USA in their quest for the first atom bomb, and Fermi's group won.

Of course, in between these two dates many other significant and historical facts were discovered pertaining to radiation. One of these was Becquerel's discovery of natural radioactivity in 1896. Marie Curie named radioactivity in 1898. Becquerel placed a crystal of uranium salt on a photographic plate, wrapped it in black paper, and left it in a drawer. Later, he discovered that the uranium had darkened the plate which revealed the phenomenon of natural radioactivity. In performing the experiment, Becquerel actually made the first autoradiograph.

Today autoradiography is a most useful tool in studying cellular function, particularly nucleic acid synthesis (see Fig. 9-6). Tritium, mostly as tritiated thymidine, is very useful for making autoradiographs of cells. Tritium emits only beta particles which travel about 1 μ. The experimental technique is not difficult. After subjecting cells, either *in vivo* or *in vitro*, to tritium they are placed on a glass slide. The slide is dipped in an emulsion, dried, and placed in a light-tight box. The slide box is stored in a refrigerator for exposure for a period of time which can vary from days to weeks. Afterwards, the slide is developed and fixed like an ordinary photographic plate. A stain is applied which will penetrate the emulsion to show cellular details. Microscopic examination shows the characteristic black dots on the film above that part of the cell from which the tritium localized to produce the radiation.

The events leading to December 2, 1942, are interesting. They show how Fermi, first in Italy and then in the USA, was able to evaluate atomic research by others, and use it to guide his own thinking about nuclear fission. In 1932 Chadwick in England discovered the neutron. Two discoveries of importance were made in 1934. Curie and Joliot discovered artificial radioactivity. They found that alpha particles produced radioactive forms of boron, magnesium, and aluminum. Following them, many workers proceeded to form various radioactive elements throughout the world. One result from these studies was by Hevesy who first used isotopes as tracer elements in chemical processes. Fermi, then in Italy, took a cue from the Curie-Joliot work, and proceeded to bombard most of the elements with neutrons and announced the discovery of elements beyond uranium in 1934.

From 1934 to 1938, Hahn, Strassmann, and Meitner in Germany repeated Fermi's work. In the summer of 1938, Meitner was forced to flee from Nazi Germany. She went to Sweden. In December 1938, Hahn and Strassmann reported that in bombarding uranium with neutrons they detected the lighter element, barium. Meitner was in communication with Hahn about this event, and she and Frisch, her nephew in Bohr's laboratory in Copenhagen, were indeed convinced that the nucleus had been split. They explained this process of division of a heavy nucleus as fission which released an enormous amount of energy.

In January 1939, Bohr came to the Princeton campus to confer with Einstein about nuclear fission and the possibility of obtaining large amounts of energy by splitting atoms as related to him by Meitner and Frisch. By this time Fermi was at Columbia University, a refugee from the Fascist regime. Later in January, he met with Bohr and Einstein and offered that neutrons might be emitted during uranium fission. Fermi knew that if fission were to be a practical source of energy, a self-sustaining reaction would have to occur. And if fission became a self-perpetuating chain reaction, enormous quantities of energy would be released. Thus an atomic bomb was possible. Finally, Fermi became convinced that military authorities ought to be informed about the bomb possibility. Consequently, he and Szilard, along with Wigner, both of whom were refugee Hungarian physicists, met with Einstein in July 1939. They suggested to Einstein that he make a direct appeal to President Roosevelt. On August 2, 1939, Einstein wrote a letter to the President informing him that atomic power was a distinct possibility, and perhaps the Nazis were farther advanced in this field of research. Einstein suggested that a staff of scientists be set up to explore the practical use of uranium. As a result of this letter, the Manhattan Project got underway at the Columbia campus by February 1940. Later the group transferred its research activity to the University of Chicago which culminated in Fermi's calm announcement under the west stands of Stagg stadium on December 2, 1942: "The chain reaction is self-sustaining."

Germany was indeed pursuing atomic research. In fact two Hamburg professors, Harteck and Groth, wrote a letter on April 24, 1939 to the German War Office stating that nuclear fission "will probably make it possible to produce an explosion many orders of magnitude more powerful than the conventional ones." A German atomic project then got underway, but it ultimately failed. An embittered rivalry began almost at once between two competing Geheimrat groups—the theoretical physicists and the experimental physicists. As a result there was a lack of a unified and centrally coordinated effort. Thus, the project failed to win the support of the Nazi government. However, until 1942, nuclear research in Germany was, at least, equal to that in the USA.

Fermi's success at Chicago elicited a search for uranium deposits, preparation of pure uranium, and construction of huge nuclear reactors. Later, at the Los Alamos, N. M., laboratory established by November 1942, experiments got underway to determine the critical mass of material needed to make a bomb. Bringing two pieces of bomb stuff together about the size of a baseball and measuring the neutron build-up was called "tickling the dragon's tail." Most unfortunately, one young scientist who had done the experiment a number of times had an accident. Somehow the two spheres came together to form a supercritical mass; the counters whined. Unhesitatingly, he threw himself forward and tore the chain-reacting mass apart with his bare hands, and prevented an atomic explosion. But the damage had been done. His body had shielded other scientists watching, but a great burst of neutrons and gamma rays penetrated him. Nine days later he died an unpleasant death from radiation sickness. Following experiments of this type were conducted by remote control.

The first atom bomb was exploded at Alamogardo, N. M., on July 16, 1945. On August 6, 1945, Hiroshima was the first city struck by an atom bomb. About 66,000 Japanese people were killed, 69,000 wounded, and two-thirds of the city's structures destroyed or damaged. Nagasaki was the next city hit by an atom bomb. On August 9, 1945, about 39,000 were killed and 25,000 wounded. Some two-fifths of the city's structures were destroyed or damaged. Thousands of the wounded in both cities died later from radiation sickness. Thus, the war initiated by Japan on December 7, 1941, with its attack on Pearl Harbor was terminated on August 14, 1945, when the Japanese accepted the Potsdam Ultimatum. In 1947 the Atomic Bomb Casualty Commission was established in Hiroshima to conduct long-range biological and medical research and to investigate radiation effects in man.

2. Ionizing Radiations

The atomic theory is not new. In the 5th century, B.C., Democritus viewed that all matter was composed of small particles he called atoms. But it was Dalton in England who started the modern atomic theory in 1803. He stated that all matter is composed of different elemental substances, and that identical atoms comprise these elements. One element varied from another only because it contains different atoms. By 1869 the Russian Mendeleyev had arranged the known elements in a periodic table which correlated their atomic weights with their chemical functions. The arrangement was so correct he predicted the discovery of elements missing in the table, and was later proved correct. In 1907, Rutherford in England advanced his nuclear theory which explained the composition of the atomic nucleus. In 1913 Bohr, visiting with Rutherford, adapted the nuclear theory by Rutherford to the quantum theory of Planck in Germany, and theorized an atomic structure which satisfied experimental findings in atomic structure.

In essence, the atom is composed of a positively-charged nucleus and one or more negatively-charged electrons orbiting in an elliptical or circular pathway around the nucleus. Although at least 32 elementary particles and antiparticles have been described in the nucleus, it may be viewed as essentially composed of uncharged neutrons, and protons each of which contain a single, positive charge.

The atomic number is placed as a subscript to the left of the symbol for the element; the atomic number represents the number of protons in the nucleus of the atom. The mass number represents the number of protons and the number of neutrons in atomic nucleus, and the mass number approximates the atomic weight. The mass number is placed to the left of the symbol for an element as a superscript. This form was recently recommended by the Joint Committee on Symbols, Units, and Notations, of the International Union of Pure and Applied Physics and the International Union of Pure and Applied Chemistry. For example, hydrogen, tritium, helium, and uranium are represented by this scheme as

$$^1_1H; \quad ^3_1H; \quad ^4_2He; \quad ^{238}_{92}U.$$

In the example of nuclides shown above, uranium is indicated as having 92 protons and 146 neutrons. The number of protons in an atomic nucleus equals the number of orbital electrons. The term nuclide is used to refer to an atomic form with a specific atomic number and specific atomic mass. A radionuclide will follow the same metabolic path as its stable isotope.

Electrons may be moved from one orbit to another unoccupied orbit. To jump to a higher energy level an electron must absorb energy. When an electron moves to a lower energy level, there is a release of energy equal to the difference in energy between the two levels. The energy is released in the form of electromagnetic radiations.

Atomic radiations are usually referred to as ionizing radiations. Essentially, there are two kinds of ionizing radiation: one kind is that caused by subatomic particles—alpha particles, beta particles, neutrons, protons, and heavy nuclei; the other kind is that produced by electromagnetic vibrations (waves)—of most importance are x-rays and gamma rays.

While in Canada in 1899, Rutherford discovered that uranium gave off one kind of rays which were stopped by a sheet of paper. He also found that uranium emitted another kind of rays which penetrated paper, but was stopped by a few sheets of aluminum. He called the former alpha rays, and the latter beta rays. These are known today as alpha particles and beta particles. Shortly thereafter, Villard in France discovered that uranium also emitted rays which were stopped by no less than several sheets of lead. Consequently, he termed these rays as gamma rays.

Alpha particles are helium nuclei. When a nucleus releases two protons and two neutrons, a loss of an alpha particle is represented. When this happens, a radioactive transformation occurs in which the atom has changed from one element to another; the change is accompanied by a release of particulate radiation. Since alpha particles are heavy and have a double charge, they travel slowly, and this makes their loss of energy high. A positively charged alpha particle loses energy by exciting and ionizing atoms of the substance through which it passes. The alpha particle has a strong attraction for a negatively charged orbiting electron. The electron may be excited, *i.e.*, pulled into a higher energy orbit, or pulled completely away to ionize the atom. The depth of penetration in a cell is less than 100 μ. Alpha particles, therefore, to produce ionization must be almost in direct contact with cells. Alpha particles traveling slowly through a cell, excite and ionize atoms in a cell. After a dissipation of their energy, an alpha particle captures two electrons in its outer orbit and becomes a helium atom.

Beta particles (beta-minus) are electrons emitted with high velocity from the nucleus of a radioactive atom. Beta-positive particles (positrons) have a similar mass to electrons, but a different charge. Both electrons and positrons transmit energy to matter in a similar way. An electron combines with a positron when it comes to rest. When this collison occurs, both particles are annihilated and mass energy is converted into two photons. High energy electrons are produced by both particle accelerators and gamma rays. No matter how they are produced, high-energy electrons have the same interactions with matter. Beta particles primarily interact with orbiting electrons. When this happens, a heavy repulsion exists between the two negative particles.

This repulsive force may excite the atom by forcing the electron to a higher energy orbit, or remove it completely to ionize the atom. If dislodged completely, this secondary electron may have acquired sufficient energy from the beta particle to cause an excitation or ionization in another atom or molecule.

A fast-moving electron may also transfer energy when it approaches an atomic nucleus. The electron is slowed by an electrical interaction, and a loss of energy results. This emission of energy is by an electromagnetic radiation. Protons may also cause ionization. This action is brought about by neutrons. Since neutrons have no charge they can penetrate a nucleus. In a series of elastic collisions with hydrogen atoms, neutrons lose some energy, and are finally slowed down to be captured by an atomic nucleus. Gamma radiation then is emitted by the unstable nucleus.

As mentioned above, electromagnetic radiation causes ionization. The electromagnetic spectrum was shown in Figure 1-1 for the approximate wave lengths of the major types of electromagnetic radiation. These are in the form of gamma rays, x-rays, ultraviolet light, visible light, infrared rays, and radiowaves. Gamma rays and x-rays produce both excitation and ionization of the substance through which they pass. Ultraviolet light transmits its energy almost entirely by excitation, not by ionization. In terms of energy delivered to a cell the most potent of all agents is by action of ionizing radiation. Although ultraviolet light is next most potent in delivered energy, it is some 50 times less effective than ionizing radiation.

Gamma rays and x-rays have wave lengths from 10^{-3} to 10 A. Gamma rays are electromagnetic radiations emitted by atomic nuclei during certain nuclear transformations. They are produced when an unstable nucleus gains stability by releasing energy. X-Rays are produced when any substance, like tungsten, is irradiated by high-energy electrons. An analogy may be made between x-ray production and throwing a rock against a wall. The sound waves emitted by the wall would be analogous to the x-rays. The high speed electrons are analogous to the rock. The electrons in the target substance are deflected from their path by the approaching high-energy electrons in the form of electromagnetic radiation. X-Ray energy, therefore, depends on the kinetic energy of the impinging electrons which produce x-rays of various energies. These x-rays and gamma radiations behave as if they were small bundles of energy traveling close to the speed of light. A bundle of this energy is called a quantum, or photon. The amount of energy carried by the photon is directly proportional to the frequency of the wave form.

Ionizing radiations caused by x-rays and gamma rays transfer their energy to the media primarily by excitation, and formation of ion pairs. When a gamma ray photon passes close to a nucleus it suddenly disappears, and an electron and positron appears in its place. Energy (E) and mass (m) are related to each other according to the Einstein formula $E = mc^2$, where c is the velocity of light. Some energy is used in pair production, the rest is transferred to the medium by excitation and ionization.

3. Measurements of Radioactivity

All living cells in the past, as well as those in the present, have been and are now exposed to background radiation. Background radiation is an ionizing radiation which includes cosmic rays, natural radioactivity, and artificial radioactivity. Cosmic rays are from outer space as first shown by Millikan in 1925. They are composed of about 79% protons, 20% alpha particles, and 1% heavy nuclei. Heavy nuclei are those which have been stripped of their electrons. When the high energy particles of cosmic rays interact with atoms in the atmosphere, electrons, neutrons, gamma rays, and mesons are formed. At sea level, mesons cause about 80% of the total radiation; electrons, the remainder. Cosmic radiation is lowest at the equator, and increases toward the poles. It also increases with latitude. Natural radiation is present in surrounding rock and soil. These contain small amounts of radionuclides. Natural radiation is the most important contributor to background radiation for sea level dwellers. Artificial radiation is usually associated with fallout from atom bomb explosions.

In order to evaluate the extent and effects of ionizing radiation on cells, the dosage must be measured and the pattern of dissipated energy in the cell system must be known. The extent of radiation damage to cells depends on the dose. There are two kinds of radioactive doses. One is exposure dose. The other is absorbed dose. The exposure dose is that amount of radiation to which the cells are exposed. The exposure dose, *i.e.*, the amount of radiation, depends primarily on the properties of the source of radiation, and the distance from the source of radiation. Only the energy from the absorbed dose has a biological effect on cells. The absorbed dose depends on two factors. One factor is the amount of irradiation. The other factor is the physical properties of the cell system irradiated.

Various units of radiation have been used to describe various quantities of radiation. Some of these radiation units in current use are defined in Table 10-1. The unit of exposure dose is the roentgen. It was adopted in 1928 as the official unit of radiation quantity. By definition, the roentgen is restricted to gamma rays and x-rays. Other types of radiation are not defined by it. Because the roentgen is a restrictive term, another unit, termed the rad, was established in 1956. Its use is not restricted like the roentgen. The rad applies to the energy absorbed from any ionizing radiation in any medium. The other units listed in Table 10-1 are self-defined.

There are various methods used in detecting and measuring radiation. An ionization chamber is the most commonly used detector of radiation exposure. No one type of ionization chamber can be used effectively to detect, or to count, all types of radiation. Luminescent dosimeters are particularly useful in tissue implantation. Certain substances normally not luminescent can be made so by exposure to ionizing radiation. For example lithium fluoride powder can be incorporated into Teflon and formed into rods, or discs. If these are inserted into a specimen, then irradiation of the specimen will show the exposure to various cells. Scintillation counters, although not generally used for measuring radiation doses, are used extensively for detecting radiation. For example, this type of counter can be used for the beta particles emitted by tritium. Calorimeters and chemical dosimeters are also useful in detecting and measuring radiation exposure. Photographic film has long been used, and is still widely used for personnel monitoring. It is most useful for visualizing gross localizations of certain radionuclides, such as tritium, as discussed previously in reference to autoradiography. Radiation causes the silver atoms in the film to ionize. These ions aggregate (see Fig. 9-6) development to produce opacity as indicated by blackening.

4. Theories of Action

The biological effects produced in cells by all types of ionizing radiation are qualitatively the same. The physical events causing these biological effects do so by excitating and ionizating atoms. Thus, excitation and ionization are the two major mechanisms whereby energy is transferred in radiation. In excitation, an electron in an atom is raised to a higher energy level. In ionization, an orbital electron is ejected

TABLE 10-1
Radiation Units

R   (Roentgen) is a unit of *exposure* of x- or gamma radiation based on the ionization that these radiations produce in air. An exposure of one roentgen results in 2.584×10^{-4} coulomb per kilogram of air, or 1 esu per cc of air at standard temperature and pressure.

Rad   is a unit of *absorbed dose* for any ionizing radiation. One rad is 100 ergs absorbed per gram of any substance.

Rem   is a unit of *dose equivalent* which is numerically equal to the dose in rads multiplied by appropriate modifying factors such as RBE (or QF) or DF.

RBE   (Relative Biological Effectiveness) is a factor expressing the relative effectiveness of radiations with differing linear-energy-transfer (LET) values, in producing a given biological effect. This unit is now limited to use in radiobiology.

QF   (Quality Factor) is another name for a linear-energy-transfer dependent factor by which absorbed doses are to be multiplied to account for the varying effectiveness of different radiations. This unit is used for purposes of radiation protection and is similar to the RBE unit used in radiobiology.

DF   (Dose Distribution Factor) is a factor expressing the modification of biological effect due to nonuniform distribution of internally deposited radionuclides.

Ci   (Curie) is a unit of *activity* of a radioactive nuclide which is equal to 3.7×10^{10} radioactive disintegrations per second.

From Casarett, A. P. 1968. *Radiation Biology*. Prentice-Hall. Englewood Cliffs, N. J. p. 32. Table 3.1.

from an atom. An atom can receive energy directly from the incident radiation or an atom may, however, receive energy indirectly by transfer from another atom. It is the indirect transfer effect which is important in aqueous solutions in that free radical formation in water occurs.

A free radical is extremely reactive. It is an atom or group of atoms containing an unpaired electron. Usually free radicals are formed in radiation as intermediates between final chemical products and ion pairs. During an ionizing radiation an electron is ejected from a water molecule as

$$H_2O \rightarrow H_2O^+ + e^-.$$

This high energy electron may be picked up by another water molecule as

$$e^- + H_2O \rightarrow H_2O^-.$$

In this way an ion pair, H_2O^+ and H_2O^-, are formed. Each ion then may, in the presence of another water molecule, form a hydrogen ion and a free radical as

$$H_2O^+ \xrightarrow{\bar{c}\ H_2O} H^+ + OH^0$$

or as

$$H_2O^- \xrightarrow{\bar{c}\ H_2O} H^0 + OH^-.$$

The H^+ and OH^- will combine to form water. The H^0 and OH^0 free radicals are very reactive. In fact, many of them can react to form H_2O_2. In cells containing catalase and peroxidases, hydrogen peroxide formation may not be significant. In the absence of such enzymes, hydrogen peroxide formation in cells may be important biologically. Free radicals may also react with oxygen to enhance the effect of radiation. Free radicals may be formed from nearly any cellular component which ionizes to contribute to the indirect effect of radiation.

B. Radiosensitivity

1. Influence of Environmental Factors

Oxygen and temperature are two environmental factors which enhance the radiosensitivity of cells. The oxygen effect is a term generally used for the interaction between radiation and oxygen. Oxygen tension seems to be the critical factor in the oxygen effect. Oxygen tension is the partial pressure of oxygen usually expressed in mm Hg as noted in Chapter 7, section B. At sea level, for example, air pressure is 760 mm Hg. The concentration of oxygen is 20.94%. Therefore, the PO_2, or oxygen tension, at sea level in the atmosphere is $760 \times 0.2094 = 159$ mm Hg assuming isothermal conditions.

In various groups of different organisms it has been shown that oxygen tension has an effect on radiosensitivity. By decreasing the partial pressure of ambient oxygen during irradiation, an organism is made hypoxic. Producing hypoxia in an organism seems the most effective manner in modifying the radiation syndrome. The oxygen tensions in most tissues are approximately 20–40 mm Hg which is similar to the partial pressure of venous blood or lymph. Tissues probably become hypoxic when the PO_2 falls below 20 mm Hg. As shown in Figure 10-1, the oxygen effect occurs between 0 and about 20 mm Hg. In other words, hypoxia provides a degree of radioprotection for cells which seems to be an almost universal effect of oxygen. It may be also seen in Figure 10-1 that a partial pressure of oxygen above about 20 mm Hg has little effect on radiosensitivity. It may be that oxygen can interact with free radicals produced in irradiation, or in chemical reactions. In doing so, the effects of radiation is increased. It may be that oxygen can also draw the free radicals into auto-oxidative chain reactions which damage the cell.

Hypothermia is related to radiosensitivity. Cooling or chilling organisms during irradiation offers a protection against radiation injury. It seems, however, that a reduction of temperature itself does not offer the radiation protection. The protection apparently comes from a resultant reduction in the partial pressure of oxygen in the tissues which in turn relates to the oxygen effect.

2. Cellular Conditions

The radiosensitivity of cells in an organism can be modified by certain biological and chemical factors. Among the biological factors are age, sex, genetic constitution, body weight, health, diurnal variation, stress, and diet. Generally, radiosensitivity decreases as an organism gets older. When old age, however, is attained the organism again becomes more sensitive to radiation. Although the sex difference is small, if present, in most species females are slightly less sensitive to radiation than males. It has been shown in many different species that a relation exists between genetic constitution and radiosensitivity. But the relation is unclear since a specimen within a species may be more resistant to radiation than another specimen within a species. Also, an organism of lesser body weight seems more susceptible to radiation than a specimen of greater body weight. Unhealthy animals are less resistant to radiation than healthy ones. The sensitivity of some animals to radiation is correlation with diurnal variation. Subjecting an animal to an additional stress besides radiation stress itself increases the magnitude of radiosensitivity. Following irradiation, those animals maintained on a well balanced diet survive better than those on a poor diet.

Cellular organelles are receiving more attention today in regard to radiation effects than they have previously. Although little information is available regarding structures such as lysosomes, ribosomes, tRNA, and soluble and bound enzymes some studies relate to mitochondria and the membrane systems of the cell. Acute cell death may occur when radiation injures the membranes including those membranes of the mitochondria. The oxidative phosphorylating system in mitochondria does not appear to be involved in primary radiation damage, however. Changes in structure after irradiation as seen in electron micro-

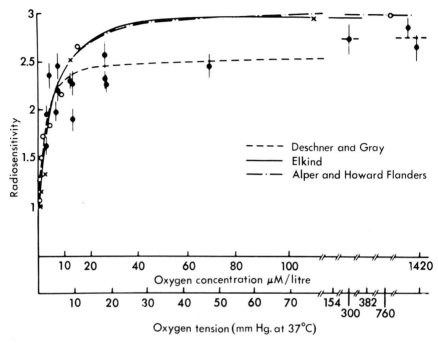

Fig. 10-1. Curves of radiosensitivity as a function of oxygen tension at the time of irradiation. Alper and Howard-Flanders, 1956; Deschner and Gray, 1959; Elkind, Swain, Alescio, Sutton and Moses, 1965. (Note: Elkind's curve drawn by Alper from his data. The three curves redrawn and superimposed by Thomlinson.) (From Churchill-Davidson, I.: The oxygen effect in radiotherapy. Oncologia 20, Suppl. 18–29, 1966.) From Pizzarello, D. J. and Witcofski, R. L. 1967. *Basic Radiation Biology*. Lea & Febiger. Philadelphia. p. 112. Fig. 11.4.

---

graphs primarily indicate mitochondrial swelling and rupture of the cristae. Dilatation of endoplasmic reticulum vesicles is also seen.

C. Radiation Effects on the Cell

1. DNA

The nucleus is more sensitive to ionizing radiation than the cytoplasm of a cell. About 40 years ago, Zirkle demonstrated this fact in plants. He irradiated the nucleus of a spore fern with alpha polonium particles. The nuclear position is eccentric, and its location is marked by a spot on the spore coat. Zirkle also irradiated spore cytoplasm in separate experiments. When the nucleus was irradiated, a lower dose was required to inhibit germinating processes than when the cytoplasm was irradiated. Nuclear sensitivity has also been demonstrated in animals. Wasp eggs of *Habrobracon* have an eccentric nucleus when newly laid. It was shown that an average of one alpha particle directly to the egg nucleus prevented hatching. To produce the same effect when the cytoplasm was bombarded, it took an average of about one million alpha particles.

A variety of biological effects may be caused by ionizing radiation. Among these various effects on the cell it is the current belief DNA is not only an important, but perhaps the principal target for radiation action. In this way nuclear sensitivity could be mediated through DNA via the chromosomes.

Since it is believed that the order and composition of the purine and pyrimidine base pairs in a DNA molecule determine the genetic code, a change in genetic composition would occur if the sequence in base pairing was altered. The purines are less sensitive to radiation than the pyrimidines. Of these, thymine is the most sensitive pyrimidine. In fact, large doses of ionizing radiation destroy thymine as well as uracil and cytosine in aqueous solutions. Ultraviolet radiation produces thymine dimers; recent evidence suggests ionizing radiation may also cause thymine dimerization.

Ionizing radiation causes a depolymerization of DNA. In doing so, it may prevent DNA replication, and halt genetic transcription. Mutagenesis or death, or both, may be the biological consequence of such damage to DNA. Depending on the nature and amount of DNA damage, complete reconstruction of the genome can occur. In some cells there exists a multienzyme mechanism which repairs DNA lesions caused not only by ionizing radiation, but also by ultraviolet irradiation, chemical mutagens, and carcinogens.

Although both ionizing radiation and ultraviolet light act fundamentally in different ways, the cell target of each are the nucleic acids. DNA absorbs most strongly at 260 mμ. The highest rate of mutations induced by ultraviolet light occurs at a wave length corresponding to maximum absorption of nucleic acids. Ultraviolet radiation is primarily absorbed by the bases which also absorb maximally at 260 mμ. Radiation destroys DNA either directly or indirectly. There are various types of direct destruction by radiation to DNA molecules. Hydrogen bonds break between chains. A base may be deleted, or changed in some way as in deamination. Single or double-chain fracture may occur. And crosslinking may occur within the DNA double helix, or with protein molecules. Indirect damage can be from peroxides, or free radicals formed from water or organic compounds by ionizing radiation to act on nucleic acids or their precursors.

Inhibition of DNA synthesis results from a direct effect. If the cell is irradiated in the S phase, there is an inhibition of DNA replication which prevents mitosis and results in cell death. However, if the cell is in mitosis, or at G_1 at the time it is irradiated, a normal replication of DNA occurs in the same cycle but mitosis is delayed, and DNA synthesis is reduced in the following cycle. Therefore, it is apparent that the DNA replication process is more radiosensitive at the S phase than at other periods in the DNA cycle.

2. Chromosome Structure

Radiation has an effect on chromosome structure. The mutagenic action of x-rays on chromosomes was reported by Muller in 1927. He found that x-irradiation increased the incidence of mutations which produced visible effects in the fruit fly *Drosophila*. The changes produced by the x-rays were the same as those caused by ordinary gene mutations. The x-rays could also cause a lethal effect in the flies. Crossover frequency was increased by x-irradiation. The x-rays caused such chromosomal aberrations as inversion, translocation, transmutation, and fragmentation.

Mutation occurs in a chromosome when the DNA present fails to replicate itself exactly. After an inexact duplication by DNA, the mutated gene is reproduced in subsequent DNA duplications. Since the genetic code is believed to reside in the sequence of base pairs of purines and pyrimidines in the DNA molecule, a mutation must be an alteration in the base sequence. The alteration may be in the form of an addition, deletion, or inversion of the bases.

There is a relation between chromosome aberrations and radiation hits. A one-hit aberration is produced by a single photon or single particle which results in a single break in a chromosome. A simple deletion is an example of a one-hit aberration. The percentage of single breaks increases linearly with the radiation dose. Two-hit aberrations result in such aberration as ring formation. Since an exchange of chromosome parts form such a figure, presumably two chromosome breaks are required. The percentage of two-hit chromosome breaks increases more rapidly than a single power of radiation dose. The rate at which x-irradiation is delivered determines the number of two-hit aberrations, but does not appreciably influence the number of one-hit aberrations. But alpha particles produce two-hit aberrations in proportion to radiation dosage.

3. Mitotic Apparatus

Structural aberrations may be produced in chromosomes by radiation at any stage of the cell division cycle, but chromosomes are more sensitive to breakage at interphase. The aberrations are most visible when the chromosomes are in metaphase or anaphase stages of mitosis. Radiation produces three types of chromosome aberrations. One type involves a breakage of the entire chromosome including the two chromatids if they are present as such. If a single break occurs, restitution can occur with complete recovery. If no restitution occurs, the fragment will remain at the equatorial plate and become lost as a terminal deletion. If two breaks occur within a single chromosome, the three fragments may form an intra-arm interchange, an interarm interchange, or an interchange between two chromosomes may result. The second type of aberration appears to come from breaks in chromatids. Radiations can produce one, two, or more breaks in one or both chromatids and various combinations can result when the parts unite. Of course, this is also true for chromosome breaks. The third type of chromosomal aberration arises as a subchromatid break, and involves only a portion of a chromatid. This subchromatid aberration may involve a change in the nucleoproteins in which the surface of the chromosome becomes sticky. In such case, chromosomes stick to each other and fail to separate at anaphase. As shown in Figure 10-2, the type of aberration is dependent on the chromosome stage in a mitotic or meiotic cycle.

An abnormal mitosis may result when a cell in any stage of cell division is exposed to ionizing radiation. The abnormality may come from the effects of radiation on the chromosomes as discussed above, or upon the spindle mechanism. The most common effect is a delay or inhibition of mitosis when a cell is irradiated during interphase. However, the mitotic delay is not visible. Essentially, mitotic delay is a temporary effect and is not related to chromosome breaks or rearrangements, or caused by damage to DNA. The delay in mitosis may be linked with inhibition of DNA synthesis, or the radiation may interfere with the oxidation-reduction of —SH compounds which is known to

326 NUCLEOCYTOPLASMIC RELATIONS

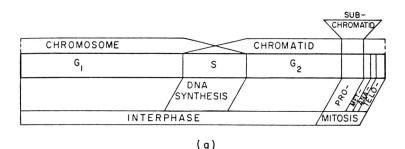

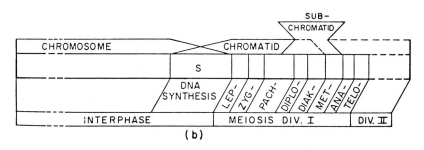

Fig. 10-2. Diagrammatic representations illustrating the relationship between the types of aberrations induced by ionizing radiations in relation to DNA synthesis and stage in development of mitotic (a) and meiotic (b) cycles. From Bartalos, M. and Baramki, T. A. 1967. *Medical Cytogenetics*. The Williams & Wilkins Co. Baltimore. p. 354. Fig. 21.1.

---

occur during mitosis. It may be also that the spindle formation is inhibited. When a cell in a stage of meiosis is exposed to ionizing radiation, it is affected in the same way as a mitotic cell.

4. Influencing Factors

No living cell is completely resistant to radiation. But several factors seem to be related to radiation effects in a cell. A clear correlation is apparent between lethal exposure and chromosome volume. Cells with the largest chromosome volumes are the most radiosensitive. Although there is a variation from one species to another in ploidy, haploid cells are more radiosensitive than diploid cells; polyploid cells are most radioresistant. There are fewer chromosomal aberrations in those cells having extra nucleoli. The position of the centromere seems important in that acrocentric chromosomes have a larger average deletion than metacentric chromosomes. It appears that the nuclear-cytoplasmic volume is an influencing factor. Those cells with a large cytoplasmic volume as compared to their nuclear volume are more radioresistant than those with a small cytoplasmic volume. Cells are also most resistant to radiation when they contain a large number of mitochondria. Younger cells of a particular tissue are more radiosensitive than mature cells. Cells with a rapid mitotic rate are less radioresistant than those cells having a low rate of mitosis and less differentiated cells seem more sensitive to radiation than highly specialized cells, but many exceptions occur.

D. Gross Effects of Radiation

1. Viruses

A virus contains either DNA or RNA surrounded by a protein coat. Ultraviolet radiation inactivates viruses and is generally a one-hit reaction. Not only does ultraviolet light destroy infectivity, but high doses destroy other properties of the virus such as immunizing capacity. The action spectrum for inactivation of viruses by ultraviolet light follow closely the absorption spectrum for nucleic acids. Both ultraviolet and ionizing radiations produce mutations in viruses.

Viruses with DNA are more sensitive to photodynamic inactivation than viruses with RNA. When viruses are exposed to a dye such as methylene blue in the presence of light, inactivation is rapid.

2. Microorganisms

Nearly all microorganisms are radioresistant. For example, the D_{37} dose, *i.e.*, the radiation dose to a cell population with a 37% survival, for human cells is about 120 rads; for sensitive bacteria, D_{37} is about 2000 rads. Some insensitive bacterial strains have a D_{37} of 30,000 rads or more. In the early history of cell radiation biology it was noted that the number of some microorganisms killed was directly proportional to the radiation dose received. This direct relation has been expressed mathematically in terms of a target theory. The target may be a whole cell, a chromosome or some other part of a cell, or a molecule such as DNA. If an event is effected in the target, the measured effect is called a hit. When the number of hits equal the number of targets, the average one-hit per target is considered a one-hit phenomenon. In a population of microorganisms, an effective event can be measured in terms of cell death, or inactivation of cell division.

However, the number of microorganisms killed is not the same at each successive dose increment. The survival curve is an exponential one and a logarithmic plot yields a straight line. In the logarithm form the equation is $\ln S = -kD$ where S equals the surviving cell fraction, D is the radiation dose, and $-k$ is the slope of the survival curve. If enough radiation is given to inactivate all cells, 63% are inactivated, and 37% (D_{37}) are protected and survive. Larger organisms do not respond to an increasing dose in a linear way. When the number of targets increases in an organism the dose at which inactivation is apparent becomes higher. The survival curve thus becomes sigmoid when the number of targets is greater than one. Radiosensitivity in microorganisms is increased in the presence of oxygen. Radiosensitivity also depends on the stage of development of the microorganism when it is irradiated.

3. Plants

Although wide differences exist among plants regarding their radiosensitivity, plants are generally more radioresistant than animals. Plants are particularly radioresistant in a dormant stage, but gradually become more radiosensitive in the early development and growth stages. When growing plants are irradiated, they may be killed, their growth inhibited, or mutations may result.

Gamma radiation has been used extensively to study radiation effects on growing plants. Modifications of roots, stems, leaves, and flowers in regard to both gross morphology and histology have been shown to occur in irradiated plants. Ionizing radiation also increases the incidence of tumor formations in plants.

4. Animals

It is an observed fact that some chemicals act to reduce the effective radiation dose. The aminothiols are perhaps the most effective group of chemical protectors against radiation. The aminothiols are characterized by —SH and —NH$_2$ groups separated by two carbon atoms, as in cysteine. Unfortunately, some of the chemicals for best protection must be administered in near lethal, or highly toxic, amounts. Just how protective a chemical may be is usually expressed by determining the dose reduction factor (DRF). The DRF may be determined as a ratio of $LD_{50\ (30)}$ for protected animals to that of $LD_{50\ (30)}$ for unprotected animals.

The term lethal dose (LD) is used in radiation studies to designate the amount of radiation required to kill all of the individuals in a large group of organisms. In practice, the medial lethal dose is used to indicate the radiation dose which kills 50% of the organisms within 30 days after exposure to $LD_{50\ (30)}$. The approximate $LD_{50\ (30)}$ in rads determined for some organisms are goat and dog, 350; guinea pig, 400; mouse, 550; pig, hamster, chicken and monkey, 600; frog and rat, 700; rabbit, 800; turtle, 1500; and goldfish, 2300. The $LD_{50\ (30)}$ for man has not been determined, but has been estimated to be somewhere between 250–450 rads.

The effect of radiation is not only mutogenic and chemogenic but is also carcinogenic. If the radiation dose is high enough to almost any part of the body, the incidence of cancer increases. But the mechanism of cancer production by radiation seems obscure. Some cancers induced by radiation seem to have a threshold, others do not. And cures of cancers by ionizing radiation seem related to the differential rates of cell population growths which occur in normal and cancer cells. It has been mentioned already that embryonic and developing tissues are generally more radiosensitive than adult tissues. These young cells belong to a population which is in a proliferating, nonsteady state. No cancer cell population is static; it is also proliferating and in a nonsteady state. The tumor cell seems to have an unlimited division potentiality in a cancer series, but the normal cell series differentiates with a finite life span.

A quantitative relation exists between the amount of radiation tolerated by different groups of organisms and their nucleic acid content. A decreasing order of sensitivity has been reported for viruses with single-stranded DNA and RNA, double-stranded DNA viruses, haploid cells of bacteria and yeast, diploid yeast cells, and avian and mammalian cells. In fact, mammals are more sensitive to radiation than birds, reptiles, amphibians, and fishes.

5. Mammalian Systems

When a body receives a lethal dose of radiation, the most sensitive tissues are those associated with bone marrow and lymphoid organs, and the lining of the intestine. The major forms of acute radiation syndrome are hemopoietic, gastrointestinal, central nervous system, and cardiovascular.

Radiation sickness severity and degree of injury by radiation is directly proportional to the dose received. As the radiation dose is increased, the latent period between exposure and onset of symptoms varies inversely with the dose. The acute radiation syndrome varies in severity and is dependent on the number of surviving cells capable of mitotic activity, unless the brain or cardiovascular system has been damaged.

Radiation casualities may be classified into three groups as indicated by Figure 10-3. In the first category recovery is probable. These individuals may be asymptomatic or develop a transient nausea which may last as long as 24 hours. They will have received a total body dose of about 125 rads or less. Individuals who have received a whole body radiation of about 250 rads or more are indicated by the second category. Here recovery is possible. These individuals suffer nausea and vomiting which should subside by 24–48 hours. No diarrhea should occur early, but may develop after a latent period of one to three weeks. In category three are those individuals exposed to a whole body irradiation of 500 to 600 rads and more. Within 1 or 2 hours after exposure nausea and vomiting begins, and progresses. An onset of explosive diarrhea develops within 48 hours. Loss of life begins to occur and recovery is improbable.

The principal threat to survival seems to be depletion of leukocytes and platelets. The lymphocyte is one of the most radiosensitive of body cells. Lymphopenia develops quickly after whole body radiation. It becomes maximal within 72 hours and serves as an early indicator of the severity of radiation damage. The erythrocytes, granulocytes, and platelets are more radioresistant than their precursor stem cells which are inhibited by radiation.

One of the chronic effects of whole body irradiation, or ionizing radiation to hemopoietic tissues, is leukemia. A significant correlation has been shown in the past between leukemia and radiologists, but is rapidly improving. An increased incidence of leukemia has been indicated in other groups such as atomic bomb survivors of Hiroshima and Nagasaki, irradiated individuals with ankylosing spondylitis (pokerback), and *in utero* children exposed in x-ray pelvimetry. However, the dose-effect relation in leukomogenic radiation is not clear, but leukocytic changes are important in diagnosing radiation sickness.

In the gastrointestinal tract, the epithelial cells lining the mucosa

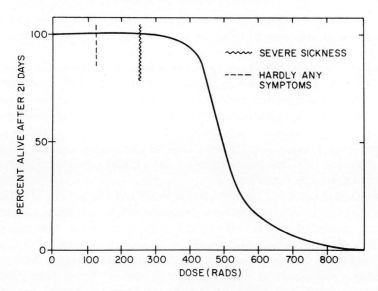

Fig. 10-3. The effect of whole body radiation on something like a mouse or a man. The number dead at 21 days is the measure of effect. Initially there is no action, and few symptoms are observed up to 125 rads. Severe sickness develops at 250 rads but no actual loss of life occurs. Then suddenly and sharply loss of life begins to occur, and then almost no one is left after 600 rads. From Pollard, E. C. 1969. The biological action of ionizing radiation. Amer. Sci. 57:211. Fig. 2.

---

turnover rate is less than for most blood cells. As a consequence radiation damage interferes with mitotic renewal of epithelial cells particularly in the small intestine. Epithelial cells lining the crypts of Lieberkuhn undergo a decrease in mitotic rate within a few minutes after radiation exposure. The villi shrink after the migrating cells slough-off the mucosa and denude the villi; these cells are not replaced. As a result, the denuded mucosa ulcerates, hemorrhages, and becomes an inflamed gangrenous tissue. Death can ensue.

Radiation damages occur to other parts of the body as well. Ionizing radiation affects the skin by injuring the basal cell layer, *i.e.*, the stratum germinitivum of the stratum malpighii, to inhibit mitosis. As a result, erythema may develop. Cells forming hair are also radiosensitive. When damaged, a loss of hair occurs. Although spermatozoa are somewhat radioresistant, spermatogonia are not and are very radiosensitive. Small amounts of radiation deplete spermatogonia quickly;

larger amounts may result in temporary or permanent sterility. Oocytes are relatively radiosensitive. Large doses of radiation injure the brain and nervous system as well as the heart and blood vessels.

Radionuclides as fallouts from nuclear explosions represent a potential hazard to biological systems. These radionuclides can be inhaled, or they can be absorbed by plants to enter the food chain for later ingestion. Of particular importance are carbon-14, iodine-131, and strontium-90; the half-life of these three radionuclides is about 5800 years, 8 days, and 28 years, respectively. The ^{14}C can readily incorporate into hydrocarbons, ^{131}I can concentrate in the thyroid gland, and ^{90}Sr can replace calcium in bone.

But radiation effects on biological systems may be beneficial. This is particularly true in medicine and dentistry where radiotherapy is used both in treatment of diseases and diagnostic procedures. For example, rapidly dividing tumor cells have an increased radiosensitivity, and selective irradiation may be applied to certain tumor tissues. And x-ray examination is an aid in dental prognosis. Iodine-131 has been used to destroy certain types of metastasizing thyroid tumors which concentrate iodine. Strontium-90, as well as cobalt-60 and x-rays, has been used as an external radiation source. Internal localized radiation to tumors, such as those of the eyelid, has been effected by seed implants containing radium-226 or gold-198. In cases like this, however, the lens must be protected so that a high irradiation dose does not cause cataract. Many treatments by radiotherapy utilize radiation dosages of hundreds or even thousands of rads. Since a single exposure could be lethal, lower radiation doses are usually given over an extended period of time which may vary from days to weeks. In any event, if the radiation dose is high enough it can kill any cell, but if radiation is used appropriately on certain cellular systems diseases may be diagnosed, abnormal body function ascertained, and malignant tumor cells treated by radiotherapy.

Suggested Reading—Chapter 10

Andrewes, C. *Viruses of Vertebrates.* 2nd edition. The Williams & Wilkins Co. Baltimore. 1967.

Andrews, J. R. *The Radiobiology of Human Cancer Radiotherapy.* W. B. Saunders Co. Philadelphia. 1968.

Bacq, Z. M. and Alexander, P. *Fundamentals of Radiobiology.* 2nd edition. Pergamon Press. New York. 1961.

Bartalos, M. and Baramki, T. A. *Medical Cytogenetics.* The Williams & Wilkins Co. Baltimore. 1967.

Baserga, R. and Kisieleski, W. E. Autobiographies of cells. Sci. Amer. *209(2):*103, 1963.
Bélanger, L. F. and Leblond, C. P. A method for locating radioactive elements in tissues by covering histological sections with a photographic emulsion. Endocrinology *39:*8, 1946.
Bloom, W. and Ozarslan, S. Electron microscopy of ultraviolet-irradiated parts of chromosomes. Proc. Nat. Acad. Sci. USA *53:*1294, 1965.
Britten, M. J. A., Halnan, K. E., and Meredith, W. J. Radiation cataract—new evidence on radiation dosage. Brit. J. Radiol. *39:*612, 1966.
Carlson, J. G. An analysis of x-ray induced single breaks in neuroblast chromosomes of the grasshopper. (*Chartophaga viridifasciata*). Proc. Nat. Acad. Sci. USA *27:*42, 1941.
Carlson, J. G. X-Ray-induced delay and reversion of selected cells in certain avian and mammalian tissues in culture. Radiat. Res. *37:*15, 1969.
Casarett, A. P. *Radiation Biology*. Prentice-Hall. Englewood Cliffs, N. J. 1968.
Dale, W. M., Meredith, W. J., and Tweedie, M. C. K. Mode of action of ionizing radiations on aqueous solutions. Nature (London) *151:*280, 1943.
Darte, J. M. M. and Little, W. M. Management of the acute radiation syndrome. Canad. Med. Ass. J. *96:*196, 1967.
Davis, G. E. *Radiation and Life*. Iowa State University Press. Ames. 1967.
Deering, R. A. Ultraviolet radiation and nucleic acid. Sci. Amer. *207(6):*135, 1962.
Eisenbud, M. *Environmental Radioactivity*. McGraw-Hill Book Co. New York. 1963.
Elkind, M. M. and Whitmore, G. F. Radiobiology of Cultured Mammalian Cells. Washington. D. C. American Institute of Biological Sciences. 1965.
Errera, M. Effects of radiations on cells. In *The Cell*. Vol. 1. Brachet, J. and Mirsky, A. E. (editors). Academic Press. New York. 1959.
Evans, H. J. Chromosome aberrations induced by ionizing radiations. Int. Rev. Cytol. *13:*221, 1962.
Fermi, E. The development of the first chain reacting pile. Proc. Amer. Phil. Soc. *90:*20, 1946.
Fermi, E. Elementary theory of the chain-reacting pile. Science *105:*27, 1947.
Ficq, A. Autoradiography. In *The Cell*. Vol. 1. Brachet, J. and Mirsky, A. E. (editors). Academic Press. New York. 1959.
Forer, A. Local reduction of spindle fiber birefringence in *Nephrotoma saturalis* (Loew) spermatocytes induced by ultraviolet microbeam irradiation. J. Cell Biol. *25:*95, 1965.
Gaulden, M. E. and Perry, R. P. Influence of the nucleolus on mitosis as revealed by ultraviolet microbeam irradiation. Proc. Nat. Acad. Sci. USA *44:*553, 1958.
Ginoza, W. The effects of ionizing radiation on nucleic acids of bacteriophages and bacterial cells. Ann. Rev. Microbiol. *21:*325, 1967.
Goldfeder, A. Radiosensitivity at the subcellular level. Laval Med. *34:*12, 1963.
Goldfeder, A. and Jelig, J. N. Radiosensitivity and biological properties of tumors. Radiat. Res. *37:*499, 1969.
Gordon, S. A. The effects of ionizing radiation of plants: biochemical and physiological aspects. Quart. Rev. Biol. *32:*3, 1957.
Grosch, D. S. *Biological Effects of Radiation*. Blaisdell Publishing Co. New York. 1965.
Hadorn, E. Transdetermination in cells. Sci. Amer. *219(5):*110, 1968.
Hahn, O. *A Scientific Autobiography*. Charles Scribner's Sons. New York. 1966.
Harbers, E., Domagk, G. F., and Müller, W. *Introduction to Nucleic Acids*. Reinhold Book Corporation. New York. 1968.
Hayflick, L. Human cells and aging. Sci. Amer. *218(3):*32, 1968.
Hollaender, A. (editor). *Radiation Biology*. Vol. 1. (Vol. 2, 1955; Vol. 3, 1956). McGraw-Hill Book Co. New York. 1954.
Hollaender, A. and Stapleton, G. E. Ionizing radiation and the cell. Sci. Amer. *201(3):*94, 1959.

Irving, D. *The German Atomic Bomb.* Simon and Schuster. New York. 1967.

Kanazir, D. T. Radiation-induced alterations in the structure of deoxyribonucleic acid and their biological consequences. Progr. Nucleic Acid Res. *9:*117, 1969.

Kopriwa, B. M. and Leblond, C. P. Improvements in the coating technique of radioautography. J. Histochem. Cytochem. *10:*269, 1962.

Lea, D. E. *Actions of Radiations on Living Cells.* 2nd edition. Cambridge University Press. Cambridge. 1955.

Leblond, C. P. and Messier, B. Renewal of chief cells and goblet cells in the small intestine as shown by radioautography after injection of thymidine-H^3 into mice. Anat. Rec. *132:*247, 1958.

Lesher, S. Radiosensitivity of rapidly dividing cells. Laval Med. *34:*53, 1963.

Loutit, J. F. Ionizing radiation and the whole animal. Sci. Amer. *201(3):*117, 1959.

Luria, S. E. and Darnell, J. E., Jr. *General Virology.* John Wiley & Sons. New York. 1967.

McLaren, A. D. and Shugar, D. *Photochemistry of Proteins and Nucleic Acids.* The Macmillan Co. New York. 1964.

Müller, A. The formation of radicals in nucleic acids, nucleoproteins, and their constituents by ionizing radiations. Progr. Biophys. *17:*99, 1967.

Muller, H. J. Artificial transmutation of the gene. Science *66:*84, 1927.

Muller, H. J. Radiation and human mutation. Sci. Amer. *193(5):*58, 1955.

Perry, R. B. Changes in ultraviolet absorption spectrum of parts of living cells following irradiation with an ultraviolet microbeam. Exp. Cell Res. *12:*546, 1957.

Pettyjohn, D. and Hanawalt, P. Evidence for repair-replication of ultraviolet damaged DNA in bacteria. J. Molec. Biol. *9:*395, 1964.

Pizzarello, D. J. and Witcofski, R. L. *Basic Radiation Biology.* Lea & Febiger. Philadelphia. 1967.

Platzman, R. L. What is ionizing radiation? Sci. Amer. *201(3):*74, 1959.

Pollard, E. C. The biological action of ionizing radiation. Amer. Sci. *57:*206, 1969.

Rogers, A. W. *Techniques of Autoradiography.* Elsevier Publishing Co. New York. 1967.

Rogers, R. W. and Von Borstel, R. C. Alpha-particle bombardment of the *Habrobracon* egg. Radiat. Res. *7:*484, 1957.

Russell, W. L. Genetic hazards of radiation. Proc. Amer. Phil. Soc. *107:*11, 1963.

Sax, K. An analysis of x-ray induced chromosomal aberrations in Tradescantia. Genetics *25:*41, 1940.

Sax, K. The effect of ionizing radiation on chromosomes. Quart. Rev. Biol. *32:*15, 1957.

Shilling, C. W. (editor). *Atomic Energy Encyclopedia in the Life Sciences.* W. B. Saunders Co. Philadelphia. 1964.

Shooter, K. V. The effects of radiations on DNA synthesis and related processes. Progr. Biophys. *17:*289, 1967.

Smyth, H. D. *Atomic Energy for Military Purposes.* Princeton University Press. Princeton, N. J. 1945.

Symposium. *Cellular Radiation Biology.* 18th Annual Symposium on Fundamental Cancer Research, University of Texas, M. D. Anderson Hospital. The Williams & Wilkins Co. Baltimore. 1965.

Turner, L. A. Nuclear fission. Rev. Mod. Phys. *12:*1, 1939.

Upton, A. C. Effects of radiation on man. Ann. Rev. Nucl. Sci. *18:*495, 1968.

Warren, S. Ionizing radiation and medicine. Sci. Amer. *201(3):*165, 1959.

Wichterman, R. Biological effects of radiations on protozoa. Bios *28:*3, 1957.

Wolff, S. *Radiation-induced Chromosome Aberrations.* Columbia University Press. New York. 1963.

Zirkle, R. E. Some effects of alpha radiation upon plant cells. J. Cell Comp. Physiol. *2:*251, 1932.

Zirkle, R. E. Partial-cell irradiation. Advances Biol. Med. Phys. *5:*103, 1957.

INDEX

Acetyl CoA (see Acetyl coenzyme A)
Acetyl coenzyme A, 173–176.
 in fatty acid synthesis, 242, 243
 in β-oxidation of fatty acids, 240, 241
 transport in mitochondria, 243
Acetyl coenzyme A carboxylase, 242
Acids, 116, 117
Acid hydrolases, and lysosomes, 40–42
Aconitase, 174, 175
Acrosomes, and lysosomes, 43
Action potential, 134
Activators, of enzymes, 164
Active transport, 128–131, 143
 of glucose, 227
Acyl carrier protein, 242
Adenosine diphosphate (ADP), formula, 149
Adenosine monophosphate (AMP), formula, 149
Adenosine triphosphatase, 130, 131
Adenosine triphosphate (ATP), 148, 169, 170, 178, 183
 and coupled reactions, 150, 151
 and free energy of hydrolysis, 149, 150
 formation of, 48, 49
 formula, 149
 in photosynthesis, 215–218
ADP (see Adenosine diphosphate)
Adsorption, 79
Aerobic metabolism (see Cellular respiration)
Aldehyde, 81
Alpha particles, discovery of, 315
Amino acid tRNA site, of ribosomes, 287–292
Amino acids
 absorption of, 244–246
 and blood transport, 246
 and membrane transport, 244
 and other foodstuffs, 234, 235
 and peptides, 90–99
 and tRNA, 287, 288–293
 deamination by liver cells, 246–249
 decarboxylation of, 249
 essential for man, 92
 formulas, 93–97
 formula weight, 93–97
 glycine and porphyrin synthesis, 249, 251, 257

 isoelectric point, 93–97, 102
 metabolism of, 249–251
 pK_a values, 93–97, 120
 synthesis of, 246–248
Amino acid systems, dissociation in acid solutions, 120
Amino acyl synthetase, 287, 288
Ammonia, fixation reactions, 246, 247
AMP (see Adenosine monophosphate)
Amphiuma tridactylum, 32
Amylopectin, 87
Amylose, 87
Anabolism, and catabolism, 226, 227
Anapleurotic pathway, in metabolism, 226, 227
Aneuploidy, 266
Antibiotic, 159
Anticodons, 286
Antimetabolite, 157–159
Antipodal molecules, and asymmetrical carbon atoms, 82, 83, 84
Apoenzyme, 163
Atomic number, 314
Atomic theory, 313, 314
ATP (see Adenosine triphosphate)
Autophagy, and lysosomes, 42, 43
Autoradiography, 26, 273, 275, 311, 318
Autotrophic cell, 143

Bacillus brevis, 98
Bacillus subtilis, 98
Background radiation, 317
Bacteriochlorophyll, 213, 214
Balbiani rings, and chromosomes, 270
Barcroft differential manometer, 187, 189, 191
Barcroft-Warburg respirometer, 185, 187, 188, 189
Barr body (see Sex chromatin)
Basal bodies
 DNA in, 54, 57
 origin of, 55–59
Basal lamina (see Basement membrane)
Base composition
 of DNA, 101–103
 of RNA, 101–103
Basement membrane, 62, 126
Base pairing, of nucleic acids, 37, 101–103, 289–293

Bases, 116, 117
Basophilia, 263
Beta particles, discovery of, 315
Bioenergetics, 143-151
Biosphere, 220, 222
Blackman reaction (see Photosynthesis, dark reactions)
Blood oxygen capacity, and vertebrate body weight, 203-209, 211
Blood volume, and vertebrate body weight, 208, 210, 211
Bohr effect, 198
Bond type, 98, 99
Brownian movement, 79
Buffers, 117-121, 196

Calorie, 183, 184
Calorimetry, 144, 183-195
Calvin cycle (see Carbon cycle, in photosynthesis)
Calypte helenae, 200
Carbamyl phosphate, 246, 247
  in ammonia fixation, 252
  in urea formation, 252
Carbohydrates
  and other foodstuffs, 232, 234, 235
  as cell constituents, 81-88
  classification of, 81
  formation and conversions, 211-222
  metabolism of, 227-235
Carbon cycle, in photosynthesis, 218-222
Carbon dioxide, production of, 185
Carbon dioxide transport, in blood, 195, 196
Carbonic anhydrase, 195, 196
Carnitine, 243
Carotenes, 110
Carotenoids, 110, 216
Catabolism, and anabolism, 226, 227
Catalysis, 151
Cathepsins, 246
Cell biology contributors
  in 19th century, 5-7
  Nobel laureates, 5-9
Cell division, 25, 275-285
Cell membrane, 55, 61-67, 123-126
Cellobiose, 70
Cell plate, 68, 69
Cell radiation
  effects on chromosome structure, 324, 325
  effects on DNA, 322-324
  effects on mitotic apparatus, 325, 326
  influencing factors, 326, 327
  ionizing radiations, 313-317
  physical aspects of, 310-320

Cells, measurement of, 19, 20
Cell size, 22
Cell surface, modification of, 67, 68
Cell theory, 4, 5
Cellulose, 70, 88
Cell wall, formation of, 68-70
Central dogma, of protein synthesis, 285, 286, 288-293
Centriole, 55-57, 60, 277, 279
Centromere, 276, 277, 279, 285
Cerebrosides, 109
Chemical energy, 148
Chiasmata, 283, 284
Chlorella, 69, 216
Chlorophyll, 103, 143, 213-217
Chlorophyll unit (see Photosynthesis, unit of)
Chloroplast, 47, 49-53, 211, 212, 215
Chromatin, 23, 25, 26, 30, 31, 99, 263, 264, 272
Chromosomes, 23, 25, 26
  aberrations of, 324-326
  morphology of, 263-275
  replication of, 279, 282, 285
Chylomicron, 235, 236, 237
Cilia, fibrils in, 54, 55
Cistron, 23, 37, 275, 286, 292
Citrate lyase, 243
Citrate synthetase, 174, 175, 243
Citric acid, 177
Citric acid cycle (see Tricarboxylic acid cycle)
Codon, 286, 293, 297
Coenocyte, 276
Coenzymes, 48, 49
Coenzyme I (see Nicotinamide adenine dinucleotide)
Coenzyme II (see Nicotinamide adenine dinucleotide phosphate)
Coenzyme Q (see Ubiquinone)
Cofactors, 162-167
Colloids, 78, 79
Competitive inhibition, 157, 158
Compound lipids, 108
Coulomb, 147
Crabtree effect, in glycolysis, 189
Cristae, in mitochondria, 45, 46, 172
Cysteine, 246
Cytochrome *c*, and imidazole conjugation, 105 106
Cytochrome oxidase, 157, 169, 170, 177
Cytochromes, 169, 170, 173
Cytokinesis, 276, 279

Dark reactions, in photosynthesis, 52, 218-222

Deoxyribonuclease, 272, 273
Deoxyribonucleic acids, 258, 259
  absorption spectrum of, 323
  and chromosomes, 264
  and cistrons, 23
  and Feulgen reaction, 263
  and histones, 264, 266, 268, 269, 271
  and hydrogen bonding, 102, 103
  and RNA synthesis, 37, 270, 273, 274
  base composition, 101–103, 285, 286
  in bacteria, 271, 275
  in basal body, 57
  in centriole, 57
  in chloroplast, 53
  in mitochondria, 47
  lac operon of, 303
  model for, 25, 26
  replication of, 275–285
  transcription to RNA, 285, 286, 289
Desmosome, 61, 67, 68
Dextrins, 88
Dialysis, 79
Differentiation, and RNA synthesis, 272–274
Diffusion, 126–128
  of glucose, 227, 228
Dipeptides, action of digestive enzymes on, 245
Diphosphopyridine nucleotide (*see* Nicotinamide adenine dinucleotide)
Diphosphoribulose carboxylase, 218
Diploid cells, and DNA content, 264, 265, 284
Disaccharides, 85, 86
  excretion of, 227, 228
DNA (*see* Deoxyribonucleic acid)
Donnan equilibrium, 131, 132
DPN (*see* Nicotinamide adenine dinucleotide)

Electrical properties
  muscle cells, 136–138
  nerve cells, 134–136
Electrochemical gradients, 131
Electromagnetic radiations, 314, 316
Electromagnetic spectrum, 13, 316
Electromotive force, 147
Electron microscopy, 10–16
  and magnification, 13, 19, 20
  and microtome, 16, 18
  and resolving power, 10, 13, 18
  procedures in, 14–16
Electron transport, 168, 169, 177–180
  in photosynthesis, 216–218
Embden-Meyerhof pathway (*see* Glycolysis)

Emeiocytosis (*see* Exocytosis)
Endergonic reactions, 145, 146, 148
Endocytosis, 132–134
Endomysium, 136
Endoneurium, 136
Endoplasmic reticulum, 28, 33, 237–239
  and enzymes in, 173
  and Golgi complex, 34
  and non-protein synthesis, 34
  and nuclear envelope, 26, 34
  and protein synthesis, 34
  and ribosomes, 39
  and zymogen, 245
Endosmosis, 127
Endothermy, 144
End-product inhibition, 159, 299
Energy of activation, 151
Enthalpy, 144
Entropy, 78, 144, 146
Enzymes, 151–162
  allosteric site, 159
  and coenzymes, 163–167
  and steady state, 151
  competitive inhibition, 157–159
  end-product inhibition, 159, 299
  kinetics, 153–157
  Lineweaver-Burk calculation, 156–158
  Michaelis-Menten hypothesis, 153–155
  nomenclature, 159–162
  noncompetitive inhibition, 157, 158
  of digestion, 244, 245
  of β-oxidation of fatty acids, 240, 241
  prosthetic groups, 163–167
  regulation of synthesis, 298–303
  substrate complex, 153
  theory of action, 152, 153
Enzyme synthesis, induction and repression of, 299–303
Epinephrine, and glycogenolysis, 229
Ergastoplasm
  and basophilia, 33
  and ribosomes, 39
Erythrocyte ghosts, 63–66
Escherichia coli, 276, 287, 295, 299
Euchromosomes, 25
Euchromatin, 25, 26, 264, 266, 280
  and DNA precursors, 26, 280
Euchromosome, 25
Evolution
  and purine degradations, 254, 256
  of microscopy, 3–5
  theory of, 5
Exergonic reactions, 145, 146, 148, 151
Exocytosis, 134
Exosmosis, 127

Exothermy, 144
External lamina, 63

FAD (see Flavin adenine dinucleotide)
Faraday, 147
Fatty acids, 105, 107
    absorption of, 235–239
    elongation of, 242, 243
    β-oxidation of, 170, 239–242
    saturated, 108
    unsaturated, 108
Fermentation, 185, 230, 231
Ferrodoxin, in photosynthesis, 216–218
Feulgen stain
    for DNA, 23
    in nucleolus, 24
Fibrous lamina, 30, 32
Flagella, fibrils in, 54, 55
Flavin adenine dinucleotide, 164–167, 174, 176, 248, 257
    and electron flow, 179
    and fatty acid degradation, 240
    and fatty acid elongation, 243
Flavin mononucleotide, 164–167
    and electron flow, 179
Flavoproteins, 165, 168, 169, 173, 180
Flavoprotein-cytochrome system, 48, 49
FMN (see Flavin mononucleotide)
Free energy, 144, 146
    of hydrolysis, 149, 150
Free radical, 319, 320
Fucoxanthin, 216
Fumarase, 174, 176

β-Galactosidase, inducible enzyme, 299
Gamma rays
    defined, 316
    discovery of, 315
Gangliosides, 109
Gene, 273, 274, 303
Genetic code, 293–298
Genetic theory, 5
Geometric isomerism, 84
Glucagon, and glycogenolysis, 229
Gluconeogenesis, 229
Glucose
    and insulin, 227
    transport system of, 227, 229
Glucose 6-phosphate, and hexoses, 232
Glucosides, 85
Glutamic acid dehydrogenase, 248
Glutamine, and ammonia conversions, 246–248
Glycine, 91
    and porphyrin synthesis, 249, 251, 257
    and purine ring, 254
    and succinyl CoA, 175
Glycocalyx (see Basement membrane)
Glycogen, 88, 146
Glycogenesis, 229, 234
Glycogenolysis, 229, 234
Glycolipids, 65, 109
Glycolysis, 173, 219, 220, 229–232
    Crabtree effect, 189
    Pasteur effect, 189
Glycosides, 85
Glyoxylic acid pathway, 230, 241
Golgi complex, 34–37
    and zymogen, 245
Gout, 254
Gramicidin, 98
Grana, 49, 51, 52, 212
GTP (see Guanosine triphosphate)
Guanosine triphosphate, 176
    in protein synthesis, 286, 287, 290, 292
    in TCA cycle, 174

Habrobracon, 323
Haploid cells, and DNA content, 264, 265, 285
Heat, 143, 144, 183
Heme
    degradation of, 105, 107
    synthesis of, 249, 251
Hemiacetal, 84
Hemoglobins
    and composition in man, 98–100
    and gas transport, 195–211
    and imidazole conjugation, 105, 106
    electropherogram of basses, 196, 197
Hemoproteins, 103, 105
Henderson-Hasselbalch equation, and pH, 118
Heterochromatin, 25, 26, 264, 266, 269, 280
Heterochromosome, 25
Heterophagy, and lysosomes, 42, 43
Heteropolymer, 80, 89
Heteropyknosis, 25
Heterosaccharides, 87
Heterotrophic cell, 143
Hexokinase, 150, 151, 152
Hexose, 82, 220
Hexose formation, in photosynthesis, 211–222
Hexose monophosphate shunt (see Phosphogluconate oxidative pathway)
High energy compounds, 148–151
Hill reaction (see also Light reactions in photosynthesis), 51

INDEX

Histones 26, 264, 266, 268, 269, 271, 272
  and embryonic development, 272
Holoenzyme, 163
Homopolymer, 80
Homosaccharides, 87
Hormones, 92, 98, 110, 111
  and peptides, 92, 98, 110
  and steroids, 110
Huygens principle, 17
Hydrogen bonds, 76, 77
Hydrogen transport, 164
Hydrostatic pressure, 126, 127
Hypertonic solution, 61, 127
Hypervitaminosis, A, D, and K, 115
Hypothermia, and radiosensitivity, 321
Hypotonic solution, 61, 127
Hypoxia, and radiosensitivity, 320, 321

Imino acids, 91, 96, 97
Induction, in protein synthesis, 299–303
Inhibition, of end-product, 159, 299
Inhibitors
  and antimetabolites, 157–159
  of enzymes, 157–159
Inosinic acid, 254, 255, 256, 257
Insulin, and glucose, 227
Intermediary metabolism, 173, 225
Interphase nucleus, 22, 25, 275, 279, 281, 284, 325
Ionic solution, 78
Ionic strength, and oxyhemoglobin affinity, 199
Isocitric acid lyase, 230, 232
Isocitric dehydrogenase, 174, 175
Isomers, 83, 84
Isoprene, 109, 110
Isothermy, 144
Isotonic solution, 127, 128
Isozymes, 162

Junctional complex, 67, 68

Karyokinesis, 276
Karyotype, of man, 266, 267
α-Ketoglutaric acid dehydrogenase, 174, 175
Ketone, 81
Ketoses, 82
Kinase, 164
Klinefelter's syndrome, 266, 269
Krebs citric acid cycle (see Tricarboxylic acid cycle)

Lactate (see Lactic acid)

Lactate dehydrogenase, 161, 162, 163
  isozymogram of, 162, 163
Lactic acid, 146, 161
Lactose, 86
Leaf, cells of, 211, 212
Lenses
  achromatic, 5
  electromagnetic, 10
  immersion, 5
Light microscopy, 5–10
  diffracted light in, 17
  measurements in, 19
  procedures in, 14, 15
Light quanta, 215
Light reactions, in photosynthesis, 52, 215–218
Light wave, characteristics of, 16. 17
Lineweaver-Burk equation, 156
Lipases, 107
Lipids, 105–111
  and other foodstuffs, 232, 234, 235
  metabolism of, 235–244
  pathways of digestion, 235–239
Lipogenesis, in cytoplasm, 242
Lipoprotein, 111, 130
Living system, essential atoms of, 75, 76, 80, 115, 116
Lolipop model, of cristal particles, 45, 171, 172
Loose junction, of cells, 67, 68
Lysosomes, 39–43
  and acrosomes, 43
  and autophagy, 42, 43

Macromolecules, 79, 80
Malic acid dehydrogenase, 174 176
Malic acid synthetase, 230, 232
Maltose, 86
Mass action, law of, 117
Mass number, 314
Matrix, physiochemical properties of, 75, 76
Meiosis, 281–285
Membrane-flow hypothesis, 35, 36
Membrane potential, development of, 128–130, 132
Membranes
  modifications of, 67, 68
  permeability of, 126
  structure of, 33, 61–67, 123–126
  transport across, 126–138
Metabolic pathways, integration of, 256, 257
Metabolic rate (Q_{O_2}), 184, 185, 201–203, 211

Metabolism, general, 225–227
Metabolite, 158
Metalloporphyrins, 103
Metaphase climax, 276, 278
Micellar membrane, model of, 66
Michaelis-Menten equation, 154
Microrespirometers
  Cartesian diver, 191–193
  Fenn, 190, 191
  Gregg, 191, 192
  Scholander, 192, 194
Microsomes, 294–295
Microtubules, 57, 60, 61
Microvilli, 67, 237–239, 246
Minerals, 115, 116
Mitochondria, 43–49
  and cloudy swelling, 171
  electron transport and oxidative phosphorylation, 44, 47, 48, 169, 177–180
  enzymes of cristae, 47, 173, 177
  enzymes of matrix, 43, 48, 173
  isolation of, 170
  lolipop particles on cristae, 171, 172
  matrix granules in, 45, 46
  membrane permeability of, 171
  outer membranes and endoplasmic reticulum, 171, 173
  pathways in, 168–173
  shuttle systems in, 176, 177
  TCA cycle in, 44, 47, 48, 169, 173–177
Mitosis, 276–281
Molecular solution, 78
Monad, 285
Monomer, 80
Monosaccharides, 82–85
  absorption of, 227, 228
Mucopolysaccharides, 88, 90
Muscle, contraction of, 138
Muscle cell, impulse transmission in, 136–138
Muton, 275
Mycoplasma laidlaivii, 22, 200
Myohematin, 170
Myon (*see* Muscle cell)
Myoneural junction, 138

NAD (*see* Nicotinamide adenine dinucleotide)
Nicotinamide adenine dinucleotide (reduced) dehydrogenase, 165, 177, 180
NADP (*see* Nicotinamide adenine dinucleotide phosphate)
Nicotinamide adenine dinucleotide phosphate (reduced), and NAD, 175
Nerve impulse, 134, 136

Neutral fats, 105, 107
Nicotinamide adenine dinucleotide (*see also* Pyridine nucleotides)
  and fermentation, 230, 231
  and glycolysis, 230, 231
  and NADPH, 175
  and oxygen, 148
  as coenzyme 163, 164, 180
  in fatty acid elongation, 242, 243
  in metabolism, 226
  in β-oxidation of fatty acids, 240, 241
Nicotinamide adenine dinucleotide phosphate (*see also* Pyridine nucleotides)
  as coenzyme, 163, 164
  in fatty acid elongation, 242, 243
  in metabolism, 226
  in photosynthesis, 215–218
Nissl bodies, 33, 38
Nitrogen cycle, 246, 247
Nuclear fission, 312
Nuclein, 258, 263
Nucleohistone (*see* Chromatin)
Nuclear envelope, 26–33
Nucleic acids, metabolism of, 251–259
Nucleolar organizer, 272, 277, 284
Nucleolar proteins, 23
Nucleoli
  extrachromosomal, 272, 273
  vacuoles in, 24
Nucleolonema, 23–25
Nucleolus, 22–25
  and protein synthesis, 38
Nucleic acids, digestion of, 252
Nucleus, 22–33
Nucleoproteins, and nucleic acids, 99–103
Nucleosides, 101
Nucleotides, 101, 102, 251, 252
Nuclides, 314

Ocular micrometer disc, in cell measurement, 19
Oligomer, 80
Oligosaccharides, 85–87
Operon, in regulation of protein synthesis, 299–303
Optical isomerism, 82
Oral mucosal smears, and sex chromatin, 265, 266
Orotidylic acid, 252, 253, 256, 257
Osmosis, 126–128
Osmotic pressure, 78, 127, 128
Oxidation-reduction potential, 147
Oxidative phosphorylation, 169, 177–180
  chemical hypothesis, 178, 179
  chemosmotic hypothesis, 178, 179

INDEX

Oxidizing agent, 147
Oxygen consumption
  methods of measuring, 185–195
  rate of, 184, 185, 189
Oxygen dissociation curves (see Oxyhemoglobin affinity curves)
Oxygen tension, and radiosensitivity, 320
Oxygen transport, in erythrocytes, 195
Oxyhemoglobin affinity
  and body weight, 209–211
  and oxygen tension, 198–201
  and pH, 199, 200
  and salt, 200
  and temperature, 199, 200
Oxyhemoglobin affinity curves
  in basses, 198, 199
  in man, 201
  in Rhesus monkey, 200
Oxysome hypothesis, 46, 47

Pachyura etrusca, 200
Pancreas
  islet alpha cells and glucagon, 229
  islet beta cells and insulin, 227
Pancreatic juice, enzymes of, 244, 245, 252
Pars amorpha, 23, 24
Pasteur effect, in glycolysis, 189
Pentose, 82
Pentose phosphate pathway (see Phosphogluconate oxidative pathway)
Pentose regeneration, 218–222
Peptide linkage, 92
Peptide tRNA site, of ribosomes, 287–292
pH, 117–121, 161
  and oxyhemoglobin affinity, 199, 200
Phagocytosis, 132–134
Phagosomes, and prelysosomes, 42, 43
Phase contrast microscopy, description of, 16–18
Phosphate transport, 164
Phosphatides (see Phospholipids)
Phosphoglucomutase, 146, 147, 229
Phosphogluconate oxidative pathway, 220, 232, 233
Phospholipids, 65, 108, 109
Phosphorolysis, 229
Photon, 52, 316–317
Photophosphorylation
  cyclic, 215–217
  noncyclic, 215–218
Photosynthesis, 211–214
  and free energy, 146
  Calvin cycle in, 218, 219
  carbon dioxide fixation, 52
  dark (Blackman) reactions, 214, 218–222
  in chloroplast, 50
  light (Hill) reactions, 51, 214–218
  photophosphorylation, redox catalysts in, 216, 217
  photophosphorylation, System I, 216, 217
  photophosphorylation, System II, 216, 217
  steps in, 52
  unit of, 52, 53
Phycocyanin, 216
Phycoerythrin, 216
Pinocytosis, 132–134, 237
Plaque membrane, model of, 125
Plasmalemma (see also Cell membrane), 123
Plasma membrane (see also Cell membrane), 123
Plasmodesmata, 68, 69
Polarization of light, 17
Polymer, 80, 88, 92
Polymerization, of macromolecules, 80, 81
Polynucleotides, 102
Polypeptides, action of digestive enzymes on, 245
Polyribosomes, 286, 287
  and protein synthesis, 39
Polysaccharides, 87–88
  excretion of, 227, 228
Porphyrins
  structure of, 103–105
  synthesis of, 175, 249, 251, 257
Postlysosomes, 42, 43
Pleuropneumonia-like organisms, 22, 200
Prelysosomes, 42, 43
Propionyl coenzyme A, in β-oxidation of fatty acids, 240, 242
Protein polysaccharide (see Mucopolysaccharides)
Proteins
  absorption of amino acids, 244–246
  and conformation, 98, 99
  as cell constituents, 88–99
  digestion of, 244–246
  histones and protamines, 25, 264
  in circulating plasma, 246
  metabolism of, 244–251
  properties of, 90
Protein synthesis
  and cell-free systems, 293–296
  and genetic code, 293–298
  and ribosomes, 285–293
  induction and repression of, 299–303
Protoplasm, inorganic composition, 80
Protoplast, 68

Pseudomonas, in photosynthesis, 50
Purines
  formula of, 101
  ring formation, 254
  synthesis and degradation of, 254–256
Pyridine nucleotides (*see also* Nicotinamide adenine dinucleotide and Nicotinamide adenine dinucleotide phosphate, 49, 164–167, 169, 170, 174, 175, 179, 180, 248
Pyridoxal phosphate, and amino acid membrane transport, 244
Pyrimidines
  formula of, 101
  ring formation, 252
  synthesis and degradation of, 252, 253
Pyruvic acid, 173, 174, 229–232, 240
Pyruvic acid dehydrogenase, 174

Q_{O_2} (*see* Metabolic rate)
Q_{10} (*see* Temperature coefficient)
Quantosomes, in photosynthesis, 53, 212

Radiation
  alpha particles, 315
  and aminothiols, 328
  beta particles, 315, 316
  by excitation, 318
  by ionization, 318, 319
  D_{37} dose, 327, 328
  historical, 310–313
  $LD_{50(30)}$, 329
  measurements of, 317, 318
  theories of action, 318–320
Radiation effects
  on animals, 328, 329
  on mammalian systems, 329–332
  on microorganisms, 327, 328
  on plants, 328
  on viruses, 327
Radiation sickness, 329, 330
Radioactivity
  artificial, discovery of, 311
  natural, discovery of, 311
Radionuclides, 332
Radiosensitivity
  and cellular conditions, 321, 322
  and environmental factors, 320, 321
Rads, defined, 318, 319
Recom, 275
Redox potential, 147, 178, 179
  in photosynthesis, 216, 217
Redox reactions, 49, 147, 148
Reducing agent, 147
Repression, in protein synthesis, 299–303

Respiration, of cells, 143, 146
Respiration rate (*see* Metabolic rate)
Respiratory chain (*see also* Electron transport), 48
Respiratory quotient (R. Q.), 183–195
Respirometer
  use with aquatic or terrestrial animals, 189, 190
  use with Winkler method, 185, 186
Riboflavin, formula of, 166
Ribonuclease, 258, 272, 273
Ribonucleic acid, 103, 256, 258
  absorption spectrum of, 38
  and the central dogma of protein synthesis, 285–293
  base composition, 53, 101–103, 286, 297, 298
  synthesis of, 268–275
Ribose 5-phosphate, and purine ring, 256
Ribosomal granules, 23, 24, 26, 30, 31
Ribosomes, 33, 36–39, 237, 295
  and protein synthesis, 285–293
  and subunits, 37, 287
  and zymogen, 245
  in animals, 38
  in bacteria, 37
  in plants, 39
Ribulose 1,5-diphosphate, 218–221
Ring structure, purine, 101, 102
RNA (*see* Ribonucleic acid)
Ribonucleic acid polymerase, 286, 289
Ribonucleic acid synthesis
  and chromosome puffs, 266, 270, 271
  and embryonic differentiation, 272–274
  on DNA, 37, 270, 273, 274
RNP granules (*see* Ribosomal granules), 24
Roentgen, defined, 318, 319

Salt, and oxyhemoglobin affinity, 200
Salt hunger, 80
Salts, 116, 117
Sarcomere, 137
Sarcoplasmic reticulum, 34, 136–138
Schindleria praematurus, 200
Sex chromatin, 265, 266, 268, 269
Sibbaldus musculus, 200
Simple lipids, 105, 107, 108
Soleplasm, 136
Solutes, 78
Soret band, of hemoproteins, 105
Spirogyra, in photosynthesis, 50
Squalus, 110
Stage micrometer, in cell measurement, 19
Starch, 86, 87

Steady state, 151, 183
Stereoisomerism, 108
Steroids, 110, 111
Succinic acid dehydrogenase, 174, 176, 177, 180
Succinic acid thiokinase, 174, 176
Succinyl coenzyme A
  and glycine, 175
  and porphyrin synthesis, 249, 257
Sucrose, 86
Sulfur, fixation of, 246, 247
Sulfur cycle, 246, 247
Suspensions, 78
Synapse, 136
Syncytium, 276

TCA cycle (*see* Tricarboxylic acid cycle)
Temperature, and oxyhemoglobin affinity, 199, 200
Temperature coefficient (Q_{10}), 153, 194
Terpenes, 109, 110
Thermodynamics
  first law, 144
  second law, 144
Thylakoids, 50, 51
Tight junction, of cells, 67, 68
Tonofibrils, 61, 67
TPN (*see* Nicotinamide adenine dinucleotide phosphate)
Transamination, of amino acids, 248
Tricarboxylic acid cycle, 48, 49, 165, 168, 169, 173–177
Triglycerides, 107, 235–237, 239
  synthesis of, 243, 244

Tritium, 311, 318
Triturus viridescens, 272, 273
Turner's syndrome, 266, 268
Tyndall effect, 79
Tyrocidin, 98

Ubiquinone (CoQ), 49, 165, 169, 177, 179, 180
Unit membrane, 51, 62, 64, 79, 125, 126
Urea, synthesis of, 246–248

Vitamin A, 112
Vitamin B complex, 114
Vitamin C, 114
Vitamin D, 112
Vitamin E, 113
Vitamin K, 113
Vitamins
  and cofactors, 164
  essential for man, 115
  fat soluble, 112, 113
  water soluble, 112, 113–115

Water, 79
  biological solvent, 76–78
  photolysis of, 52, 214, 217
Waxes, 105, 108

X-rays
  defined, 316
  discovery of, 310

Zwitterion, 92
Zymogen, and digestive enzymes, 245

polyribosomes

chloroplast

cilium

centrioles

mitochondrion

Golgi complex

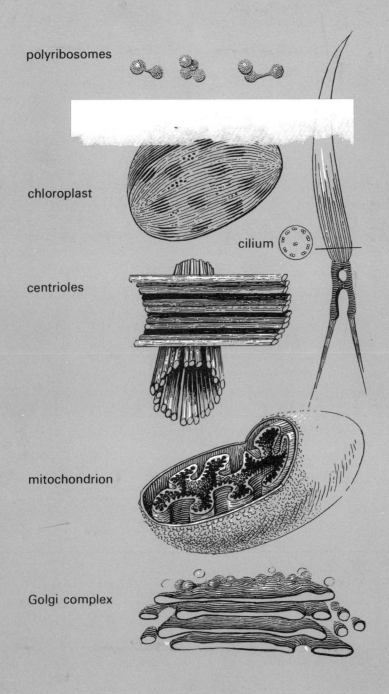